T0260164

PLANTS IN MESOZOIC TIME

Life of the Past

James O. Farlow, editor

PLANTS IN MESOZOIC TIME

Morphological Innovations, Phylogeny, Ecosystems

Edited by Carole T. Gee

Indiana University Press
Bloomington & Indianapolis

This book is a publication of

Indiana University Press
601 North Morton Street
Bloomington, Indiana 47404-3797 USA

www.iupress.indiana.edu

Telephone orders   800-842-6796
Fax orders   812-855-7931
Orders by e-mail   iuporder@indiana.edu

Manufactured in the United States of America
Library of Congress Cataloging-in-Publication Data

Plants in Mesozoic time : morphological innovations, phylogeny, ecosystems / edited by Carole T. Gee.
      p. cm. — (Life of the past)
  Includes bibliographical references and index.
  ISBN 978-0-253-35456-3 (cloth : alk. paper) 1. Paleobotany. 2. Paleontology—Mesozoic. I. Gee, Carole T.
  QE905.P54 2010
  561'.116—dc22
                                                                        2009044852

1 2 3 4 5 15 14 13 12 11 10

Ted Delevoryas Dedication Volume

CONTENTS

C

PREFACE, DEDICATION, AND ACKNOWLEDGMENTS

P

The Mesozoic was a curious time. On the surface, it resembled today's world, but instead of large herbivores like elephants, there were sauropods; instead of large carnivores like lions, there were tyrannosaurs; instead of trees with flowers and broad leaves, there were pteridosperms and bennettitaleans with naked seeds and fernlike fronds. The Mesozoic was thus a wondrous time. It was an era of giant dinosaurs, flying reptiles, and crown-tufted plants. It also encompassed a huge amount of time, over 185 million years, making the existence of the Mesozoic fauna and flora on earth an incredibly long success story.

This is perhaps why there has been a huge resurgence in interest in the Mesozoic biota in the last few decades, although much less attention has been focused on the flora than on the fauna. Yet the Mesozoic plants were more than just stage props for the dinosaurs and their daily activities, as they formed and inhabited a world of their own making. In fact, they were the basic building blocks of landscapes that the world had never seen before and will likely never see again. To describe just a few uniquely Mesozoic plant communities, there were lush forests of ginkgoes, araucarias, stout-trunked bennettitaleans, and low-growing ferns and cycads; there were vast meadows of ferns with scattered groves of seed ferns and woody gymnosperms; there were fleshy-leaved cheirolepidiaceous conifer trees growing on saline soils; there were massive tree-forming ferns with delicate angiosperm vines twining around their trunks. We know all this and more through the efforts of decades of dedicated research by Mesozoic paleobotanists, one of whom—Ted Delevoryas—will be appearing again in this connection.

Although the all-encompassing term *Mesozoic flora* is being used here, the Mesozoic flora was by no means uniform in space or time. Like the modern-day flora, different plants preferred different habitats and had various growth habits, and there were many major plant groups in the Mesozoic that do not have any modern counterpart or living relatives today, such as the enigmatic seed ferns or the bennettitaleans. However, some taxa did survive to this day and can even be recognized to genus level. These living fossils include the sphenophyte *Equisetum*, the maidenhair tree *Ginkgo*, and the southern conifer *Araucaria*, just to name a few.

In regard to time, the Mesozoic flora was not static, but evolved and developed over the course of 185 million years. The Triassic, for instance, saw the advent of the Mesozoic flora, the shift from the fern- and lycopod-dominated Paleozoic to a largely arborescent assemblage of gymnosperms. The Jurassic was not only the Age of the Dinosaurs but also the Age of the Cycadophytes. Indeed, the gymnosperms ruled the global flora for most of the Mesozoic, relinquishing their dominance only in the mid-Cretaceous when the radiation of the angiosperms tipped the scales in favor of plants with flowers.

Despite being a large, diverse plant group, the gymnosperms had in common several major evolutionary advantages over the ferns and fern allies. One was the development of secondary tissues, such as wood, which enabled gymnosperms to grow into tall, solid trees. Another key innovation was that of the seed, a unit of reproduction and dispersal consisting of a plant embryo and food reserves in a protective seed coat—or, as Ted Delevoryas used to explain to his general botany students, a baby plant in a box with its lunch.

Ted Delevoryas is today professor emeritus at the University of Texas at Austin and is considered one of the leading paleobotanists in America. Ted began his career in 1954, when he graduated with a Ph.D. in botany at the University of Illinois at Urbana-Champaign. Afterward, he took on a postdoctoral position at the Museum of Paleontology, University of Michigan, for one year. His first real job, as an assistant professor of botany at Michigan State University, followed in 1955. Ted then taught at Yale University from 1956–1972, except for two years, 1960–1962, at Illnois, before finally moving on to U.T. Austin in the 1970s.

Most of Ted's work has focused on the morphology and evolution of plants in two general areas: first, Paleozoic ferns and seed ferns such as *Psaronius*, *Medullosa*, and *Glossopteris*, and second, Triassic and Jurassic cycads and bennettitaleans, such as *Leptocycas*, *Cycadeoidea*, and *Williamsonia*. In actuality, though, Ted is a morphologist par excellence of the entire plant kingdom. With regard to research, Ted is meticulous and scholarly, and his enthusiasm for all plants—both fossil and living—is boundless. His groundbreaking work in paleobotany not only formed a solid foundation for further studies (including chapters in this volume by Osborn and Taylor [chapter 3], Rothwell and Stockey [chapter 4], and Crepet and Stevenson [chapter 10]), but also served as an inspiration to others in other areas of paleobotany.

Ted is also a highly skilled draftsman and used his talent to reconstruct beautiful and detailed images of whole-body fossil plants, some of which adorn the cover of this book. Anyone who was lucky enough to have had a class with him will certainly remember his spectacular, synchronous, two-handed chalkboard drawings of an apical meristem.

Ted was my doctoral advisor, or, as we called it, my major professor, at U.T. Austin in the 1980s. However, here in Germany, where I now live and work, a Ph.D. supervisor is called a *Doktorvater*. Although this may strike some as too familiar a term for what should be a strictly professional relationship between mentor and student, I find that it is quite an apt designation in many cases. For it is the *Doktorvater*, or *Doktormutter* for that matter, who takes responsibility for the academic development of the young scientist—advising, encouraging, and guiding—and imbues the fledging scientist with his or her vision of science, research, and

Preface Figure 1. Ted Delevoryas digging for Early Permian plants in north-central Texas on an unusually warm day in January 1982, here with Chuck Daghlian in the foreground. Ted loves prospecting and collecting in the field and is good at it. *Photo by Carole T. Gee.*

academic life. As my *Doktorvater*, Ted was easygoing but watchful over my scholarly development, and he always had a kind and encouraging word for me. Through his teaching and scientific work, he passed on to me his passion for plant morphology and anatomy. And like a good academic parent, he introduced me to his world, taking me along like a tag-along kid on a number of field trips and collecting trips throughout Texas (Preface Fig. 1). On these trips, I got to see and appreciate paleobotany through his eyes.

Most importantly, Ted taught me to understand that plant fossils are not just patterns in the rock record—objects like postage stamps to be collected and sorted and labeled—but parts of once growing, photosynthesizing, transpiring, reproducing living beings. This is perhaps the most precious inheritance from my academic dad: the realization and appreciation that paleobotanical fossils simply represent plants that lived in the very distant past. To look at them in that way is to be able to visualize them with deep respect as actual trees, shrubs, herbs, and vines that grew and greened the earth hundreds of millions of years ago.

During my doctoral work, I was in the lucky position of being able to give new names to several species, but I was too embarrassed then to name any of this scrappy material after my major professor. Now, however, some 20 years later, I have the opportunity of naming another fossil plant—this time, a Late Jurassic *Araucaria* complete with seed cones, cone scales, seeds, pollen cones, leaves, twigs, and branches (see Gee and Tidwell, chapter 5 in this volume), a once-living, thriving, whole plant—after my *Doktorvater* in this Festschrift on the happy occasion of his eightieth birthday in July 2009.

This volume is thus dedicated to Ted Delevoryas as a tribute to his manifold and valuable contributions to our knowledge of the Mesozoic flora. The 14 chapters in this book, Ted's biography, and this preface are offered here with gratitude, respect, and affection for Ted from his academic offspring and his extended family of colleagues—over 30 of them in this book alone—botanists, paleobotanists, paleontologists. During his 40-plus years in academia, Ted has sired several generations of paleobotanists, producing a robust academic family tree (Preface Fig. 2).

Although Ted was equally at home in the Paleozoic, we are focusing here in this volume on his greatest interest in later years, the Mesozoic flora. In view of the excellent textbooks on paleobotany available these days, we decided not to duplicate their efforts, but to complement these general works by showcasing scholarly paleobotanical research on the Mesozoic. Each chapter thus highlights the latest research on a particular plant, plant group, or topic. This is reflected as state-of-the-art reviews on plant groups such as the Late Triassic ginkgoleans of North America (Ash, chapter 8), and the Mesozoic cycadophytes (Cúneo et al., chapter 9) and Early Cretaceous conifers (Archangelsky and Del Fueyo, chapter 11) of Argentina. There are also several chapters elucidating key innovations that show up in the Mesozoic flora, namely in the leaf crowns of Mesozoic forests (Sussex et al., chapter 1), sphenophytes (Schwendemann et al., chapter 2), bennettitaleans (Osborn and M. Taylor, chapter 3; Rothwell and Stockey, chapter 4), conifers (Gee and Tidwell, chapter 5), and flowering plants (Dilcher, chapter 6; D. Taylor, chapter 7). One chapter applies phylogenetic methods to the recurrent burning issue of bennettitalean and angiosperm relationships (Crepet and Stevenson, chapter 10). Even studies on evolutionary developmental biology ("evo-devo") are no longer the sole domain of those working on living organisms, because fossils can, surprisingly, reveal quite a bit about how development and evolution interact (Sussex and Kerk, chapter 1; D. Taylor; chapter 7).

Most of the topics on this book involve gymnosperms, but also included here are in-depth studies on other Mesozoic plant groups such as the sphenophytes (Schwendemann et al., chapter 2) and angiosperms (Dilcher, chapter 6; D. Taylor, chapter 7; Tidwell et al., chapter 12), including the first woody dicot vine known from the fossil record (Tidwell et al., chapter 12). From a geographic point of view, paleobotanical research on the Western Interior of North America remains as active as ever—for example, on the Late Triassic Chinle Formation (Ash, chapter 8), the Late Jurassic Morrison Formation (Gee and Tidwell, chapter 5; Tidwell et al., chapter 12; Hotton and Baghai-Riding, chapter 13; Sander et al., chapter 14) and the Cretaceous Cedar Mountain, Burro Canyon, Mojado, and Dakota Sandstone Formations (Tidwell et al., chapter 12).

Theodore Delevoryas
Ph.D. University of Illinois at Urbana-Champaign 1954

Lila M. Cohen
M.S. Yale University 1958

Donald A. Eggert
Ph.D. Yale 1960, Postdoc Univ. Illinois Urbana-Champaign 1960-1961 and Yale University 1964-1965

David L. Dilcher
Ph.D. Yale University 1964

Thomas N. Taylor
Post-doc Yale University 1964-1965

Gar W. Rothwell
M.S. University of Illinois at Chicago Circle 1969

Ruth A. Stockey
M.S. Ohio University 1974, Ph.D. Ohio State University 1977

Edith L. Taylor
M.S. 1978, Ph.D. Ohio State University 1983

N. Ruben Cúneo
Postdoc Ohio State University 1990-1991

Ignacio Escapa
Ph.D. Comahue University, Argentina 2009

Jeffrey M. Osborn
Ph.D. Ohio University 1991

Mackenzie L. Taylor
B.A. Truman State University 2005

Georgina M. Del Fueyo
Postdoc Ohio State University 1992-1994

Michael Krings
Postdoc University of Kansas 1999-2000, 2002-2003

Andrew B. Schwendemann
Ph.D. candidate University of Kansas

Raj K. Jain
Post-doc Yale University 1965-1966

Rodney Gould
Post-doc Yale University 1969-1971

William L. Crepet
Ph.D. Yale University 1973

David Winship Taylor
Ph.D. University of Connecticut Storrs 1987

Christopher P. Person
Ph.D. University of Texas at Austin 1976

Shyam Chandra Srivastava
Postdoc University of Texas at Austin 1976-1977

Charles P. Daghlian
Ph.D. University of Texas at Austin 1977

Bruce S. Serlin
Ph.D. University of Texas at Austin 1980

Carole T. Gee
Ph.D. University of Texas at Austin 1987

John A. Stanley
M.A. University of Texas at Austin 1988

Sharon L. Fergusson
M.A. University of Texas at Austin 1988

John M. Mendenhall
M.A. University of Texas at Austin 1995

Nina L. Baghai
Ph.D. University of Texas at Austin 1996

Preface Figure 2. Genealogy of Ted Delevoryas's academic family. The F_1 generation is shown in its entirety; members of the F_2 and F_3 generations are listed only if they are authors in this volume. For clarity, the family tree is depicted here as a linear series of academic relationships, although in reality it is an intricate, close-knit, weblike network.

And to illustrate the pivotal role that plants play within an ecosystem, as well as their intimate relationship with the giant, plant-dependent reptiles of the Mesozoic, this book includes a chapter on Mesozoic plants and dinosaur herbivory (Sander et al., chapter 14).

Many paleobotanists helped to put this book together, and I am greatly indebted to Ted's many friends and colleagues who have contributed to this book, especially to P. Martin Sander, David Winship Taylor, and David L. Dilcher for their valuable advice and assistance. Many thanks should also go to the reviewers of the individual chapters for their selfless work, some of whom reviewed more than one chapter. This includes Nan Crystal Arens, N. Rúben Cúneo, James Farlow, Patricia G. Gensel, Carol L. Hotton, Hans Kerp, Harufumi Nishida, Gar W. Rothwell, Ruth A. Stockey, David W. Taylor, William D. Tidwell, Alfred Traverse, and Zhiyan Zhou.

I thank the talented and hard-working technical staff at the University of Bonn, Steinmann Institute, Division of Paleontology: technical illustrator Dorothea Kranz for her splendid artwork on the book's cover, as well to Georg Oleschinki for his excellent refinement of the photos in this book. My gratitude also extends to Robert Sloan and James Farlow at Indiana University Press for their support for this project, the first volume on paleobotany in the Life of the Past series. This book was edited during the tenure of a grant from DFG Research Unit 533 on the Biology of the Sauropod Dinosaurs: The Evolution of Gigantism, which is gratefully acknowledged here. My heartfelt thanks go to Volker Mosbrugger, director of the Senckenberg Museum, for serving as principal investigator of my project on Food Ecology. This book is DFG Research Unit 533 contribution no. 56.

Carole T. Gee
University of Bonn
Bonn, Germany

THE CAREER OF TED DELEVORYAS: APPRECIATION AND PUBLICATIONS

Thomas N. Taylor, Edith L. Taylor,
and Charles P. Daghlian

It is impossible to define a career in an introduction to a book, especially one as distinguished as the career of Ted Delevoryas. Any feat of this magnitude would have the writer begging for more space, adding additional superlatives to the prose, and always questioning whether the appropriate tenor has been achieved or whether the honoree will approve of his career as summarized by others. The preceding might be appropriate for most individuals, but not for Ted Delevoryas. His raison d'etre has always been the celebration of the success of others, his students and colleagues, rather than even remotely noting his own accomplishments, which, as all in paleobotany know, are numerous and in many ways have redefined the discipline. We also know that our dear friend Ted will fuss a little about having any type of celebration that in any way might make reference to his career and certainly would have objected to the preparation of this volume, had his opinion been sought. And of course that is exactly the reason why his opinion was not solicited! There is also another, perhaps more significant reason why we have ignored his opinion on the preparation of this volume. Those of us who love Ted Delevoryas also want him to know that he instilled in us the importance of sometimes ignoring the advice and counsel of our teacher and friend, and doing what we believed was important and right. Sorry, Ted, but we were paying attention to your lessons on life as we listened to your lectures on *Medullosa*, *Psaronius*, *Cycadeoidea*, and even trace departure in coenopterid ferns!

Ted's career and scientific accomplishments are many and extraordinary. Among these are a couple of areas that we wish to highlight. Some paleobotanists working with permineralized fossil plants realized that these cells and tissue systems represented a unique level of inquiry that had the potential to answer questions about fossil plant development, but Ted, and his then graduate student Don Eggert, shifted the paradigm by using coal-ball plant organs to discuss ontogeny. Eggert's classic papers on tissue development in arborescent lycopsids and sphenopsids are two studies that were supervised by Ted. In fact, Delevoryas's own dissertation on *Medullosa* and colleague Jeanne Morgan's study of *Psaronius* were early approaches that used numerous specimens of the same morphospecies to document changes in organ development. Studies that involve an understanding of ontogenetic stages and developmental patterns of fossil plants have become the norm in paleobotany as a result of this early work. Whether the questions focus on species descriptions or phylogenetic analyses, the stimulus to address such hypotheses can be traced to Ted's creative approach of viewing the fossils as once dynamic organisms. Perhaps more than anyone before him, Ted viewed collections of abundant specimens as new sources of information in which each of the specimens might well reveal a slightly younger or more mature organism that could be understood within the context of ontogeny.

This developmental approach in the study of fossil plants was also incorporated into Ted's studies of fossil Bennettitales. Using the rich collection of silicified *Cycadeoidea* trunks and cones at the Peabody Museum at Yale University, which were initially collected by George W. Wieland, Delevoryas initiated a new research program that focused on cycadeoid biology and reproduction. Several of his early studies made significant contributions to understanding the vascular anatomy of the trunks of these plants, and these in turn provided new data and hypotheses on the homologies and evolution of both vegetative and reproductive organs. Biologists of varying research specialties have always been fascinated by questions relating to the origin of angiosperms. Thus, when Wieland published his two-volume study of silicified cycadeoids and suggested that the reproductive organs appeared much like an open flower at maturity, many at the time regarded these plants as the progenitors to the angiosperms. By using material from the Wieland Collection, in addition to preparing countless new petrographic thin sections of *Cycadeoidea* cones, Ted produced a series of outstanding papers that more accurately reflected how the cones actually grew and illustrated the relationship between the pollen and ovule-bearing parts. As a result of these papers, together with those of graduate student Bill Crepet, we now know a great deal about the structural organization of the reproductive parts of these plants, including, not surprisingly, their ontogeny and pollination biology, and the fact that they have little to do with angiosperms.

The study of Mesozoic plants is yet another dimension of Ted's career that has had a profound impact on paleontology. Whether these studies are papers on Triassic ferns, conifers, and cycads from North Carolina, Jurassic bennettitaleans from Mexico, or glossopterids from Australia, the work is creative, precise, and always intellectually stimulating. His book on paleobotany, *Morphology and Evolution of Fossil Plants*, his book on plant evolution, *Plant Diversification*, and his collaborations with Harold C. Bold and Constantine Alexopoulos on the *Morphology of Vascular Plants and Fungi* will serve to enlighten students of the plant sciences well into the future.

No one is capable in these few lines of summarizing a career that has spanned many decades and been responsible for an incredible body of insightful research, worldwide collaboration, and personal and professional friendships—certainly not us. But we do know that those who have been taught, counseled, or have in some way interacted with Ted Delevoryas during his career, like us, possess an extraordinary degree of admiration and affection that cannot be expressed on any printed page. It is with that caveat that this volume honors Ted Delevoryas.

Stewart, W., and T. Delevoryas. 1952. Bases for determining relationships among the Medullosaceae. *American Journal of Botany* 39: 505–516.

Morgan, J., and T. Delevoryas. 1952. *Stewartiopteris singularis:* a new psaroniaceous fern rachis. *American Journal of Botany* 39: 479–497.

Morgan, J., and T. Delevoryas. 1952. An anatomical study of *Stipitopteris. American Journal of Botany* 39: 474–478.

Delevoryas, T., and J. Morgan. 1952. *Tubicaulis multiscalariformis:* a new American coenopterid. *American Journal of Botany* 39: 160–166.

Delevoryas, T. 1953. A new male cordaitean fructification from the Kansas Carboniferous. *American Journal of Botany* 40: 144–150.

Delevoryas, T. 1954. The Medullosae—structure and relationships. Ph.D. diss., University of Illinois, Urbana-Champaign.

Morgan, J., and T. Delevoryas. 1954. An anatomical study of a new coenopterid and its bearing on the morphology of certain coenopterid petioles. *American Journal of Botany* 41: 198–203.

Delevoryas, T., and J. Morgan. 1954. A new pteridosperm from Upper Pennsylvanian deposits of North America. *Palaeontographica*, Abt. B, 96: 12–23.

Delevoryas, T., and J. Morgan. 1954. A further investigation of the morphology of *Anachoropteris clavata. American Journal of Botany* 41: 192–203.

Delevoryas, T., and J. Morgan. 1954. Observations on petiolar branching and foliage of an American *Botryopteris. American Midland Naturalist* 52: 374–387.

Delevoryas, T. 1955. A *Palaeostachya* from the Pennsylvanian of Kansas. *American Journal of Botany* 42: 481–488.

Delevoryas, T. 1955. The Medullosae—structure and relationships. *Palaeontographica*, Abt. B, 97: 115–167.

Stewart, W. N., and T. Delevoryas. 1956. The Medullosan pteridosperms. *Botanical Review* 22: 45–80.

Delevoryas, T. 1956. The shoot apex of *Callistophyton poroxyloides. Contributions from the Museum of Paleontology, University of Michigan* 12: 285–299.

Delevoryas, T. 1957. Anatomy of *Sigillaria approximata. American Journal of Botany* 44: 654–660.

Delevoryas, T. 1958. A fossil stem apex from the Pennsylvanian of Illinois. *American Journal of Botany* 45: 84–89.

Delevoryas, T. 1959. Investigations of North American cycadeoids: *Monanthesia. American Journal of Botany* 46: 657–666.

Cohen, L. M., and T. Delevoryas. 1959. An occurrence of *Cordaites* in the Upper Pennsylvanian of Illinois. *American Journal of Botany* 46: 545–549.

Delevoryas, T. 1960. Cycadeoidales. In *McGraw-Hill Encyclopedia of Science and Technology*, 689. New York: McGraw-Hill Book Co.

Delevoryas, T. 1960. Book review: The Morphology and Anatomy of American Species of the Genus *Psaronius*, by J. Morgan. *Bulletin of the Torrey Botanical Club* 87: 225–226.

Delevoryas, T. 1960. Book review: Anatomy of Seed Plants, by Katherine Esau. John Wiley & Sons, New York. 1960. *AIBS Bulletin* 10 (3): 40.

Eggert, D. A., and T. Delevoryas. 1960. *Callospermarion*—a new seed genus from the Upper Pennsylvanian of Illinois. *Phytomorphology* 10: 131–138.

Delevoryas, T. 1960. Investigations of North American cycadeoids: trunks from Wyoming. *American Journal of Botany* 47: 778–786.

Delevoryas, T. 1961. Book review: Studies in Paleobotany, by Henry H. Andrews, John Wiley & Sons, New York and London, 1961. *The Quarterly Review of Biology* 36: 212.

Delevoryas, T. 1961. Book review: Principles of Paleobotany, by William C. Darrah, Ronald Press Co., New York, 1960. *The Quarterly Review of Biology* 36: 212.

Delevoryas, T. 1961. Notes on the cone of *Cycadeoidea*. *American Journal of Botany* 48: 540.

Delevoryas, T. 1962. Book review: The Yorkshire Jurassic Flora. I. Thallophyta-Pteridophyta, by Thomas Maxwell Harris, British Museum (Natural History), London, 1961. *The Quarterly Review of Biology* 37: 181–182.

Delevoryas, T. 1962. Book review: Fossile Wandstrukturen. Handbuch der Pflanzenanatomie. Abteilung Cytologie, Band III, Teil 5, by Erich Hennes, Gebrüder Borntraeger, Berlin, 1959. *The Quarterly Review of Biology* 37: 182.

Delevoryas, T. 1962. *Morphology and Evolution of Fossil Plants*. New York: Holt, Rinehart and Winston.

Delevoryas, T. 1963. Investigations of North American cycadeoids: cones of *Cycadeoidea*. *American Journal of Botany* 50: 45–52.

Delevoryas, T. 1964. The role of palaeobotany in vascular plant classification. In V. H. Heywood and J. McNeill (eds.), *Phenetic and Phylogenetic Classification*, pp. 26–36. London: Systematics Association.

Delevoryas, T. 1964. Origin and evolution of ferns. *Memoirs of the Torrey Botanical Club* 21 (5): 1–2.

Delevoryas, T. 1964. Ontogenetic studies of fossil plants. *Phytomorphology* 14: 299–314.

Taylor, T. N., and T. Delevoryas. 1964. Paleozoic seed studies: a new Pennsylvanian *Pachytesta* from southern Illinois. *American Journal of Botany* 51 (2): 189–195.

Delevoryas, T. 1964. Two petrified angiosperms from the Upper Cretaceous of South Dakota. *Journal of Paleontology* 38: 584–586.

Delevoryas, T. 1964. A probable pteridosperm microsporangiate fructification from the Pennsylvanian of Illinois. *Palaeontology* 7: 60–63.

Delevoryas, T. 1965. Investigations of North American cycadeoids: microsporangiate structures and phylogenetic implications. *The Palaeobotanist* 14: 89–93.

Delevoryas, T. 1966. Book review: Spores. Ferns, Microscopic Illusions Analyzed, by Clara S. Hires, Mistaires Laboratories, Millburn, N.J., 1965. *Bulletin of the Torrey Botanical Club* 93: 460.

Delevoryas, T. 1966. Book review: The Morphology of Gymnosperms. The Structure and Evolution of Primitive Seed-Plants, by K. R. Sporne, Hillary House Publishers, New York, 1965. *The Quarterly Review of Biology* 41: 422.

Delevoryas, T. 1966. *Plant Diversification*. New York: Holt, Rinehart and Winston.

Ostrom, J. H., and T. Delevoryas. 1966. *The Age of Reptiles: A Guide to the Rudolph Zallinger Mural in the Peabody Museum, Yale University*. New Haven, Conn.: Yale University Press.

Delevoryas, T. 1966. Hunting fossil plants in Mexico. *Discovery* 2 (1): 7–13.

Delevoryas, T. 1966. Investigations of North American cycadeoids: microsporangiate structures and phylogenetic implications. *The Palaeobotanist* 14: 89–93.

Eggert, D. A., and T. Delevoryas. 1967. Studies of Paleozoic ferns: *Sermaya*, gen. nov. and its bearing on filicalean evolution in the Paleozoic. *Palaeontographica*, Abt. B, 120: 169–180.

Jain, R. K., and T. Delevoryas. 1967. A Middle Triassic flora from the Cacheuta Formation, Minas de Petroleo, Argentina. *Palaeontology* 10: 564–589.

Delevoryas, T. 1967. Further remarks on the ontogeny of certain Carboniferous plants. *Phytomorphology* 17: 330–336.

Delevoryas, T. 1968. Book review: Traité de Paléobotanique. Tome II. Bryophyta, Psilophyta, Lycophyta, by E. Boureau, Masson et Cie, 1964. *Bryologist* 71: 157–158.

Delevoryas, T. 1968. Some aspects of cycadeoid evolution. *Botanical Journal of the Linnean Society* 61: 137–146.

Delevoryas, T. 1968. Paleobotanical investigations. *Yale Scientific Monthly* 42 (8): 4–9, 21.

Delevoryas, T. 1968. Investigations of North American cycadeoids: structure, ontogeny and phylogenetic considerations of cones of *Cycadeoidea*. *Palaeontographica*, Abt. B, 121: 122–133.

Delevoryas, T. 1969. Paleobotany, phylogeny, and a natural system of classification. *Taxon* 18: 204–212.

Delevoryas, T., and T. N. Taylor. 1969. A probable pteridosperm with eremopterid foliage from the Allegheny Group of northern Pennsylvania. *Postilla* 133: 1–14.

Delevoryas, T. 1969. Glossopterid leaves from the Middle Jurassic of Oaxaca, Mexico. *Science* 165: 895–896.

Delevoryas, T. 1970. Book Review: Xylotomy of the Living Cycads with a Description of Their Leaves and Epidermis, by Pal Greguss, Akademiai Kiado, Budapest, 1968. *Economic Botany* 24: 108.

Delevoryas, T. 1970. Book review: The Yorkshire Jurassic Flora. III: Bennettitales, by Thomas Maxwell Harris, British Museum (Natural History), London, 1969. *The Quarterly Review of Biology* 45: 383.

Delevoryas, T. 1970. Book review: A Short History of Botany in the United States, by Joseph Ewan, Ed. Hafner, New York, 1969. *Science* 169: 1194.

Delevoryas, T. 1970. Book review: The Fossil Flora of the Drywood Formation of Southwestern Missouri, by P. W. Basson, University of Missouri Press, Columbia, 1969. *BioScience* 20: 251.

Delevoryas, T. 1970. Book review: Anatomy of the Monocotyledons IV. Juncales, by D. F. Cutler, ed. by C. R. Metcalf, Oxford University Press, New York, 1969. *BioScience* 20: 831.

Delevoryas, T. 1970. Biotic provinces and the Jurassic–Cretaceous floral transition. In E. L. Yochelson (ed.), *Proceedings of the North American Paleontological Convention*, vol. 2, pp. 1660–1674. Lawrence, Kans.: Allen Press.

Delevoryas, T. 1970. Plant life in the Triassic of North Carolina. *Discovery* 6 (1): 15–22.

Delevoryas, T., and R. E. Gould. 1971. An unusual fossil fructification from the Jurassic of Oaxaca, Mexico. *American Journal of Botany* 58: 616–620.

Delevoryas, T. 1971. Cycadeoidales. In *McGraw-Hill Encyclopedia of Science and Technology*, 3rd ed., p. 689. New York: McGraw-Hill Book Co.

Delevoryas, T., and R. C. Hope. 1971. A new Triassic cycad and its phyletic implications. *Postilla* 150: 1–21.

Delevoryas, T. 1971. Biotic provinces and the Jurassic–Cretaceous floral transition. In K. M. Waage (ed.), *North American Paleontological Convention, Proceedings, Part L*, pp. 1660–1674. Lawrence, Kans.: Allen Press.

Crepet, W. L., and T. Delevoryas. 1972. Investigations of North American cycadeoids: early ovule ontogeny. *American Journal of Botany* 59: 209–215.

Delevoryas, T. 1973. Postdrifting Mesozoic floral evolution. In B. J. Meggers, E. S. Ayensu, and W. D. Duckworth (eds.), *Tropical Forest Ecosystems in Africa and South America: A Comparative Review*, pp. 9–19. Washington, D.C.: Smithsonian Institution Press.

Delevoryas, T., and R. C. Hope. 1973. Fertile coniferophyte remains from the Late Triassic Deep River Basin, North Carolina. *American Journal of Botany* 60: 810–818.

Delevoryas, T., and R. E. Gould. 1973. Investigations of North American cycadeoids: williamsonian cones from the Jurassic of Oaxaca, Mexico. *Review of Palaeobotany and Palynology* 15: 27–42.

Delevoryas, T. 1973. Book review: Fossilium Catalogus; II, Plantae; Pars 79, Gymnospermae (Ginkgophyta et Coniferae) I, W. J. Jongmans and S. J. Dijkstra, 1971; Gymnospermae (Ginkgophyta et Coniferae) II, W. J. Jongmans and S. J. Dijkstra, 1971, Dr. W. Junk's-Gravenhage, The Hague, 1971. *Journal of Paleontology* 47: 595.

Delevoryas, T. 1973. Biological nomenclature. *Systematics Association Publication* 6: 29–36.

Delevoryas, T. 1975. Book review: Bryophytes of the Pleistocene: The British Record and its Chorological and Ecological Implications, by J. H. Dickson, Cambridge University Press, Cambridge, 1973. *The Quarterly Review of Biology* 50: 75.

Delevoryas, T. 1975. Mesozoic cycadophytes. In K. S. W. Campbell (ed.), *Gondwana Geology*, pp. 173–191. Canberra, Australia: Australian National University Press.

Delevoryas, T., and C. P. Person. 1975. *Mexiglossa varia* gen. et sp. nov., a new genus of glossopteroid leaves from the Jurassic of Oaxaca, Mexico. *Palaeontographica*, Abt. B, 154: 114–120.

Delevoryas, T., and R. C. Hope. 1975. *Voltzia andrewsii*, n. sp., an Upper Triassic seed cone from North Carolina, U.S.A. *Review of Palaeobotany and Palynology* 20: 67–74.

Skog, J. E., P. G. Gensel, T. L. Phillips, T. Delevoryas, S. H. Mamay, A. T. Cross, A. E. Kasper, and H. T. Andrews. 1975. Preface to special issue on Henry N. Andrews Jr. *Review of Palaeobotany and Palynology* 20: 2.

Skog, J. E., P. G. Gensel, T. L. Phillips, T. Delevoryas, S. H. Mamay, A. T. Cross, A. E. Kasper, and H. T. Andrews (eds.). 1975. Special issue dedicated to Henry N. Andrews Jr. on his retirement. *Review of Palaeobotany and Palynology* 20: 2–131.

Skog, J. E., P. G. Gensel, T. L. Phillips, T. Delevoryas, S. H. Mamay, A. T. Cross, A. E. Kasper, and H. T. Andrews. 1975. Henry N. Andrews Jr.: a biographical sketch. *Review of Palaeobotany and Palynology* 20: 3–11.

Delevoryas, T., and R. C. Hope. 1976. More evidence for a slender growth habit in Mesozoic cycadophytes. *Review of Palaeobotany and Palynology* 21: 93–100.

Delevoryas, T. 1977. *Plant Diversification*, 2nd ed. New York: Holt, Rinehart and Winston.

Ostrom, J. H., and T. Delevoryas. 1977. *The Age of Reptiles*. New Haven, Conn.: Yale University Press.

Gould, R. E., and T. Delevoryas. 1977. The biology of *Glossopteris*: evidence from petrified seed-bearing and pollen-bearing organs. *Alcheringa* 1: 387–399.

Delevoryas, T., and R. C. Hope. 1978. Habit of the Upper Triassic *Pekinopteris auriculata*. *Canadian Journal of Botany* 56: 3129–3135.

Delevoryas, T. 1978. *Diversificacao nas Plantas*. São Paulo: Pioneira.

Taylor, T. N., D. L. Dilcher, and T. Delevoryas. 1979. Plant reproduction in the fossil record—introduction. *Review of Palaeobotany and Palynology* 27: 211–212.

Taylor, T. N., D. L. Dilcher, and T. Delevoryas (eds.). 1979. Plant reproduction in the fossil record. *Review of Palaeobotany and Palynology* 27: 211–358.

[Papers presented at a symposium of Paleobotanical Section, Botanical Society of America, August 1977, at Michigan State University.]

Ball, H. W., H. W. Borns Jr., B. A. Hall, H. K. Brooks, F. M. Carpenter, and T. Delevoryas. 1979. Biota, age, and significance of lake deposits, Carapace Nunatak, Victoria Land, Antarctica. In B. Laskar and C. S. R. Rao (eds.), *Fourth International Gondwana Symposium: Papers*, pp. 166–175. New Delhi: Hindustan Publishing.

Delevoryas, T. 1980. Polyploidy in gymnosperms. In W. H. Lewis (ed.), *Polyploidy: Biological Relevance*, pp. 215–218. New York: Plenum Press.

Daghlian, C. P., W. L. Crepet, and T. Delevoryas. 1980. Investigations of Tertiary angiosperms: a new flora including *Eomimosoidea plumosa* from the Oligocene of eastern Texas. *American Journal of Botany* 67: 309–320.

Delevoryas, T., and R. C. Hope. 1981. Conifer diversity in the Upper Triassic. *BioScience* 31: 326–328.

Serlin, B. S., T. Delevoryas, and R. Weber. 1981. A new conifer pollen cone from the Upper Cretaceous of Coahuila, Mexico. *Review of Palaeobotany and Palynology* 31: 241–248.

Delevoryas, T., and S. C. Srivastava. 1981. Jurassic plants from the department of Francisco Morazan, central Honduras. *Review of Palaeobotany and Palynology* 34: 345–357.

Delevoryas, T. 1981. *Diversificación Vegetal*. Mexico City: CECSA.

Delevoryas, T., and R. C. Hope. 1981. More evidence for conifer diversity in the Upper Triassic of North Carolina. *American Journal of Botany* 68: 1003–1007.

Person, C. P., and T. Delevoryas. 1982. The Middle Jurassic flora of Oaxaca, Mexico. *Palaeontographica*, Abt. B, 180: 82–119.

Delevoryas, T. 1982. A new *Coniopteris* from the Middle Jurassic of Tecomatlan Puebla, Mexico. In D. D. Nautiyal (ed.), *Studies on Living and Fossil Plants*, pp. 71–76. Allahabad, India: Society of Plant Taxonomists.

Delevoryas, T. 1982. Perspectives on the origin of cycads and cycadeoids. *Review of Palaeobotany and Palynology* 37: 115–132.

Taylor, T. N., and T. Delevoryas (eds.). 1982. Gymnosperms: Paleozoic and Mesozoic. *Review of Palaeobotany and Palynology* 37: 1–154. [Papers presented at a symposium held at the 13th International Botanical Congress, Sydney, Australia, 21–28 August 1981.]

Taylor, T. N., and T. Delevoryas. 1982. Introduction to gymnosperms: Paleozoic and Mesozoic. In T. N. Taylor and T. Delevoryas (eds.), *Review of Palaeobotany and Palynology* 37: 1–5.

Smoot, E. L., T. N. Taylor, and T. Delevoryas. 1985. Structurally preserved fossil plants from Antarctica. I. *Antarcticycas*, gen. nov., a Triassic cycad stem from the Beardmore Glacier area. *American Journal of Botany* 72: 1410–1423.

Bold, H. C., C. J. Alexopoulos, and T. Delevoryas. 1986. *Morfología de las Plantas y los Hongos*. Barcelona, Spain: Ediciones Omega.

Ostrom, J. H., L. J. Hickey, and T. Delevoryas. 1987. *The Age of Reptiles*, 2nd ed. New Haven, Conn.: Peabody Museum of Natural History, Yale University.

Taylor, T. N., T. Delevoryas, and R. C. Hope. 1987. Pollen cones from the Late Triassic of North America and implications on conifer evolution. *Review of Palaeobotany and Palynology* 53: 141–149.

Bold, H. C., C. J. Alexopoulos, and T. Delevoryas. 1987. *Morphology of Plants and Fungi*, 5th ed. New York: Harper and Row.

Laudon, T. S., D. J. Lidke, T. Delevoryas, and C. T. Gee. 1987. Sedimentary rocks of the English coast, eastern Ellsworth Land, Antarctica. In G. D. McKenzie (ed.), *Gondwana Six: Structure, Tectonics, and Geophysics*, pp. 183–189. Washington, D.C.: American Geophysical Union.

Delevoryas, T., and R. C. Hope. 1987. Further observations on the Late Triassic conifers *Compsostrobus neotericus* and *Voltzia andrewsii*. *Review of Palaeobotany and Palynology* 51: 59–64.

Delevoryas, T. 1990. Comments on the role of cycadophytes in Antarctic fossil floras. In T. N. Taylor and E. L. Taylor (eds.), *Antarctic Paleobiology— Its Role in the Reconstruction of Gondwana*, pp. 173–178. New York: Springer-Verlag.

Delevoryas, T. 1991. Investigations of North American cycadeoids: *Weltrichia* and *Williamsonia* from the Jurassic of Oaxaca, Mexico. *American Journal of Botany* 78: 177–182.

Gustavson, T. C., and T. Delevoryas. 1992. *Caulerpa*-like marine alga from Permian strata, Palo Duro Basin, west Texas. *Journal of Paleontology* 66: 160–161.

Delevoryas, T., T. N. Taylor, and E. L. Taylor. 1992. A marattialean fern from the Triassic of Antarctica. *Review of Palaeobotany and Palynology* 74: 101–107.

Delevoryas, T. 1993. Book review: Jurassic and Cretaceous Floras and Climates of the Earth, by V. A. Vakhrameev, ed. by N. F. Hughes; translated by J. V. Litvinov, Cambridge University Press, Cambridge and New York, 1991. *The Quarterly Review of Biology* 68: 96–97.

Delevoryas, T. 1993. Origin, evolution, and growth patterns of cycads. In D. W. Stevenson and K. J. Norstog (eds.), *The Biology, Structure, and Systematics of the Cycadales*, pp. 236–245. Milton, Australia: Palm and Cycad Societies of Australia.

Axsmith, B. J., T. N. Taylor, T. Delevoryas, and R. C. Hope. 1995. A new species of *Eoginkgoites* from the Upper Triassic of North Carolina, U.S.A. *Review of Palaeobotany and Palynology* 85: 189–198.

Delevoryas, T., and J. E. Mickle. 1995. Upper Cretaceous magnoliaceous fruit from British Columbia. *American Journal of Botany* 82: 763–768.

PLANTS IN MESOZOIC TIME

PART 1

Morphological Innovations in Mesozoic Plants

©Marlene Hill Donnellly

ARCHITECTURAL INNOVATION AND DEVELOPMENTAL CONTROLS IN SOME MESOZOIC GYMNOSPERMS, OR, WHY DO THE LEAF CROWNS IN MESOZOIC FORESTS LOOK TUFTED?

1

Ian Sussex, Nancy Kerk, and Carole T. Gee

The concept of plant architecture was first formalized by Hallé, Oldeman, and Tomlinson in 1978 and was based on patterns observed in forest-forming trees. Plant architecture is dependent primarily on the function of meristems and the growth patterns of shoots and roots. There has been an evolutionary progression in architecture throughout the history of plant life, and Mesozoic gymnosperms occupy an intermediate position between the earliest and the latest patterns. We shall discuss the molecular, genetic, and physiologic bases of advanced plant patterns and then deduce which changes have occurred in evolutionary terms that make the shoot architecture of Mesozoic gymnosperms different from modern forest-forming trees. This may be largely the result of an evolutionary progression in shoot growth patterns, such as internodal extension, which was, in the Mesozoic, likely a hormone-regulated process and possibly under the developmental control of polar auxin flow, as it is today.

Fig. 1.1. Reconstruction of a Late Triassic forest in Greenland, showing the tufted appearance of Mesozoic forest tree crowns. *Image courtesy of Marlene Donnelly and Jennifer McElwain.* © *by Marlene Hill Donnelly.*

Introduction

Many Mesozoic gymnosperms occupy a unique place in plant evolution, demonstrating as they do the transitions between the relatively simple morphologies of the ferns, lycophytes, and early seed ferns to the much more complex morphologies, including floral morphologies, of the angiosperms. In this era, which spanned some 180 million years, the most common plants were cycads, bennettitaleans (also known as cycadeoids), conifers, and ginkgophytes. Plants of these groups were the dominant members of Mesozoic forests, and they shared a common feature in possessing stems with little or very restricted internodal extension, although this was less a feature of the Ginkgoales. The consequence of this growth habit was that Mesozoic forests were composed of several major plant groups with tree crowns that were remarkably tufted and with stems that were extensively covered and often protected by scaly leaf bases (Fig. 1.1; Plate 1). They looked dramatically different from the forests, particularly the tropical forests, of the present. They also differed from Paleozoic coal swamp forests in which only one plant group—the arborescent lycophytes—had tufted crowns. What were the changes in growth characteristics that underlay the transition from the tuft-crowned Mesozoic gymnosperms to the modern trees of the present-day forests? The huge differences in morphology of the vascular plants make this seem like a daunting task. However, examination of the architecture of plants provides a method for categorizing and systematizing the events of this transition.

Shoot Architecture

The concept of plant architecture, which was first formalized by Hallé et al. (1978), is based primarily on earlier descriptions of tropical forest trees. Most attention has been paid to shoot architecture, although there are some descriptions of root architecture (Atger and Edelin 1994; Osmont et al. 2007). Because of the paucity of preserved root systems of Mesozoic plants, we shall not consider these in this analysis.

The architectural analysis of shoot systems as originally proposed by Hallé et al. (1978) conceived the plant as conforming through its growth cycle to an architectural model that is an abstraction for the deterministic genetic blueprint on which the construction of the plant is based. Under ideal conditions of growth, the development and architecture of the plant would conform strictly to the model. Hallé et al. identified 23 different architectures that described essentially all of the vascular plant growth habits, as well as those of 24 fossil plants, including several members of the Bennettitales, Cycadales, Cordaitales, and Coniferales. The fact that plants of the same species differ in appearance was attributed to interactions between the environment and the development of the individual organism. They termed these changes opportunistic, so that the developmental morphology of a plant represented a balance between deterministic, genetic and opportunistic, environmental effects (Hallé 1999). The architectural models were considered to be indicative of the growth and developmental processes of a particular phase of the plant's life cycle, and they found that as a plant passed through various phases of its life cycle, such as transitions involved in phase change, aging, heteroblastic development, or flowering, the model might change to another one indicative of different growth processes.

Vascular plants are modular organisms that develop by the repetition of certain basic entities, albeit with some modification in the expression of these, during the phases of development. To a large extent, architectural models can be described in terms of the behaviors of meristems and the phytomers that they produce: whether apical meristems are active continuously or intermittently; whether they undergo a fate change such as vegetative to reproductive; whether branching is terminal or axillary; whether branches grow orthotropically (vertically), plagiotropically (horizontally), or with an intermediate orientation; and whether phytomers elongate equally, unequally, or insignificantly (Sussex and Kerk 2001). These aspects of meristem and phytomer behavior, which are the basis of the architectural models, are increasingly becoming the subject of physiologic, genetic, and molecular analyses that are likely to lead to new ways to regulate meristem function throughout development, and perhaps to the generation of new architectural models (Sussex and Kerk 2001; Barthélémy and Caraglio 2007). In addition, these approaches should allow us to identify connections between current architectural models and thus put them into an evolutionary context.

The shoot system of vascular plants is composed of a series of repeating units, termed phytomers or modules. These originate through the activity of an apical meristem that produces a succession of leaf primordia in a phyllotactic sequence characteristic of the species. Each phytomer consists of a leaf (sometimes several) attached to a node on the stem. The nonnodal part of the stem is an internode, which may be elongated or not. An axillary bud is usually located in the axil of the leaf. It repeats the organization of the terminal bud and, through its growth, produces an axillary branch that may then repeat the branching process. Variations in the extent of internodal elongation and axillary branch development characterize many of the architectural differences between a number of Mesozoic gymnosperm trees and modern forests. Efforts to understand the basis of these differences have included physiological, genetic, and molecular analyses.

The most extreme shoot morphology is that of the Mesozoic bennettitaleans and cycads in which the vegetative stem has the appearance of a persistent short shoot with closely packed leaves and no internodal extension (Fig. 1.2) (Delevoryas 1971; Stewart and Rothwell 1993; Taylor et al. 2009). This ancient growth habit persists today in some cycads and palms.

The ancient conifers possessed shoot morphologies that were not unlike those of modern representatives. Leaves were generally closely spaced along the stem because of the short internodes. Many modern conifers (*Pinus, Larix, Cedrus, Pseudolarix*) possess dimorphic shoots (Fig. 1.3), although this is not a feature of araucariaceous or podocarpaceous genera. The major axes, called long shoots, have leaves that are separated from one another by internodal elongation, although the internodes are quite short, usually only a few millimeters long. The axillary shoots, called short shoots or spur shoots, typically are determinate in growth and produce a characteristic number of leaves that serve as the major photosynthetic organs of the plant. The extent of major axis growth in an annual growth cycle in *Pinus taeda*, for example, depends on several phenotypic, environmental, and genetic factors. These cause differences in the number of phytomers produced and the extent of internodal elongation (Bridgwater 1990; Parisi 2006).

Mesozoic ginkgophytes possessed a more pronounced differentiation between the main axes and the axillary branches that they produced. The main axes were long shoots with extended internodes, whereas axillary branches were short shoots in which the internodes were not extended, resulting in closely packed leaves (Fig. 1.4). The physiological basis of shoot dimorphism in the single extant member of the Ginkgoales was

Organization of the Vascular Plant Shoot

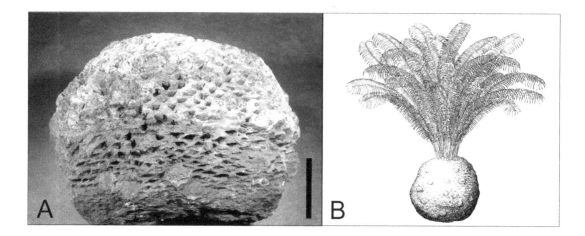

Fig. 1.2. *Cycadeoidea dacotensis* (MacBride) Ward from the Early Cretaceous of North America. (A) Fossilized stem showing the closely spaced leaf bases and absence of internodal extension. Scale bar = 10 cm. *Original photo courtesy of Ted Delevoryas.* (B) Reconstruction of the whole plant showing its squat stem with crowded, persistent leaf bases and a tuft of pinnate fronds at its crown. *Reconstruction by Ted Delevoryas.*

investigated by Gunkel and Thimann (1949), who found that buds of both long and short shoots of *Ginkgo biloba* produce diffusible auxin as they turn green and begin to expand early in the growth cycle. Auxin production decreases as the bud opens. In the short shoots, auxin content remains low, but as long shoots begin to elongate, there is a second rise in auxin yield in the stem to a level considerably above the previous peak that is correlated with elongation of the internodes. Interconversion between long shoots and short shoots of *Ginkgo* occurs spontaneously when a terminal bud is converted from the long shoot habit to that of a short shoot. Experimental conversion of short shoots to long shoots can be induced by decapitation of a terminal long shoot bud, but is inhibited by exogenous application of the plant hormone indole acetic acid to the cut surface (Gunkel et al. 1949). A similar auxin control of shoot dimorphism was also demonstrated in *Cercidiphyllum japonicum*, an angiosperm with distinctive long and short shoots (Titman and Wetmore 1955).

The involvement of several hormones in axillary bud development has been demonstrated in *Lupinus angustifolius*, where the relative concentrations of cytokinin, auxin, and abscisic acid controlled axillary bud outgrowth. The balance between these hormones changed during development, resulting in minimal expansion of axillary shoots in some regions of the plant, thus mimicking the long shoot/short shoot habit (Emery et al. 1998). The existence of mutants that affect hormonal status and branching in *Arabidopsis* plants permits experimental examination of these events. In wild-type *Arabidopsis*, axillary shoots are inhibited in outgrowth and remain suppressed in the rosette phase of development. Mutant *axr1* plants are highly branched because axillary buds are not suppressed and grow out in the rosette phase. The axillary buds of *axr1* plants are resistant to topically applied auxin, which suppresses the outgrowth of buds in wild-type plants (McSteen and Leyser 2005). Bennett et al. (2006) have suggested an alternative manner by which axillary shoot branching may be regulated in *Arabidopsis*. On the basis of mutant analysis, they have concluded that gene products moving upward in the plant from the roots suppresses the activity of auxin efflux proteins in the stem, thus decreasing polar auxin transport and allowing increased polar transport of axillary bud auxin and concomitant bud growth.

Shoot architecture provides a means of comparing and contrasting developmental programs across the vascular plant assemblage. As we have shown in this chapter, one of the most striking differences in architecture is the degree of internodal extension, from none in the bennettitaleans, to slight in the present-day conifers and presumably in their Mesozoic ancestors, to longer internodes in *Ginkgo* and its fossil relatives and in many present-day angiosperms. Internodal elongation in present-day plants is regulated by polar auxin flow. Evidence for polar auxin flow in fossil plants as old as 375 million years was inferred by Rothwell and Lev-Yadun (2005), who observed in the wood of the Late Devonian progymnosperm *Archaeopteris* the differentiation of circular tracheary elements, which in modern species is symptomatic of polar auxin flow blockage. This opens up the possibility that polar auxin flow may have been an underlying cause of internodal elongation in fossil plants too. However, it is noteworthy that plants lacking vegetative internodal extension exist today in the cycads (presumably a primitive feature) and palms. Internodal extension is a hormone-regulated developmental process, so that evolutionary changes in the level of critical hormones or the genes involved in their synthesis or expression may underlie the changes in internodal extension.

Because some plants can alter their growth habit from one architectural model to another, it is tempting to think that such changes may be under the control of regulatory genes and that the number of such genes might be small. An example of this is the phase change from juvenile to adult in *Arabidopsis*. Several genes and mutants have been identified that either accelerate or retard this transition. Willmann and Poethig (2005) describe several genes and small RNAs that help us understand the vegetative phase changes and that suggest a link between vegetative phase change and RNA silencing. By means of genetic approaches to understand developmental pathways that control heteroblasty in angiosperms, Hunter et al. (2006) found genes that caused premature expression of genes that were associated with adult vegetative traits. The mutant phenotypes were attributed to "the aberrant expression of genes normally repressed by these small RNAs." Furthermore, they showed that the precocious phenotypes that influenced the pace of heteroblasty could be attributed to "sensitivity of leaf primordia to a temporal signal rather than by serving as a component of a developmental clock" (Hunter et al. 2006: 2973). Gene regulation by microRNAs is becoming a topic for increasing attention. Floyd and Bowman (2004: 485) have shown that 2 microRNAs have been "conserved in homologous sequences of all lineages of land plants, including bryophytes, lycopods, ferns and seed plants," indicating that these microRNAs may mediate gene regulation in nonflowering and flowering plants over periods dating back more than 400 million years.

Discussion

Fig. 1.3. Shoot tip of *Pinus* sp. showing parts of two growth cycles. The second year axis (below) consists of a long shoot (although the internodes are quite short) bearing nonphotosynthetic scale leaves. In the axil of each scale leaf is an axillary short shoot bearing the photosynthetic needle leaves. The enlarging first-year axis is terminal on the stem. Each scale leaf has a short shoot in its axil with elongating needle leaves.

Fig. 1.4. A shoot of *Ginkgo biloba* showing a long shoot with extended internodes and axillary short shoots with nonextended internodes sheathed by leaf bases. Leaves are borne on the long shoot only in its first year of growth. Leaves are borne on short shoots each year that they are viable. The pendulous, berry-like fructifications are borne on stalks on the short shoots as well.

A more comprehensive understanding of the role of regulatory genes in evolutionary changes in morphology is that of the MADS-box genes (Theissen et al. 2000). Genes in this small family act as homeotic selector genes that determine floral organ identity and floral meristem identity in flowering plants. These genes are also present and have different regulatory roles in ferns and gymnosperms. It seems likely, therefore, that changes in regulatory genes, albeit by structure, function, expression, or interaction with iRNA or by other mechanisms, have been instrumental in morphological and architectural changes from the Mesozoic to the present. Identification and study of comparable gene families that regulate development of the vegetative shoot system in present-day plants of different affinities may allow us to understand more fully the evolutionary differences in shoot morphology between Mesozoic gymnosperms and trees of today.

Conclusions

In this chapter, we have focused on the architectural innovations that distinguish between Mesozoic and modern plants, paying particular attention to axis development. An alternative approach to these questions is that developed by Sanders et al. (2007) who have examined the evolutionary developmental biology of leaf development. Together, these two studies provide a broader perspective on the evolutionary and developmental differences between Mesozoic and modern plants.

Acknowledgments

We thank Ted Delevoryas for his many years of guiding us through the thickets of Mesozoic plants, Gary Peter for discussions on the dynamics of pine shoot growth, Vivian Irish for discussions about evolution and development, and David Winship Taylor and an anonymous reviewer for paleobotanical insight. This is contribution no. 59 of the DFG Research Unit 533 on the Biology of the Sauropods: The Evolution of Gigantism.

References Cited

Atger, C., and C. Edelin. 1994. Premières données sur l'architecture comparée des systèmes racinaires et caulinaires. *Canadian Journal of Botany* 72: 963–975.

Barthélémy, D., and Y. Caraglio. 2007. Plant architecture: a dynamic, multilevel and comprehensive approach to plant form, structure and ontogeny. *Annals of Botany* 99: 375–407.

Bennett, T., T. Sieberer, B. Willett, J. Booker, C. Luschnig, and O. Leyser. 2006. The *Arabidopsis* MAX pathway controls shoot branching by regulating auxin transport. *Current Biology* 16: 553–563.

Bridgwater, F. E. 1990. Shoot elongation patterns of loblolly pine families selected for contrasting growth potential. *Forest Science* 36: 641–656.

Delevoryas, T. 1971. Cycadeoidales. In *McGraw-Hill Encyclopedia of Science and Technology*, 3rd ed., p. 689. New York: McGraw-Hill Book Co.

Emery, R. J. N., N. E. Longnecker, and C. A. Atkins. 1998. Branch development in *Lupinus angustifolius* L. II. Relationship with endogenous ABA, IAA and cytokinins in axillary and main stem buds. *Journal of Experimental Botany* 49: 555–562.

Floyd, S. K., and J. L. Bowman. 2004. Ancient microRNA target sequences in plants. *Nature* 428: 485–486.

Gunkel, J. E., and K. V. Thimann. 1949. Studies of development of long shoots and short shoots of *Ginkgo biloba* L. III. Auxin production in shoot growth. *American Journal of Botany* 36: 145–151.

Gunkel, J. E., K. V. Thimann, and R. H. Wetmore. 1949. Studies of development of long shoots and short shoots of *Ginkgo biloba* L. IV. Growth habit, shoot expression and the mechanism of its control. *American Journal of Botany* 36: 309–318.

Hallé, F. 1999. Ecology of reiteration in tropical trees. In M. H. Kurmann and A. R. Hemsley (eds.), *The Evolution of Plant Architecture*, pp. 147–158. London: Royal Botanic Gardens, Kew.

Hallé, F., R. A. A. Oldeman, and P. B. Tomlinson. 1978. *Tropical Trees and Forests: An Architectural Analysis*. Berlin: Springer-Verlag.

Hunter, C., M. R. Willmann, G. Wu, M. Yoshikawa, L. Gutierrez-Nava, and S. R. Poethig. 2006. Trans-acting siRNA-mediated repression of ETTIN and ARF4 regulates heteroblasty in *Arabidopsis*. *Development* 133: 2973–2981.

McSteen, P., and O. Leyser. 2005. Shoot branching. *Annual Review of Plant Biology* 56: 353–374.

Osmont, K. S., R. Sibout, and C. S. Hardtke. 2007. Hidden branches: developments in root system architecture. *Annual Review of Plant Biology* 58: 93–113.

Parisi, L. M. 2006. Shoot elongation patterns and genetic control of second-year height growth in *Pinus taeda* L. using clonally replicated trials. M.S. diss., School of Forest and Conservation, University of Florida.

Rothwell, G. W., and S. Lev-Yadun. 2005. Evidence of polar auxin flow in 375 million-year-old fossil wood. *American Journal of Botany* 92: 903–906.

Sanders, H., G. W. Rothwell, and S. Wyatt. 2007. Paleontological context for the developmental mechanisms of evolution. *International Journal of Plant Science* 168: 719–728.

Stewart, W. N., and G. W. Rothwell. 1993. *Paleobotany and the Evolution of Plants*, 2nd ed. New York: Cambridge University Press.

Sussex, I. M., and N. M. Kerk. 2001. The evolution of plant architecture. *Current Opinion in Plant Biology* 4: 33–37.

Taylor, T. N., E. L. Taylor, and M. Krings. 2009. *Paleobotany: The Biology and Evolution of Fossil Plants*, 2nd ed. San Diego: Academic Press.

Theissen, G., A. Becker, A. Di Rosa, A. Kanno, J. T. Kim, T. Munster, K.-U. Winter, and H. Saedler. 2000. A short history of MADS-box genes in plants. *Plant Molecular Biology* 42: 115–149.

Titman, P. W., and R. H. Wetmore. 1955. The growth of long and short shoots in *Cercidiphyllum*. *American Journal of Botany* 42: 364–372.

Willmann, M. R., and R. S. Poethig. 2005. Time to grow up: the temporal role of small RNAs in plants. *Current Opinion in Plant Biology* 8: 548–552.

Fig. 2.1. Map showing the location of Fremouw Peak (arrow) within the Queen Alexandra Range of the Transantarctic Mountains. Inset shows the relative position of the collecting site in Antarctica.

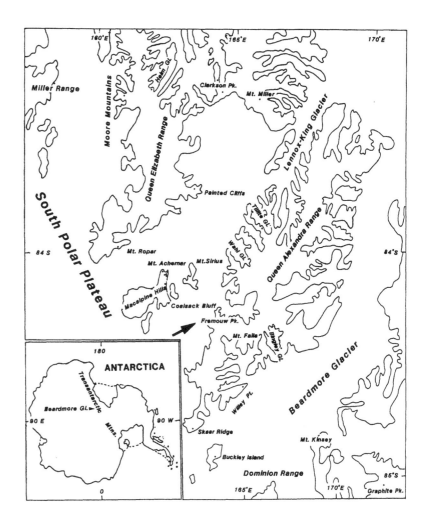

MODERN TRAITS IN EARLY MESOZOIC SPHENOPHYTES: THE *EQUISETUM*-LIKE CONES OF *SPACIINODUM COLLINSONII* WITH IN SITU SPORES AND ELATERS FROM THE MIDDLE TRIASSIC OF ANTARCTICA

2

Andrew B. Schwendemann, Thomas N. Taylor,
Edith L. Taylor, Michael Krings, and Jeffrey M. Osborn

Structurally preserved cones of the early Middle Triassic sphenophyte *Spaciinodum collinsonii* have been discovered within permineralized peat from Fremouw Peak, Antarctica. Cones consist of whorls of peltate sporangiophores bearing approximately 10 sporangia each. Spores have a perispore and four elaters with spatulate ends, making *Spaciinodum* the earliest known Triassic sphenophyte with elater-bearing spores. These equisetalean cones occur alongside the vegetative stems, leaves, and dormant buds of *Spaciinodum collinsonii*. This close association and the absence of other sphenophytes at the locality indicate that the various plant organs comprise a single species. On the basis of information that has recently become available for *Spaciinodum collinsonii*, this taxon can today be regarded as one of the best-understood Mesozoic sphenophytes. The morphology and anatomy of *S. collinsonii* correspond to those of the modern *Equisetum* subgenus *Equisetum*, which suggests that the origin of the extant subgenus *Equisetum* dates as far back as 240 million years ago. Fossil remains previously described as reproductive axes of *Spaciinodum* are reinterpreted as vegetative axes with a fungal infection. The species diagnosis is therefore emended on the basis of the discovery of new reproductive axes that differ significantly from those previously defined.

Introduction

The Sphenophyta are a phylum of pteridophytes with a rich fossil history dating back to the Devonian, making it one of the most ancient lineages of plants. The group reached its peak diversity during the Late Carboniferous (Pennsylvanian), a time when most sphenophytes were arborescent. Only herbaceous species, however, survived into the Mesozoic (or perhaps not, see Schweitzer et al. 1997) and persist to the present. Much of our understanding concerning the evolutionary history of this group is due to permineralized specimens found in Carboniferous coal balls being correlated with the compression and impression fossils. The ability to characterize the Carboniferous sphenophytes on the basis of their morphology and anatomy has led to increased understanding of their taxonomy and ecology (e.g., DiMichele and Phillips 1994; DiMichele et al. 2005). In contrast, although several Mesozoic sphenophytes have been described, relatively little is known about their phylogenetic relationships, how they evolved from Paleozoic forms, and how they gave rise to the extant representatives of the group (Taylor et al. 2009). This gap in our understanding is largely due to the lack of permineralized sphenophytes from the Mesozoic. To date, only a single anatomically preserved sphenophyte has been described from this era based exclusively on permineralized specimens, *Spaciinodum collinsonii*.

Spaciinodum collinsonii is a permineralized sphenophyte from the lower Middle Triassic of the Fremouw Formation in Antarctica (Osborn and Taylor 1989; Osborn et al. 2000). The species was originally described from aerial stems and rhizomes, and the taxon is characterized by jointed and ribbed stems with diagnostic pith canals, carinal canals, and vallecular canals that are restricted to nodes (Osborn and Taylor 1989). Osborn et al. (2000) later described reproductive remains in organic association with the vegetative stems. The reproductive axes were described by Osborn et al. (2000) as having a vascular system consisting of 31 to 33 collateral vascular bundles that are continuous through successive nodes and internodes. This vascular condition is found in extant *Equisetum* cones (Browne 1912, 1915, 1920, 1933, 1941; Barratt 1920; Page 1972), but not in the vegetative axes (Golub and Wetmore 1948a, 1948b; Bierhorst 1959; Page 1972). Sporangia of *S. collinsonii* were described as occurring in a single whorl attached to the axis in association with cortical chambers, not as occurring on peltate sporangiophores, as in extant *Equisetum* and the majority of Mesozoic sphenophytes. Cell layers of the sporangial wall were unidentifiable as a result of preservation; however, the remains of a tapetal membranelike layer were suggested. Sporangia were reported as containing abundant spores averaging 10 μm in diameter. Spores were described as spheroidal, with rugulate surface ornamentation and a sporoderm averaging 1.0 μm in thickness.

New evidence on the anatomy of dormant buds and vegetative axes of *Spaciinodum* (Ryberg et al. 2008), as well as additional recently discovered reproductive specimens have cast doubt on the interpretation of Osborn et al. (2000). The objective of our study is to describe recently discovered reproductive axes and spores from the Fremouw Peak locality that are comparable to those of extant *Equisetum* and that are associated with the vegetative axes of *Spaciinodum*. Additionally, the taxonomic status of the previously described cone of *S. collinsonii* is discussed.

Stratigraphy and Specimen Preparation

Cones of *Spaciinodum* are preserved in permineralized peat collected from the Fremouw Peak locality in the Queen Alexandra Range of the Transantarctic Mountains (84°17'41"S, 164°21'48"E; Fig. 2.1; Barrett and Elliot 1973). The peat is dated as early Middle Triassic on the basis of the palynomorph assemblage and vertebrate fossils (Farabee et al. 1990; Hammer et al. 1990). Peat blocks were sectioned and the polished surface etched with 49% hydrofluoric acid (HF) for 1–5 minutes. Cellulose acetate peels (Galtier and Phillips 1999) were made from the prepared surface; some peels were subsequently mounted on slides with Eukitt mounting medium (O. Kindler GmbH, Freiburg, Germany). Slides are housed in the Paleobotany Division of the Natural History Museum and Biodiversity Research Center, University of Kansas, Lawrence, under accession numbers 23015–23138, 26339, and 26343–26364. Peels and slides of the 13 cones were made from blocks 11277 A, 11277 B$_{top}$, 11277 B2$_{side\ top}$, and 11277 B3$_{side\ bot}$, 10017 C$_{bot}$, and 10160 D1S2.

Microscopy

For light microscopy, all specimens were photographed with a Leica DC500 digital camera attachment on a Leica DM 5000B compound microscope and a Leica MZ 16 dissecting microscope. Digital images were processed by Adobe Photoshop CS, version 8.0 (1999–2003, Adobe Systems Incorporated). High magnification (>640×) images were taken under oil immersion.

For scanning electron microscopy (SEM), two methods of spore isolation were used. In one method, spores were macerated directly from the slab with 49% HF that had been pipetted within an elevated wax well surrounding *Spaciinodum* sporangia containing spores. The fresh HF was allowed to react with the slab for 2 minutes, after which it was pipetted into a container and diluted with distilled water. The spores were allowed to settle and the supernatant was then pipetted out. After this, fresh distilled water was added to the test tube. These steps were repeated until the solution attained a neutral pH. Subsequently, the mixture was pipetted onto stubs coated with conductive putty. Once the water had evaporated, stubs were sputter-coated with gold and imaged with a Leo 1550 scanning electron microscope at 5 kV. In the second method, spores were recovered from acetate peels by excising a portion of the peel containing sporangia and dissolving the peel with several changes of acetone. Spores were pipetted onto stubs coated with conductive putty, which were then sputter-coated and imaged as described above. SEM stubs are housed in the Paleobotany Division of the Natural History Museum and Biodiversity Research Center, University of Kansas, Lawrence, under accession numbers AS(1–9) 08.

Phylum Sphenophyta
Order Equisetales Dumortier
Family Equisetaceae Michaux ex DeCandolle
Genus *Spaciinodum* Osborn et T. N. Taylor

SPACIINODUM COLLINSONII OSBORN ET T. N. TAYLOR EMEND.
SCHWENDEMANN ET AL.

FIGS. 2.2 AND 2.3; PLATES 2 AND 3

Emended diagnosis: Stem with ribbed and furrowed surface and jointed organization representing distinct nodal and internodal regions, in longitudinal section internodal regions completely condensed to partially elongated and becoming progressively taller basally. Nodes characterized by a solid pith, absence of vallecular (cortical) and carinal (protoxylem) canals, and large vascular bundles that form a nearly continuous ring. Internodal regions characterized by vallecular canals separated by uniseriate cellular partitions (fimbrils) occurring in a continuous series for the circumference of the cortex. Position of vascular bundles offset in successive internodes, bundles collateral, xylem endarch. Internodal anatomy differing with degree of maturation of the internode; relatively immature internodes with vallecular canals filled with cells (although the future canal boundaries and uniseriate fimbrils are clearly visible), carinal canals absent; more mature (elongate) internodes characterized by open vallecular and carinal canals; most mature internodes with large pith canal; sclerenchyma and distinct stem ribs rarely observed, possibly only developing at the periphery of the cortex in very mature stems or rarely preserved. Leaves occurring in whorls at nodes and enclosing and overarching the apex of the young stem, fused proximally and free and tapering distally, each vascularized by a single bundle with scalariform xylem thickenings, free distal portions rectangular in transverse section, mesophyll unifacial. Branches arising within the cortex of the stem, alternating with the leaves; branches organized internally much like larger stems, vascular bundles collateral, inner cortex with developing vallecular canals, outer cortex cellular; branch-borne leaves given off at the nodes and overarching the apices of the branch primordia. Stomata superficial. Buds consist of a telescoped stem with jointed organization, stem apex overarched by leaves, bud attached to additional tissue basally that may represent a root or rhizome. **Fertile structures** composed of whorls of sporangiophores attached to a central axis and aggregated into a cone, either terminal on the main aerial stem or a lateral branch. Vascular system of 9–12 bundles that occasionally fuse with adjacent bundles; bundles bounded by the endodermis; six sporangiophores in each whorl with approximately 10 sporangia per sporangiophore; sporangiophores alternate at each successive whorl, and sporangia prolate with a sporangial wall 1 cell layer thick and with conspicuous spiral thickenings. **Spores** spheroidal, trilete, averaging 48 μm in diameter, spore wall of a perispore layer attached to spore body at a single point with four elaters terminating in spatulate ends.

Several incomplete cones of *Spaciinodum collinsonii* have been found dispersed throughout the silicified peat in various states of decay. All of the 13 specimens identified appear to have undergone microbial degradation, and all cones are typically closely associated with a diverse assemblage of apparently saprotrophic fungi and funguslike organisms. Most abundant in the vicinity of these reproductive structures is *Combresomyces cornifer* (Peronosporomycetes), which is thought to be a saprotroph (Dotzler et al. 2008; Schwendemann et al. 2009). This high degree of fungal degradation could account for the lack of cellular preservation within most of the tissues of *Spaciinodum* cones. However, the overall morphology of the cone axis, as well as the cone axis vasculature, sporangiophores, sporangia, and spores are adequately preserved, and collectively these characters afford the opportunity to provide a detailed description of the structure.

Cone Morphology

The cones consists of whorls of peltate sporangiophores (Fig. 2.2A; Plate 2A), which each originate from the central axis at approximately 90 degrees. The central axis is approximately 1.0 mm in diameter, and the entire cone is approximately 3.0 mm in diameter. A hollow pith (medullary canal) continuing through the nodes and internodes characterizes the axis. In some sections, remains of tissue can be seen associated with the pith (Fig. 2.2B; Plate 2B). The number of sections with this tissue is small, and it is unclear whether the tissue represents the remains of a cellular pith, a nodal diaphragm, or an unrelated structure. Nine to 12 collateral vascular bundles appear throughout the nodes and internodes. More bundles are likely present because the vasculature rarely seems to be preserved for the entirety of the vascular cylinder when viewed in transverse section. Bundles are typically positioned at the base of a sporangiophore located in the same level or at the base of a sporangiophore located at the levels above and below (Fig. 2.2C; Plate 2C). Unlike the stem of *Spaciinodum*, vascular bundles in the cone do not form a complete ring at nodes, but neighboring bundles occasionally fuse, and bundles have also been observed diverging. The periphery of each bundle is tightly appressed to the cortex, with both sides of each bundle bounded by a thin layer that circumscribes the pith cavity (Fig. 2.2C; Plate 2C); this is likely the remains of the endodermis. Tracheids have annular secondary thickenings (Fig. 2.2D; Plate 2D), although a few annular–helical and helical tracheids have also been found. A lack of good preservation obscures the divergence pattern of xylem from the cone axis to the sporangiophores and the branching of xylem to the sporangia within the sporangiophores. In the cones examined, there are typically six sporangiophores per whorl. Most sections, however, show only four or fewer intact sporangiophores, with bases indicating the positions of degraded sporangiophores (Fig. 2.2B; Plate 2B). Sporangiophores alternate in position from node to node. The length of the sporangiophore from

the point where the sporangiophore stalk connects the cone axis to the outermost portion of the sporangiophore head ranges from 0.8 to 1.2 mm in a median longitudinal section. The diameter of the sporangiophore heads ranges from 1.0 to 1.3 mm. Sporangiophore heads are tightly appressed to neighboring heads, making the boundaries of the individual sporangiophores difficult to define in poorly preserved specimens. In some cases, heads from adjacent sporangiophores appear to be fused, creating a structure that encompasses most of the circumference of the cone axis (Fig. 2.2E; Plate 2E). On the basis of the alternation of sporangiophore position from node to node and the possible geometry of this arrangement, the sporangiophore heads are interpreted as hexagonal or rhomboidal in shape.

On the abaxial surface of each sporangiophore head is a single whorl of approximately 10 sporangia. Sporangia are 0.59–0.86 mm long and 0.28–0.40 mm wide. Fragments of sporangia, indicated by the remains of sporangial walls, are found more frequently than intact sporangia attached to the sporangiophore head (Fig. 2.2F; Plate 2F). Sporangia are tightly appressed to the cone axis, pressing against concave indentations on the outer surface of the cone axis. This close association often suggests that sporangial walls are attached to the cone axis. Closer inspection reveals this to not be the case (Fig. 2.2F; Plate 2F). The sporangial wall consists of a single layer of longitudinally elongated cells with spiral thickenings. The thickness of this layer ranges from 26.3 to 58.1 µm, with a mean of 38.2 µm (n = 23). No cell layers are found interior to the sporangial wall (Fig. 2.2G; Plate 2G).

Spore Morphology

Within each intact sporangium are numerous spherical, slightly immature spores surrounded by the remnants of the tapetal plasmodium (Fig. 2.2H, Plate 2H). The spore wall is made up of numerous layers, including the spore body proper, a perispore (Fig. 2.3A; Plate 3A) (=middle layer of Uehara and Kurita 1989), and elaters on some spores. The spore body ranges from 19.7 to 47.9 µm in diameter with a mean of 29.3 µm (n = 38). The perispore is attached to the spore body proper at a single point, possibly near the aperture, as in spores of extant *Equisetum*. In most peels, there appears to be a significant gap between the spore body and the perispore (Fig. 2.3A; Plate 3A). This may be the actual condition of *Spaciinodum* spores, or alternatively a result of spore body shrinkage during diagenesis associated with fossilization. When measurements of spore diameter include the perispore, the spores range from 29.3 to 64.5 µm with a mean of 48.1 µm (n = 38).

There are several structures observed with light microscopy and SEM that represent elaters. Figure 2.3B and Plate 3B show a coiled structure found inside a sporangium. This structure, which is not associated with any spore, consists of coiled bands that measure about 1.5 µm in

Fig. 2.2. Cone of *Spaciinodum collinsonii* from the lower Middle Triassic of Antarctica. (A) Transverse section of cone showing a single peltate sporangiophore with two empty sporangia (arrows); slide no. 23017. Scale bar = 250 µm. (B) Transverse section of cone axis with several intact sporangiophores (arrows), sporangiophore stumps (arrowheads), and tissue (T) in the pith canal; slide no. 23019. Scale bar = 1 mm. (C) Higher magnification of (B) showing vascular bundles and encompassing endodermis at the base of sporangiophores; slide no. 23019. Scale bar = 200 µm. (D) Longitudinal section of cone tracheids showing annular and annular–helical secondary thickenings; slide no. 23020. Scale bar = 50 µm. (E) Transverse section of cone axis with fused sporangiophore heads and unfused sporangiophore stalks (arrows); slide no. 23018. Scale bar = 1 mm. (F) Transverse section of cone axis showing sporangia appressed against concave indentations in the cone axis; slide no. 23068. Scale bar = 200 µm. (G) Longitudinal section through sporangial wall showing the thickening pattern of the wall buttresses; slide no. 23017. Scale bar = 30 µm. (H) Longitudinal section through sporangia displaying spores surrounded by the remains of the tapetal plasmodium; slide no. 23069. Scale bar = 150 µm.

thickness; however, at the edge of the structure (Fig. 2.3B; Plate 3B), the coil appears to flatten. This flattened area measures 4.4 µm at the widest point and may represent the spatulate end of an elater. Another coiled structure was found with bands 9.1 µm thick and associated with a spore (Fig. 2.3C; Plate 3C). In the region of the arrow in Figure 2.3C and Plate 3C, the bands appear to form a spatulate end. Information based on transmitted light microscopy can be correlated with data obtained through SEM of spores macerated directly from *Spaciinodum* sporangia. Figure 2.3D and Plate 3D show a highly fractured spore with apparent elaters surrounding the spore body. The thinner strands measure 8.3 µm thick, while the thicker portions that are more spatulate are approximately 12.5 µm thick. Figure 2.3E and Plate 3E show at least three possible elaters originating from a common point on a fractured spore body. Thinner strands are about 3.1 µm thick with the spatulate end approximately 12.5 µm thick. One additional specimen bearing elaters was observed (Fig. 2.3F; Plate 3F). In this specimen, the elaters are thinner and apparently coiled. The strands measure approximately 2.9 µm in thickness and are terminated by two larger spatulate ends measuring approximately 10.3 µm in thickness. Only a single spore with an unequivocal trilete suture was observed (Fig. 2.3G; Plate 3G); however, many spores have apparent trilete sutures that could also be interpreted as folds of the spore wall. It is highly probable that the haptotypic mark is obscured by the perispore in most specimens.

Discussion

Although no organic connection between the vegetative axes of *Spaciinodum collinsonii* and the cones described here have been observed, it appears that this cone type was borne on vegetative axes of *Spaciinodum collinsonii*. To date, only a single species of sphenophyte has been discovered in the permineralized peat at Fremouw Peak (Osborn and Taylor 1989; Ryberg et al. 2008), and numerous specimens of these vegetative shoots occur throughout all of the blocks where the cones have been found. Additionally, the diameter of the cones described here (3.0 mm) is consistent with the described diameter of vegetative *Spaciinodum* axes (1.8–3.0 mm; Osborn and Taylor 1989). This is especially significant given the unusually small size of *Spaciinodum* vegetative axes (Osborn and Taylor 1989). Moreover, the vegetative and reproductive axes share several anatomical features in common, such as a similar number of vascular bundles, vascular bundles surrounded by a double endodermis, and the presence of annular tracheids that grade into helical tracheids (Osborn and Taylor 1989; Ryberg et al. 2008).

On the basis of information available, it appears that the overall organization and structure of the cone of *Spaciinodum* is nearly indistinguishable from that of extant *Equisetum*. This is not entirely unexpected because the vegetative remains of *Spaciinodum* share many features with extant *Equisetum*, particularly species within the subgenus *Equisetum* (Osborn and Taylor 1989; Ryberg et al. 2008). The apparent alliance of

Spaciinodum with subgenus *Equisetum* is of particular interest because subgenus *Hippochaete* was originally thought to have originated in Gondwana, whereas subgenus *Equisetum* was thought to have evolved in Laurasia. These biogeographic hypotheses are based on distributions of extant taxa and the assumption that large stem size and bisexual gametophytes were ancestral in *Equisetum* (Schaffner 1930; Hauke 1963, 1978). Recent molecular phylogenies, however, are at odds with these assumptions, and the molecular studies report that these character states are derived in extant *Equisetum* (Des Marais et al. 2003; Guillon 2004, 2007). The presence of *Spaciinodum*, a relatively small sphenophyte, in Gondwana during the Triassic adds some fossil support to these recent phylogenies.

Features of the cone are relatively conserved throughout the extant species, whereas morphology and anatomy of the stem are variable and are therefore the primary characters traditionally used to differentiate extant species (Hauke 1963, 1978). Reproductive characters that may have taxonomic significance—such as position of the cone or cones on the plant, whether or not the axis of the cone is extended above the uppermost whorl of sporangiophores, and whether stems of the species are dimorphic or monomorphic—are inapplicable in this study because the *Spaciinodum* cone specimens are incomplete. The incompleteness of the specimens is not surprising because of the rapid decomposition seen in extant *Equisetum* (Marsh et al. 2000) and the abundance of decomposers in the same peat blocks in which *Spaciinodum* occurs. Marsh et al. (2000) have also demonstrated that relative to its biomass, extant *Equisetum* absorbs a disproportionate amount of phosphorus, potassium, and calcium by extending its rooting system into the C soil horizon. If these characteristics were shared by *Spaciinodum*, a significant amount of otherwise trapped nutrients could have been rapidly recycled within the plant community and would have allowed *Spaciinodum* to play a significant role in the structuring of ancient plant communities.

Cone Morphology in Comparison to Other Mesozoic Sphenophytes

To date, *Spaciinodum collinsonii* represents the only Mesozoic sphenophyte known only from permineralized remains. Comparisons with Mesozoic compression specimens are difficult because of the few characters that are comparable between the two types of preservation. However, numerous compressed sphenophyte cones have been described from the Mesozoic, and a comparison between them could help to correlate the morphological features of the compressions with the anatomy of *Spaciinodum*.

Equisetites fertilis from the Upper Triassic of Argentina (Frenguelli 1944) is not comparable; although it is considered to have reproductive structures, they are not aggregated into cones as in *Spaciinodum*. Several additional cones can also be dismissed on the basis of the size of the specimens. The *Spaciinodum* cone is fairly small, measuring approximately

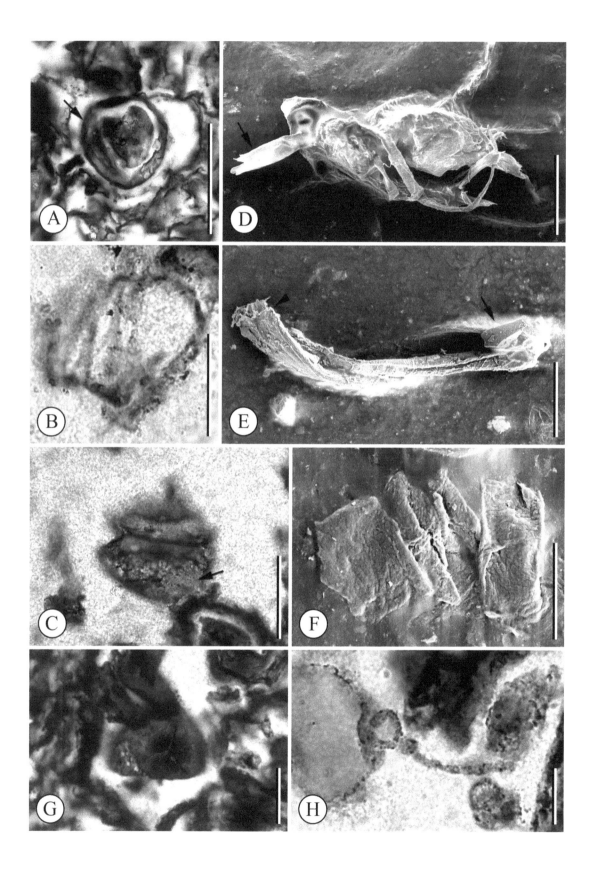

3.0 mm in diameter. Cones such as *Equisetites arenaceus*, *Equisetites mougeotii*, *Equisetites quindecimdentata*, *Neocalamostachys takahashii*, *Equisetostachys suecicus*, *Equisetum muensteri*, and *Equisetum columnare* all have cones that measure at least 20 mm in diameter, and their cones bear sporangiophores of a size considerably larger than those of *Spaciinodum* (Halle 1908; Menéndez 1958; Kon'no 1962, 1972; Boureau 1964; Barnard 1967; Harris 1978; Kelber and van Konijnenburg-van Cittert 1998; Kustatscher et al. 2007). The cone axes of *Equisetites nagatensis* and *Equisetites naitoi* are less substantial, but still measure over 10 mm (Kon'no 1962).

Although some *Spaciinodum* specimens contain remains of the tapetal plasmodium (Fig. 2.2H; Plate 2H), which indicates that the cone was not fully mature, the sporangial wall in all specimens is only a single layer thick, indicating that the cone was close to maturity at the time of fossilization (Bierhorst 1971). It is possible, however, that the cone fragments described here may have comprised the uppermost or lowermost portions of the cone and therefore represent the narrowest parts of the cone. Although possible, it appears unlikely that each of the 13 cones discovered to date represents only these portions of *Spaciinodum* cones.

Several Mesozoic sphenophyte compressions bearing cones have been found with cone sizes that are more comparable to *Spaciinodum* (≤10 mm). Of these, *Equisetum boureaui* (Upper Triassic of Cambodia) differs from *Spaciinodum* in that it has only four sporangia per sporangiophore (Vozenin-Serra and Laroche 1976). In the Upper Triassic *Equisetostachys verticillata*, the length of the sporangia (up to 1.5 mm) and the sporangiophore stalk (up to 3 mm) extend beyond the size range of *Spaciinodum* (Grauvogel-Stamm 1978). Unlike *Spaciinodum*, *Equisetostachys nathorstii* has at least 16 sporangiophores per whorl (Halle 1908), and *Equisetites woodsii* has been described as bearing 50 to 60 sporangia per sporangiophore (Jones and de Jersey 1947). Disarticulated sporangiophores of *Equisetites lyellii* have been described by Watson (1983) and Watson and Batten (1990); these specimens differ most obviously from *Spaciinodum* in having rounded sporangia, as opposed to the elongated shape of a prolate spheroid. Additionally, *E. lyellii* has circular sporangiophore heads with 24 surface ribs radiating from its surface (Watson and Batten 1990). No surface ribs were detected in *Spaciinodum*, and the heads of the sporangiophores are thought to be hexagonal. *Spaciinodum* differs from *Equisetites pusillus* from the Aptian of Patagonia in having sporangiophores that are helically arranged on the cone axis (Villar de Seoane 2005). *Neocalamostachys pedunculatus* has fertile whorls composed of 20 sporangiophores, with sporangiophore stalks twice as long as those found in *Spaciinodum* (Kon'no 1962, 1972; Boureau 1964). Similarly, the sporangiophore stalks of *Equisetites asaensis* are nearly four times as long as those of the cone described here (Kon'no 1962). *Equisetites bracteosus* is an interesting cone from the Upper Triassic of Japan with several whorls of sporangiophores that are occasionally interrupted by a single whorl of leaves (Kon'no 1962), but no leaf whorls are

Fig. 2.3. Spores of *Spaciinodum collinsonii* from the lower Middle Triassic of Antarctica. (A) Spore with perispore (arrow) surrounding the spore body proper; slide no. 23021. Scale bar = 50 μm. (B) Helically coiled elaters detached from spore and found within *Spaciinodum* cone; slide no. 23111. Scale bar = 20 μm. (C) Spore showing coiled elaters tightly appressed to the spore. Arrow marks the position of spatulate end; slide no. 23070. Scale bar = 30 μm. (D) Spore surrounded by elaters terminating in spatulate ends (arrow); SEM no. AS1 08. Scale bar = 60 μm. (E) Elaters with spatulate ends (arrowhead) attached to a spore at a single point (arrow); SEM no. AS5 08. Scale bar = 25 μm. (F) Coiled elaters detached from spore with spatulate ends at top and bottom of image; SEM no. AS2 08. Scale bar = 15 μm. (G) Spore without perispore showing a prominent trilete suture; slide no. 23021. Scale bar = 20 μm. (H) Spore with germinating hypha from "sporangia" described by Osborn et al. (2000); slide no. 15651. Scale bar = 10 μm.

known to occur in the cone of *Spaciinodum*. *Equicalastrobus chinleana* and *Equisetites aequecaliginosus* are Triassic sphenophytes that have the general structure of an equisetalean cone, but with an additional lanceolate, leaflike structure attached to the center of the sporangiophore head that is not found in *Spaciinodum* (Grauvogel-Stamm and Ash 1999; Weber 2005). *Equisetites laterale* has a comparable cone width to that of *Spaciinodum*, but the preservation of *E. laterale* inhibits any further comparison (Gould 1968).

The inability to correlate structurally preserved remains of *Spaciinodum* with equisetalean compressions says more about the diversity of equisetalean cones during the Mesozoic than it does about the preservation quality of the specimens. Equisetaleans, although prevalent throughout most of the Paleozoic, apparently did not reach a cosmopolitan distribution until the Late Permian and the Triassic (Escapa and Cúneo 2005; Cúneo and Escapa 2006). The array of equisetalean fossils found during the Mesozoic suggests an increase in the diversity of herbaceous equisetalean plants, perhaps the largest diversity of the group since the decline of the arborescent habit within the group. Discovery of additional structurally preserved cones from the Mesozoic would facilitate a more accurate quantification of equisetalean diversity at this important geological time and help elucidate the evolutionary history of one of the most ancient plant lineages.

Spores, Elaters, and Comparisons to Other Elater-Bearing Spores

Spores of *Spaciinodum collinsonii* are comparable to those of modern *Equisetum* as well as to those of the Mesozoic sphenophytes. Spores of extant *Equisetum* generally range from 35 to 65 μm in diameter (Hauke 1963, 1978; Duckett 1970), whereas spore diameter in Mesozoic taxa ranges from 28 to 68 μm (e.g., Halle 1908; Barnard 1967; Gould 1968; Vozenin-Serra and Laroche 1976; Grauvogel-Stamm 1978; Harris 1978; Watson 1983; Watson and Batten 1990; Kelber and van Konijnenburg-van Cittert 1998; Villar de Seoane 2005); the majority of these taxa have spore diameters from 40 to 50 μm. The range of diameters and the mean diameter of *Spaciinodum* spores fits well within the ranges reported for related spores.

The seemingly large range of spore diameters in *Spaciinodum* may reflect natural variation in spore size, but it also may reflect immaturity of some spores. Although a few spores were found with attached elaters (Fig. 2.3C–E; Plate 3C–E), most have shown no evidence of elaters. There are several reasons why elaters may have not been reported in more in situ spores. One is that the spores were still immature and the elaters had not yet been deposited on most spores. Evidence for this exists in the form of a preserved structure interpreted as the remains of the tapetal plasmodium (Fig. 2.2H; Plate 2H). In extant *Equisetum*, the tapetal plasmodium is responsible for the deposition of the spore wall layers, including the elaters, which are deposited after the perispore (Lugardon 1969; Uehara and Kurita 1989; Uehara and Murakami 1995). The continued presence of the tapetal plasmodium in the fossil after the deposition of the perispore (Fig. 2.2F, H; Plate 2F, H) indicates that the elaters were in the process of being deposited when fossilization occurred. Another scenario involves the elaters being lost during the fossilization process. Elaters are delicate structures (Halle 1908; Kedves et al. 1991) and may not have survived fossilization. Although the darkened nature of the preserved tissue indicates that the material may have been exposed to heat during fossilization, the preservation of the tapetal plasmodium implies that the first scenario is perhaps more accurate. Another possibility is that the structures that are interpreted here as elaters are not actually elaters, but some form of detritus or contamination with an elater-like appearance.

The similarity in both morphology and size of the elaters discussed in this chapter with those of extant *Equisetum* support the belief that the *Spaciinodum* structures are elaters. The width of elaters in extant *Equisetum* is approximately 2.5 μm (Uehara and Kurita 1989), and these widths can range up to 5 μm (Lugardon 1969), similar to several of the elater-like structures we describe (Fig. 2.3B–F; Plate 3B–F). The widths of elaters in Figures 2.3B, 2.3E, and 2.3F (see also Plate 3B, E, F) are comparable to those of extant *Equisetum*, whereas those in Figures 2.3C and 2.3D (Plate 3C, D) appear to be much wider. The elater-like structures in Figures 2.3C and 2.3D (Plate 3C, D) are within the natural variation of *Spaciinodum* elater width, but with such a small sample size, it is difficult to determine whether this is actually the case. It is possible that the structures in Figure 2.3C and Plate 3C are actually parts of a ripped perispore that give the impression of being elaters. In Figure 2.3D and Plate 3D, the elater-like structures are also considerably larger than those of extant *Equisetum*, but they do not appear to be the remains of a torn perispore. This structure could represent contamination or a natural variation in elater width.

Other similarities with spores of *Equisetum* are the spatulate ends of the elaters. In Figure 2.3E and Plate 3E, the elaters are attached to a common point on the spore. This position of attachment is also found in spores of *Equisetum* (Lugardon 1969; Uehara and Kurita 1989) and *Elaterites triferens* (Wilson 1943; Kurmann and Taylor 1984). Spores of extant *Equisetum* have four narrow elaters, which are spirally coiled around the spore and terminate in spatulate ends, while each of the three broad elaters of *Elaterites triferens* are circinately coiled against themselves, and those coils rest against the spore and terminate in a blunt tip (Wilson 1943; Kurmann and Taylor 1984). The evolution of elater number from three to four likely occurred as a result of more efficient packing. Of the few *Spaciinodum* spores found with attached elaters, the largest number of elaters found on any specimen is three. With such a small sample of elater-bearing spores, it is difficult to determine whether this is the actual number of elaters borne by spores of *Spaciinodum*. The narrow width and spatulate ends of its elaters indicate that *Spaciinodum* spores had four elaters; a calculation of surface area by using the available images of *Spaciinodum* elaters indicates that four elaters would be required to cover the surface area of typical *Spaciinodum* spore. Additional specimens will have to be discovered to answer these questions unequivocally.

Elaters are thought to function in both mechanical ejection from the interior of the sporangia and in dispersal away from the sporangia. Intertwining elaters of neighboring spores create a spore mass, which is able to disperse farther than would be the case with individual spores. This phenomenon can be explained by the increase in surface area accompanied by a proportionally smaller increase in mass (Niklas 1992; Vogel 1994; Schwendemann et al. 2007). Spore masses also allow gametophytes to develop in close association with each other, making fertilization more likely. This functional aspect of elaters may carry extra significance for *Spaciinodum* because of its occurrence at a high paleolatitude. The Fremouw Peak flora was located at 70–75°S latitude (Kidder and Worsely 2004), greatly limiting the amount of photosynthetic active radiation (PAR) available for photosynthesis in species living underneath the canopy of larger plants. The amount of PAR available is known to affect the sex ratios in gametophyte populations of extant *Equisetum* (Guillon and Fievet 2003). Guillon and Fievet (2003) demonstrated that fewer female gametophytes develop at lower PAR than at higher PAR, and that the gametophyte fitness of both sexes is reduced at lower levels of PAR. Less fit gametophytes with fewer females to fertilize could have deleterious effects on *Spaciinodum* populations if the individual gametophytes were dispersed at significant distances from one another. These effects could also have been avoided if *Spaciinodum* grew in an unshaded habitat where the PAR was not further reduced by a canopy. The occurrence of *Spaciinodum* alongside the broad-leaved conifer *Notophytum kraeuselii* (Meyer-Berthaud and Taylor 1991; Axsmith et al. 1998) indicates that this was not the case and underscores the importance of elaters in this paleoecosystem.

The presence of a trilete mark sets the spores of *Spaciinodum* apart from those of extant *Equisetum* (Lugardon 1969; Uehara and Kurita 1989) and from several other Mesozoic sphenophytes (Halle 1908; Barnard 1967; Gould 1968; Vozenin-Serra and Laroche 1976; Grauvogel-Stamm 1978; Harris 1978; Watson 1983; Watson and Batten 1990; Kelber and van Konijnenburg-van Cittert 1998; Villar de Seoane 2005). Extant *Equisetum* was long thought to be inaperturate, but subsequent studies that used electron microscopy indicated a small circular preformed aperture on the proximal surface of the spore too small to see with light microscopy (Lugardon 1969; Uehara and Kurita 1989). A few Mesozoic sphenophytes have been identified as trilete (e.g., *Equisetites arenaceus*: Kelber and van Konijnenburg-van Cittert 1998; *Equisetostachys nathorstii*: Halle 1908; *Equisetostachys verticillata*: Grauvogel-Stamm 1978), but most have been described as alete. The situation with alete Mesozoic taxa is likely similar to that of extant *Equisetum*, in which their small aperture size resulted in them not being identified. Spores of *Spaciinodum* may represent a transitional state between spores with elaters and germination through a trilete mark and spores with elaters and a specialized circular aperture. Evidence for this is confounded by the presence in younger strata of spores lacking elaters and described as alete. It is quite possible that these geologically younger spores once bore elaters that were destroyed during fossilization or preparation.

Status of Previously Described Reproductive Structures

Osborn et al. (2000) originally described reproductive remains of *Spaciinodum* mainly on the basis of two features: the anatomy of the axis and the presence of spores found in a single whorl of the axis. Recent work on *Spaciinodum collinsonii* dormant buds by Ryberg et al. (2008) has resulted in a reinterpretation of the axes described by Osborn et al. (2000). Ryberg et al. (2008) have concluded that the axis is actually a dormant bud and that the anatomy of the axis is similar to that of vegetative axes. In fact, Osborn et al. (2000: 233) did not completely rule out the possibility that the specimens were not reproductive: "Another interpretation of the new *Spaciinodum* fossils is that they are not fertile apices, but vegetative apices that are infected by fungal spores produced either by a saprophytic or a parasitic organism." As noted by Osborn et al. (2000), the fungus *Paleofibulus* (Osborn et al. 1989) is present in the "reproductive" specimens of *Spaciinodum collinsonii*, but the spores found in the *Spaciinodum* axis are much smaller than those of *Paleofibulus* or any other Antarctic fossil fungus described up to that time. However, the spores described by Osborn et al. (2000) do fit the size range and general morphology of zoospore cysts of a peronosporomycete. Moreover, peronosporomycete oogonia have recently been discovered in silicified peat from Fremouw Peak (Schwendemann et al. 2009) and are also present in close association with the "reproductive" axes figured in Osborn et al. (2000). Reexamination of the spores and associated axes described in Osborn et al. (2000) was undertaken for this current study. A small number of "spores" in the "sporangia" of the *Spaciinodum* axis have been found with a germinating hypha (Fig. 2.3H; Plate 3H), leading to the conclusion that the previously described reproductive axis is most likely a dormant bud harboring spores of a funguslike organism.

Conclusions

A great diversity of equisetalean cones can be found throughout Mesozoic sediments. The new Triassic sphenophyte cone described in this chapter adds more depth to the evolutionary history of this group. *Spaciinodum collinsonii* is an equisetalean plant displaying affinities with *Equisetum* subgenus *Equisetum*, suggesting an origin of this subgenus by the Triassic. Spores of *S. collinsonii* bear elaters of the equisetalean type, marking the first appearance of these structures in the fossil record. Reexamination of material previously described as the reproductive axes of *S. collinsonii* has led to the conclusion that those axes were most likely dormant buds of *S. collinsonii* harboring spores of a peronosporomycete.

Acknowledgments

This chapter is dedicated to Ted Delevoryas, friend, mentor, and colleague, who has greatly influenced our personal lives and professional careers—and to whom we are all indebted. This study was supported in part by funds from the National Science Foundation (ANT-0635477 to Edith L. Taylor and Thomas N. Taylor, and EAR-0542170 to Thomas N. Taylor and Michael Krings).

References Cited

Axsmith, B. J., T. N. Taylor, and E. L. Taylor. 1998. Anatomically preserved leaves of the conifer *Notophytum krauselii* (Podocarpaceae) from the Triassic of Antarctica. *American Journal of Botany* 85: 704–713.

Barnard, P. D. W. 1967. Flora of the Shemshak Formation, part 2. Liassic plants from Shemshak and Ashtar. *Rivista Italiana di Paleontologia e Stratigrafia* 71: 1123–1168.

Barratt, K. 1920. A contribution to our knowledge of the vascular system of the genus *Equisetum. Annals of Botany* 34: 201–235.

Barrett, P. J., and D. H. Elliot. 1973. Reconnaissance geologic map of the Buckley Island Quadrangle, Transantarctic Mountains, Antarctica. Antarctic Map, USGS Series, no. 3. Washington, D.C.: United States Geological Survey.

Bierhorst, D. W. 1959. Symmetry in *Equisetum. American Journal of Botany* 46: 170–179.

———. 1971. *Morphology of Vascular Plants.* New York: Macmillan.

Boureau, É. 1964. *Traité de Paléobotanique: Tome III—Sphenophyta, Noeggerathiophyta.* Paris: Masson et Cie.

Browne, I. 1912. Anatomy of the cone and fertile stem of *Equisetum. Annals of Botany* 26: 663–703.

———. 1915. A second contribution to our knowledge of the anatomy of the cone and fertile stem of *Equisetum. Annals of Botany* 29: 231–264.

———. 1920. Phylogenetic considerations on the internodal vascular strands of *Equisetum. New Phytologist* 19: 11–25.

———. 1933. A fifth contribution to our knowledge of the anatomy of the cone and fertile stem of *Equisetum. Annals of Botany* 47: 459–475.

———. 1941. A sixth contribution to our knowledge of the anatomy of the cone and fertile stem of *Equisetum. Annals of Botany* 5: 425–453.

Cúneo, N. R., and I. Escapa. 2006. The equisetalean genus *Cruciaetheca* nov. from the Lower Permian of Patagonia, Argentina. *International Journal of Plant Sciences* 167: 167–177.

Des Marais, D. L., A. R. Smith, D. M. Britton, and K. M. Pryer. 2003. Phylogenetic relationships and evolution of extant horsetails, *Equisetum*, based on chloroplast DNA sequence data (*rbcL* and *trnL-F*). *International Journal of Plant Sciences* 164: 737–751.

DiMichele, W. A., and T. L. Phillips. 1994. Paleobotanical and paleoecological constraints on models of peat formation in the Late Carboniferous of Euramerica. *Palaeogeography, Palaeoclimatology, Palaeoecology* 106: 39–90.

DiMichele, W. A., R. A. Gastaldo, H. W. Pfefferkorn. 2005. Plant biodiversity partitioning in the Late Carboniferous and Early Permian and its implications for ecosystem assembly. *Proceedings of the California Academy of Sciences* 56: 32–49.

Dotzler, N., M. Krings, R. Agerer, J. Galtier, and T. N. Taylor. 2008. *Combresomyces cornifer* nov. gen. et sp., an endophytic peronosporomycete in *Lepidodendron* from the Carboniferous of central France. *Mycological Research* 112: 1107–1114.

Duckett, J. G. 1970. Spore size in the genus *Equisetum*. *New Phytologist* 69: 333–346.

Escapa, I., and R. Cúneo. 2005. A new equisetalean plant from the Early Permian of Patagonia, Argentina. *Review of Palaeobotany and Palynology* 137: 1–14.

Farabee, M. J., E. L. Taylor, and T. N. Taylor. 1990. Correlation of Permian and Triassic palynomorph assemblages from the central Transantarctic Mountains, Antarctica. *Review of Palaeobotany and Palynology* 65: 257–265.

Frenguelli, J. 1944. Contribuciones al conocimiento de la flora del Gondwana superior en la Argentina XXIV. *Equisetites fertilis* n. comb. (*Equisetites scitulus* y *Macrotaenia fertilis*). *Notas del Museo de la Plata* 9: 501–509.

Galtier, J., and T. L. Phillips. 1999. The acetate peel technique. In T. P. Jones and N. P. Rowe (eds.), *Fossil Plants and Spores: Modern Techniques*, pp. 67–70. London: Geological Society.

Golub, S. J., and R. H. Wetmore. 1948a. Studies of development in the vegetative shoot of *Equisetum arvense* L. I. The shoot apex. *American Journal of Botany* 35: 755–767.

———. 1948b. Studies of development in the vegetative shoot of *Equisetum arvense* L. II. The mature shoot. *American Journal of Botany* 35: 767–781.

Gould, R. E. 1968. Morphology of *Equisetum laterale* Phillips, 1829, and *E. bryanii* sp. nov. from the Mesozoic of south-eastern Queensland. *Australian Journal of Botany* 16: 153–176.

Grauvogel-Stamm, L. 1978. La flore du Grès à *Voltzia* (Buntsandstein supérieur) des Vosges du Nord (France). *Sciences Géologiques, Université Louis Pasteur de Strasbourg, Institut de Géologie, Memoir* 50: 1–225.

Grauvogel-Stamm, L., and S. R. Ash. 1999. "*Lycostrobus*" *chinleana*, an equisetalean cone from the Upper Triassic of the southwestern United States and its phylogenetic implications. *American Journal of Botany* 86: 1391–1405.

Guillon, J.-M. 2004. Phylogeny of horsetails (*Equisetum*) based on the chloroplast *rps4* gene and adjacent noncoding sequences. *Systematic Botany* 29: 251–259.

———. 2007. Molecular phylogeny of horsetails (*Equisetum*) including chloroplast *atpB* sequences. *Journal of Plant Research* 120: 569–574.

Guillon, J.-M., and D. Fievet. 2003. Environmental sex determination in response to light and biased sex ratios in *Equisetum* gametophytes. *Journal of Ecology* 91: 49–57.

Halle, T. G. 1908. Zur Kenntnis der mesozoischen Equisetales Schwedens. *Kungliga Svenska Vetenskapsakademiens Handlingar* 43: 1–36.

Hammer, W. R., J. W. Collinson, and W. J. Ryan III. 1990. A new Triassic vertebrate fauna from Antarctica and its depositional setting. *Antarctic Science* 2: 63–167.

Harris, T. M. 1978. A reconstruction of *Equisetum columnare* and notes on its elater bearing spores. *Palaeobotanist* 25: 120–125.

Hauke, R. L. 1963. A taxonomic monograph of the genus *Equisetum* subgenus *Hippochaete*. *Nova Hedwigia* 8: 1–123.

———. 1978. A taxonomic monograph of the genus *Equisetum* subgenus *Equisetum*. *Nova Hedwigia* 30: 385–455.

Jones, O. A., and N. J. de Jersey. 1947. Fertile Equisetales and other plants from the Brighton Beds. *University of Queensland, Department of Geology Papers* 3 (4): 1–16.

Kedves, M., A. Toth, and E. Farkas. 1991. High temperature effect on the spores of *Equisetum arvense* L. *Plant Cell Biology and Development* 1: 8–14.

Kelber, K.-P., and J. H. A. van Konijnenburg-van Cittert. 1998. *Equisetites arenaceus* from the Upper Triassic of Germany with evidence for reproductive strategies. *Review of Palaeobotany and Palynology* 100: 1–26.

Kidder, D. L., and T. R. Worsley. 2004. Causes and consequences of extreme Permo-Triassic warming to globally equable climate and relation to the Permo-Triassic extinction and recovery. *Palaeogeography, Palaeoclimatology, Palaeoecology* 203: 207–237.

Kon'no, E. 1962. Some species of *Neocalamites* and *Equisetites* in Japan and Korea. *Science Reports, Tohoku University, Sendai, Japan,* 2nd ser. (Geology). Special Volume 5: 21–47.

———. 1972. Some Late Triassic plants from the southwestern border of Sarawak, east Malaysia. *Geology and Palaeontology of Southeast Asia* 10: 125–178.

Kurmann, M. H., and T. N. Taylor. 1984. Comparative ultrastructure of the sphenophyte spores in *Elaterites* and *Equisetum. Grana* 23: 109–116.

Kustatscher, E., M. Wachtler, and J. H. A. van Konijnenburg-van Cittert. 2007. Horsetails and seed ferns from the Middle Triassic (Anisian) locality Kühwiesenkopf (Monte Prà della Vacca), Dolomites, northern Italy. *Palaeontology* 50: 1277–1298.

Lugardon, B. 1969. Sur la structure fine des parois sporales d'*Equisetum maximum* Lamk. *Pollen et Spores* 11: 449–474.

Marsh, A. S., J. A. Arnone III, B. T. Bormann, and J. C. Gordon. 2000. The role of *Equisetum* in nutrient cycling in an Alaskan shrub wetland. *Journal of Ecology* 88: 999–1011.

Menéndez, C. A. 1958. *Equisetites quindecimdentata* sp. nov. del Triasico superior de Hilario, San Juan. *Revista de la Asociación Geológica Argentina* 13: 5–14.

Meyer-Berthaud, B., and T. N. Taylor. 1991. A probable conifer with podocarpacean affinities from the Triassic of Antarctica. *Review of Palaeobotany and Palynology* 67: 179–198.

Niklas, K. J. 1992. *Plant Biomechanics: An Engineering Approach to Plant Form and Function.* Chicago: University of Chicago Press.

Osborn, J. M., C. J. Phipps, T. N. Taylor, and E. L. Taylor. 2000. Structurally preserved sphenophytes from the Triassic of Antarctica: reproductive remains of *Spaciinodum. Review of Palaeobotany and Palynology* 111: 225–235.

Osborn, J. M., and T. N. Taylor. 1989. Structurally preserved sphenophytes from the Triassic of Antarctica: vegetative remains of *Spaciinodum,* gen. nov. *American Journal of Botany* 76: 1594–1601.

Osborn, J. M., T. N. Taylor, and J. F. White Jr. 1989. *Palaeofibulus* gen. nov., a clamp-bearing fungus from the Triassic of Antarctica. *Mycologia* 81: 622–626.

Page, C. N. 1972. An assessment of inter-specific relationships in *Equisetum* subgenus *Equisetum. New Phytologist* 71: 355–369.

Ryberg, P. E., E. J. Hermsen, E. L. Taylor, and T. N. Taylor. 2008. Development and ecological implications of dormant buds in the high-paleolatitude Triassic sphenophyte *Spaciinodum* (Equisetaceae). *American Journal of Botany* 95: 1443–1453.

Schaffner, J. H. 1930. Geographic distribution of the species of *Equisetum* in relation to their phylogeny. *American Fern Journal* 20: 89–106.

Schweitzer, H.-J., J. H. A. van Konijnenburg-van Cittert, and J. van der Burgh. 1997. The Rhaeto–Jurassic flora of Iran and Afghanistan. 10.

Bryophyta, Lycophyta, Sphenophyta, Pterophyta—Eusporangiatae and Protoleptosporangiatae. *Palaeontographica*, Abt. B, 243: 103–192.

Schwendemann, A. B., T. N. Taylor, E. L. Taylor, M. Krings, and N. Dotzler. 2009. *Combresomyces cornifer* from the Triassic of Antarctica: evolutionary stasis in the Peronosporomycetes. *Review of Palaeobotany and Palynology* 154: 1–5.

Schwendemann, A. B., G. Wang, M. L. Mertz, R. T. McWilliams, S. L. Thatcher, and J. M. Osborn. 2007. Aerodynamics of saccate pollen and its implications for wind pollination. *American Journal of Botany* 94: 1371–1381.

Taylor, T. N., E. L. Taylor, and M. Krings. 2009. *Paleobotany: The Biology and Evolution of Fossil Plants*. 2nd ed. San Diego: Academic Press.

Uehara, K., and S. Kurita. 1989. An ultrastructural study of spore wall morphogenesis in *Equisetum arvense*. *American Journal of Botany* 76: 939–951.

Uehara, K., and S. Murakami. 1995. Arrangement of microtubules during spore formation in *Equisetum arvense* (Equisetaceae). *American Journal of Botany* 82: 75–80.

Villar de Seoane, L. 2005. *Equisetites pusillus* sp. nov. from the Aptian of Patagonia, Argentina. *Revista del Museo Argentino de Ciencias Naturales* 7: 43–49.

Vogel, S. 1994. *Life in Moving Fluids: The Physical Biology of Flow*. Princeton, N.J.: Princeton University Press.

Vozenin-Serra, C., and J. Laroche. 1976. Présence d'une Equisetacée dans les formations de Chres—Cambodia occidental. *Palaeontographica*, Abt. B, 159: 158–166.

Watson, J. 1983. Two new species of *Equisetum* found in situ. *Acta Palaeontologica Polonica* 28: 265–269.

Watson, J., and D. J. Batten. 1990. A revision of the English Wealden flora, II. Equisetales. *Bulletin of the British Museum (Natural History), Geology* 46: 37–60.

Weber, R. 2005. *Equisetites aequecaliginosus* sp. nov., ein Riesenschachtelhalm aus der spättriassischen Formation Santa Clara, Sonora, Mexiko. *Revue de Paléobiologie, Genève* 24: 331–364.

Wilson, L. R. 1943. Elater-bearing spores from the Pennsylvanian strata of Iowa. *American Midland Naturalist* 30: 518–523.

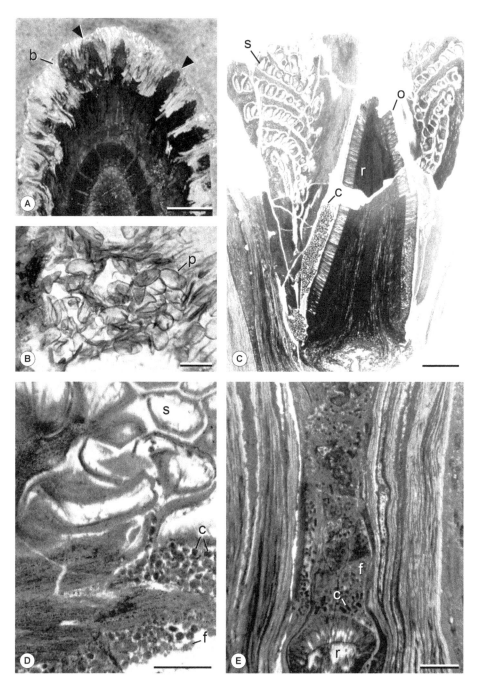

Fig. 3.1. Trunk and cone morphology, *Cycadeoidea*. (A) Transverse section through a trunk showing several cones (arrowheads) surrounded by a thick ramentum of helically arranged bracts (b). Scale bar = 3 cm. (B) Transverse section of a synangium showing a single pollen sac containing in situ pollen grains (p); LM. YPM 5083. Scale bar = 50 mm. (C) Longitudinal section through a bisporangiate cone showing the fleshy microsporangiate region with multiple synangia (s) surrounding a conical ovulate receptacle (r) containing many ovules (o). Note the abundant coprolites (c) in the space between the microsporophyll and the ovulate receptacle, as well as the bore tunnels within the microsporophyll; LM. YPM 5061. Scale bar = 5 mm. (D) Detail of a microsporophyll showing synangia (s) and insect tunnels containing coprolites (c) and frass (f); LM. YPM 5107. Scale bar = 1 mm. (E) Longitudinal section through an immature cone with disarticulated plant material, coprolites (c), and frass (f) surrounding the dome-shaped ovulate receptacle (r); LM. Crepet slide, NS 136 (no YPM number). Scale bar = 5 mm.

POLLEN AND COPROLITE STRUCTURE IN *CYCADEOIDEA* (BENNETTITALES): IMPLICATIONS FOR UNDERSTANDING POLLINATION AND MATING SYSTEMS IN MESOZOIC CYCADEOIDS

3

Jeffrey M. Osborn and Mackenzie L. Taylor

The Mesozoic seed plant group Bennettitales is most comprehensively known from permineralized fossils of *Cycadeoidea* from the Early Cretaceous of North America. Cones were bisporangiate and have been interpreted as remaining closed at maturity. Consequently, self-pollination has been hypothesized in *Cycadeoidea* through developmental disintegration of the fused cones. Additionally, entomophily has been hypothesized as a secondary pollination syndrome because extensive beetle tunnel and gallery systems occur within reproductive tissues. In what is to our knowledge the first study of the structure and contents of coprolites and frass pellets extracted from beetle tunnels within *Cycadeoidea* cones, pollen grains and pollen wall fragments were found to be absent. New data on direct insect interactions, combined with data on pollen ultrastructure, are discussed here as they relate to interpretations of pollination mechanisms and mating system biology in *Cycadeoidea* and the Bennettitales. Because pollination biology determines the frequency and diversity of mating opportunities, the two pollination syndromes would potentially have had different consequences for mating systems in *Cycadeoidea*. Self-pollination alone would have almost certainly resulted in obligate selfing, whereas occasional to predominant insect pollination would have likely resulted in a mixed mating system

The Mesozoic seed plant group Bennettitales is most comprehensively known from permineralized fossils of *Cycadeoidea* from the Lower Cretaceous of North America. Fossils have been collected from a wide range of localities, and many taxonomic species have been described. However, the majority of data on *Cycadeoidea* are derived from investigations of specimens collected in the Black Hills of South Dakota and Wyoming, U.S.A. (Wieland 1906, 1916; Crepet 1974).

The reproductive structure and pollination system of *Cycadeoidea* have received much attention over the years. Wieland's (1906) early

Introduction

35

anatomical and morphological work characterized the reproductive organ of *Cycadeoidea* as flowerlike, and he reconstructed the plant with open "flowers" protruding from the surface of the stem. At maturity, each open "flower" was interpreted to consist of pinnately compound, pollen-bearing microsporophylls surrounding a centrally positioned ovulate receptacle (Wieland 1906). Such an open structure would have permitted wind pollination, as has also been hypothesized for some members of the Williamsoniaceae, the second and more ancient (Late Triassic–Late Cretaceous) family within the Bennettitales (e.g., Watson and Sincock 1992; Labandeira et al. 2007; Taylor et al. 2009).

Wieland's (1906) interpretation and reconstruction were subsequently shown to be incorrect. Upon further investigation of Wieland's specimens, as well as of other unstudied fossils, Delevoryas (1968) and Crepet (1974) demonstrated that all *Cycadeoidea* cones were bisporangiate and remained permanently closed at maturity. The cones were shown to have a highly complex ontogenetic structure. The pinnate microsporophylls did not open but rather recurved during development, and the distal tips became ontogenetically fused near the base of the ovulate receptacle (Delevoryas 1968; Crepet 1974). Consequently, self-pollination has been hypothesized in *Cycadeoidea* through developmental disintegration of the fused cones (Crepet 1972, 1974). In addition, entomophily has been hypothesized as a secondary pollination syndrome because extensive beetle tunnel and gallery systems occur within reproductive tissues of *Cycadeoidea* cones (Crepet 1972, 1974).

Labandeira et al. (2007) have recently proposed a generalized plant–insect association between *Cycadeoidea* and beetles (likely belonging to suborder Polyphaga, or its subclade Phytophaga) analogous to the association that characterizes extant cycads (Stevenson et al. 1998) and perhaps extinct cycads as well (Klavins et al. 2005). The model of Labandeira et al. (2007) is based on cone structure and other anatomical characters for *Cycadeoidea*, a range of data on fossil beetles, and the commonality of insect damage to bennettitalean cones described from a range of geographically and geologically disjunct localities (Reymanówna 1960; Bose 1968; Saiki and Yoshida 1999; Stockey and Rothwell 2003). However, there is actually very little direct information about the fossil insects associated with *Cycadeoidea* or other bennettitaleans (see Labandeira et al. 2007 and references therein).

The objective of the present study is to investigate the structure and contents of coprolites and frass pellets extracted from beetle tunnels within *Cycadeoidea* cones, as well as to review and reexamine the pollen structure of *Cycadeoidea*. The new data on insect evidence of phytophagy, combined with the pollen data, are then discussed as they relate to interpretations of pollination mechanisms and mating system biology in *Cycadeoidea* and the Bennettitales.

The genus *Cycadeoidea* ranges in age from Late Triassic to Early Cretaceous, and a number of species have been described from a variety of localities (Taylor et al. 2009). Several key Early Cretaceous sites in the Black Hills of South Dakota and Wyoming in North America have yielded the majority of specimens (see Wieland 1906 and Crepet 1974 for historical synopses and locality details). The pollen grains, coprolites, and frass pellets described here were isolated from *Cycadeoidea* fossils collected from these North American sites, and the specimens are housed in the Paleobotanical Collections of the Peabody Museum of Natural History, Yale University, New Haven, Conn. The light micrographs illustrated in Figure 3.1 were photographed from the original specimens and thin sections produced and studied by G. R. Wieland (1906, 1916), T. Delevoryas (1963, 1965, 1968), and W. L. Crepet (1972, 1974). These thin sections, along with the silicified trunk pieces from which the sections were made, are housed in Yale's Peabody Museum (YPM). The pollen grains described and studied here were isolated from cones of *Cycadeoidea dacotensis* (trunk 213) and *Cycadeoidea* sp. (trunk CF 78), and the coprolites and frass pellets were isolated from cones of *Cycadeoidea wielandii* (trunk 77) and *Cycadeoidea* sp. (trunk CF 78). In addition, *Cycadeoidea* pollen from the fossils studied by Taylor (1973) was reexamined; these specimens are housed in the Paleobotanical Collections of the University of Kansas, Lawrence.

For electron microscopy, in situ pollen, coprolites, and frass were isolated directly from synangia and insect tunnels within the cones by building elevated wax wells directly around the areas of interest on the surfaces of the silicified trunk slabs. Pollen grains, coprolites, and frass pellets were then pipette macerated with 48% hydrofluoric acid and washed several times with water. For scanning electron microscopy (SEM), pollen, coprolites, and frass were pipetted directly onto aluminum stubs, sputter-coated with gold–palladium, and imaged with Hitachi S-500 and JEOL JSM-840 scanning electron microscopes at accelerating voltages of 15–20 kV. For transmission electron microscopy (TEM), pollen, coprolites, and frass were pipetted directly onto cellulose filters under suction. The filters were coated on both sides with agar, dehydrated in a graded ethanol series, and then transferred to 100% acetone to dissolve the filters. The agar-embedded specimens were gradually infiltrated with Spurr's epoxy resin and then embedded in flat aluminum pans. Once polymerized, the resin was removed from the aluminum pans and examined with a stereo light microscope to identify specific microfossils of interest. To obtain single-pollen grains, coprolites, and frass pellets in desired orientations for ultramicrotomy, small resin "blocks" surrounding the microfossils were cut from the resin disks with a jeweler's saw. The specimens were then sectioned on a Reichert ultramicrotome with a diamond knife. Thin sections (70–90 nm) were collected and dried on formvar-coated slot grids, stained with 1% potassium permanganate (3–12 minutes), 1% uranyl acetate (6–12 minutes), and lead citrate (3–6

Materials and Methods

minutes), and then imaged with a Zeiss EM-10 transmission electron microscope at 60–80 kV.

Results and Discussion

Cone Morphology

Trunks of *Cycadeoidea* bear cones on short shoots that are embedded within a thick ramentum of leaf bases and bracts; these cones only partially extend beyond the trunk surface (Fig. 3.1A). All cones of *Cycadeoidea* are bisporangiate, producing both ovules and pollen during the course of cone development (Fig. 3.1B, C). Cones have recurved, pinnate microsporophylls, each bearing numerous synangia. Synangia are reniform and multiloculate, typically consisting of 22 elongate locules, and the locules contain abundant in situ pollen grains (Fig. 3.1B–D). After pollen dehiscence from the synangia and disintegration of the microsporangiate organs, the receptacular organs remain intact and continue development (Crepet 1974), and the overall reproductive structure then resembles a monosporangiate ovulate cone (Fig. 3.1A).

Although hundreds of sectioned trunks and cones were examined in the present study to identify specimens with pollen, coprolites, and frass, relatively few *Cycadeoidea* trunks were found with cones bearing microsporangiate organs (i.e., in a bisporangiate developmental stage). The paucity of specimens with microsporangiate organs, also noted by Delevoryas (1968) and Crepet (1974), led early workers to conclude that *Cycadeoidea* bore monosporangiate cones. Critical investigation of the cones with pollen-bearing organs, however, later contributed to the recognition that *Cycadeoidea* actually produced closed, bisporangiate cones with a highly complex ontogenetic structure. The anatomical, ontogenetic, and phylogenetic aspects of cone structure were described in detail by Delevoryas (1968) and Crepet (1974).

Pollen Morphology

Pollen grains of *Cycadeoidea* are typically elliptic to prolate in shape and average 25 μm in length and 12 μm in width (Fig. 3.1B and Fig. 3.2A, C). Spheroidal grains have also been observed (Osborn and Taylor 1995); however, most pollen grains exhibit significant folding of the exine, which contributes to a range of other pollen shapes (Fig. 3.1B). In fact, early reports of the preservation of cellular microgametophyes in *Cycadeoidea* pollen (Wieland 1906) were shown by Taylor (1973) to actually represent exine folds. *Cycadeoidea* pollen is monosulcate, and the relatively broad width of the aperture (seen in section in Fig. 3.2B) causes this region of the pollen wall to become highly folded. This condition has made detailed observation of the external surface of the aperture difficult to achieve (see Osborn and Taylor 1995). The nonapertural pollen surface exhibits a punctate to psilate ornamentation pattern (Fig. 3.2C, E).

The pollen wall is well preserved at the ultrastructural level, averages 0.73 μm in overall thickness, and consists of an electron-translucent sexine and an electron-dense nexine (Fig. 3.2B, D, F). The outer sexine consists of two well-defined layers, a homogenous tectum and a granular infratectum (Fig. 3.2D, F). The tectum has intermittent thin zones corresponding to surface punctuations (Fig. 3.2D, F). In many grains, the tectum and infratectum are not as clearly defined from each other structurally because the granules are so densely packed within the infratectum and appear to evenly grade into the homogeneous tectum (Fig. 3.2D). This structural condition led the first report of pollen wall architecture in *Cycadeoidea* to characterize the exine as homogeneous and lacking ultrastructural detail (Taylor 1973). However, in other pollen grains, individual granules within the infratectum are well preserved, exhibit uniform spacing and diameters, and are clearly delimited from the overlying homogeneous tectum (Fig. 3.2F).

The electron-dense nexine is uniform in thickness throughout the pollen wall, including both apertural and nonapertural regions of the grain (Fig. 3.2B). Endexine lamellations are not discernible in most *Cycadeoidea* pollen grains because the lamellae have become highly compressed during exine ontogeny (Osborn and Taylor 1995; Osborn 2000), a developmental phenomenon also observed in extinct and extant gnetalean pollen (Osborn et al. 1993; Doores et al. 2007). Endexine lamellae may become detectable under certain preservational conditions when fossil pollen grains become stretched or compacted (Osborn and Taylor 1994, 1995; Osborn 2000).

The apertural wall consists of a thin tectal layer, resulting from lateral thinning of both the tectum and infratectum, and a uniformly thick, electron-dense nexine (Fig. 3.2B).

In addition to a range of vegetative and reproductive characters (cf. Rothwell et al. 2009; Crepet and Stevenson 2010; Rothwell and Stockey 2010), several structural characters of the pollen wall from *Cycadeoidea* have been important in evolutionary considerations of the genus, as well as in interpreting the phylogenetic position of the Bennettitales relative to other seed plants. In particular, the granular infratectum and the uniform thickness of the nexine in both nonapertural and apertural regions have played a role in phylogenetic hypotheses about relationships among Bennettitales, Gnetales, and several other gymnospermous taxa, as well as basal lineages of angiosperms (e.g., Osborn 2000; Doores et al. 2007; Taylor et al. 2008).

Coprolite and Frass Morphology

Both coprolites and frass pellets are present in the insect tunnels within *Cycadeoidea* cones. Coprolites are typically oval to spherical in shape and have well-defined outer margins (Fig. 3.3A). The contents of coprolites include a heterogeneous assemblage of plant tissue and cell wall

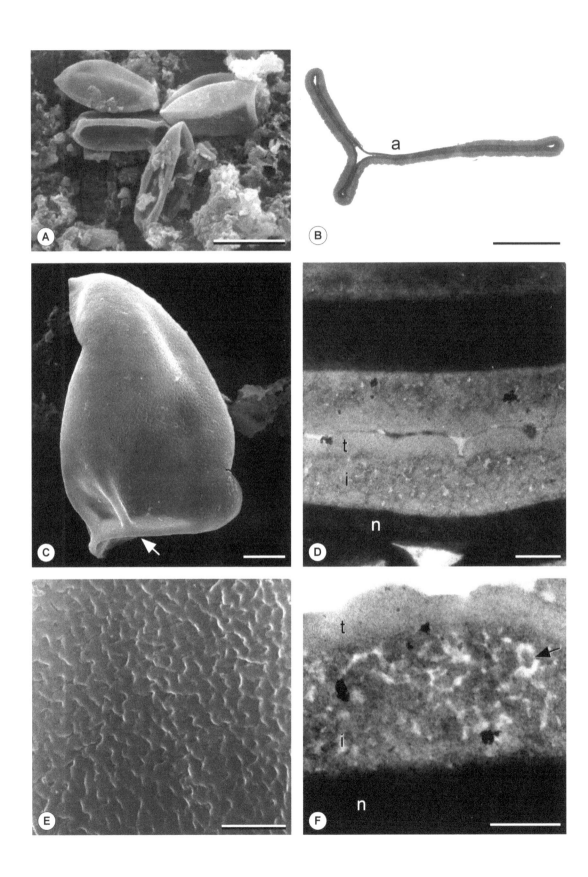

fragments of varying sizes (Fig. 3.3A–C). The tissue and cell wall fragments are often branched and have an amorphous ultrastructure (Fig. 3.3B, C).

The frass pellets isolated from the same insect boring tunnels are smaller in size, irregular in shape, and have a markedly different ultrastructure (Fig. 3.3D). In comparison to coprolites, the frass pellets consist of plant tissue that is more homogeneous in organization and likely represents chewed plant material that has become disarticulated in chunks (Fig. 3.3D, E).

Labandeira et al. (2007) have proposed a generalized life history pattern of insect herbivory, ovule predation, and pollination in *Cycadeoidea*. Five phases of tissue consumption, associated pollination, and formation of tunnel/gallery systems by small, robust beetles have been hypothesized (Labandeira et al. 2007). The coprolites and frass pellets investigated in the current study were extracted from tunnels within the microsporangiate organs of cones bearing mature pollen, and they therefore represent the fourth and fifth phases of the beetle life history of Labandeira et al. (2007). Consequently, if the beetles were consuming pollen grains, we would expect to find either entire pollen grains or resistant pollen wall layers within the coprolites isolated from the beetle tunnels. In analogous taxa and systems, intact and well-preserved pollen grains have been identified both within the gut contents of several fossil insects (Krassilov et al. 1998; Krassilov and Rasnitsyn 1999) and within the contents of coprolites produced by several fossil beetles (Middle Triassic cycads, Klavins et al. 2005; Late Cretaceous angiosperms, Lupia et al. 2002). However, in the current study, neither entire pollen grains nor pollen walls/exine layers, which are structurally distinctive and exhibit differential staining patterns, were found within the coprolites extracted from *Cycadeoidea* cones (e.g., cf. Fig. 3.2B, D, and F with Fig. 3.3A–E).

Pollination Biology and Mating Systems

Current reconstructions of cycadeoid cones as closed bisporangiate structures indicate that it is highly likely that *Cycadeoidea* exhibited a high rate of selfing. The degree of selfing, which would have depended on the efficiency of insect visitors as pollinators and on post-pollination processes, would potentially have had a profound effect on the ability of cycadeoids to persist and diversify over evolutionary time (Stebbins 1970). On the basis of reconstructions of cone morphology and development, evidence of insect visitors, and comparisons with extant gymnosperms, we consider two major possibilities for mating systems in *Cycadeoidea*: obligate selfing and facultative outcrossing.

OBLIGATE SELFING

Self-pollination likely occurred in *Cycadeoidea* through autonomous selfing, the transfer of self-pollen without the assistance of an outside vector.

Fig. 3.2. Pollen morphology and ultrastructure, *Cycadeoidea*. (A) Four elongate pollen grains, in which each grain exhibits substantial folding of the exine; SEM. Scale bar = 25 μm. (B) Transverse section through a single pollen grain with a distinctly bilayered exine and thinning in the apertural region (a); TEM. Scale bar = 5 μm. (C) Proximal view of a single pollen grain with slightly folded exine (arrow); SEM. Scale bar = 5 μm. (D) Detail of the pollen walls from two adjacent pollen grains, showing the electron-translucent sexine and the electron-dense nexine (n). The outer sexine consists of two well-defined layers, a homogenous tectum (t) with intermittent thin zones corresponding to surface punctuations, and a granular infratectum (i); TEM. Scale bar = 0.25 mm. (E) Detail of the punctate pollen grain surface; SEM. Scale bar = 2 μm. (F). Detail of pollen wall ultrastructure showing homogeneous tectum (t), infratectum (i) composed of distinct granules (arrow), and electron-dense nexine. Endexine lamellations are not discernible in the nexine (n) of this pollen grain; TEM. Scale bar = 0.25 mm. *All specimens illustrated are from trunk 213.*

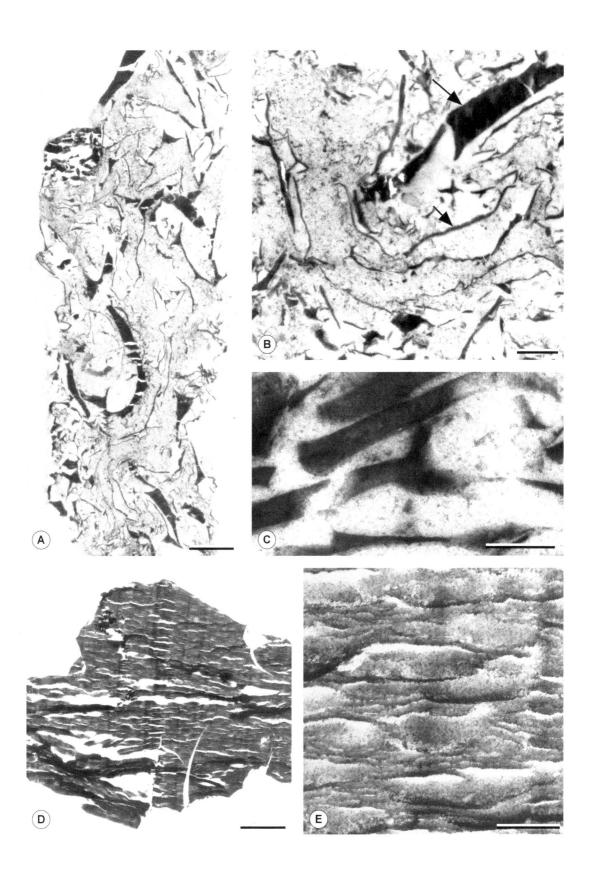

Many fertilized cones have been found that exhibit no evidence of insect boring, providing support for the hypothesis that insect visitation was not necessary for pollination and fertilization (Crepet 1974). Moreover, the morphology of bisporangiate cones, with the synangia positioned above the ovulate receptacle and pollen maturing before the ovules, would have facilitated autonomous selfing. Several stages of bisporangiate cones have been described in which the synangia are in various degrees of disintegration, and the remnants of the synangia have come into contact with the ovulate receptacle (Crepet 1972, 1974). This would have facilitated successful self-pollination if pollen contacted the ovules and remained viable until ovules and female gametophytes became developmentally receptive.

In extant gymnosperms, pollen germination and subsequent pollen tube growth are quite slow, and in extant conifers, there is often a period of dormancy between pollination and fertilization (e.g., Fernando et al. 2005). Preserved pollen tubes in *Williamsonia bockii* have been described as nearly identical to extant araucarian conifers, specifically *Agathis australis* (Stockey and Rothwell 2003). In *A. australis*, pollination and fertilization are separated by a year or more, which results from a 3-month delay in pollen germination, slow pollen tube growth, and pollen tube dormancy during the winter months (Owens et al. 1995). It is probable that the Bennettitales, including *Cycadeoidea*, exhibited similar pollen tube growth patterns that allowed for ovule development between pollination and fertilization.

Autonomous selfing is beneficial in cases where pollinators or mates are limited because such selfing provides assurance of seed set in the absence of these (Lloyd 1992). The ability to autonomously self would also have been beneficial if there were strong selection favoring the evolution of closed cones. Open cones that exhibited cross-pollination via either wind or insects are hypothesized to be plesiomorphic in the Bennettitales (e.g., Crepet 1974; Labandeira et al. 2007; Taylor et al. 2009). Crepet (1974) suggested that the evolution of fused microsporophylls, and possibly the protective position of the cones within the trunk, in *Cycadeoidea* might have been in response to beetle predation. A transition from a predominantly outcrossing mating system to an obligate or predominantly selfing mating system may have accompanied cone closure.

The primary evolutionary consequence of a transition to selfing is a loss of heterozygosity and therefore a reduction in genetic diversity. This reduces the ability of a population to evolve in response to selective pressures, especially in the face of a changing environment. Because of this, obligate selfing is generally considered an evolutionary dead end (Stebbins 1970). Crepet (1974) and Taylor et al. (2009) suggested that a loss of genetic diversity due to self-fertilization may have made the Cycadeoidales less able to evolve in a changing environment in the Late Cretaceous and thus may have contributed to their decline.

In *Cycadeoidea* cones, damage from insect boring, as well as the fact that coprolites and frass are found both near the pollen-containing

Fig. 3.3. Coprolite and frass ultrastructure extracted from *Cycadeoidea* cones. (A) Section through a single coprolite showing heterogeneous contents of plant cell walls and tissue fragments, as well as the well-defined outer margins of the coprolite; TEM. Scale bar = 25 μm. (B) Coprolite contents showing plant tissue fragments of varying sizes (electron-dense cell walls; arrows) and having an amorphous ultrastructure. Note the absence of whole pollen grains, as well as structurally distinctive and/or differentially stained exine layers; TEM. Scale bar = 0.5 mm. (C) Detail of coprolite contents showing amorphous tissue fragments and cell walls. Note that one tissue fragment is branched, and that the fragments lack the structural characters and differential staining of *Cycadeoidea* pollen walls and are also thinner than the exine; TEM. Scale bar = 0.25 μm. (D) Section of disarticulated plant tissue found among frass pellets showing likely chewed plant material. Compare the overall morphology, ultrastructure, and staining pattern of this frass pellet to the sectioned coprolite shown in (A); TEM. Scale bar = 2.5 μm. (E) Detail of frass pellet in (D) showing homogenous nature of the plant tissue; TEM. Scale bar = 0.5 mm. All specimens illustrated are from trunk CF 78.

synangia and near the ovulate receptacle, indicate that insects were often present in cycadeoid cones and may have transferred pollen. Thus, the Cycadeoidaceae may have also exhibited facilitated selfing, in which beetles facilitated the movement of pollen from the synangia to the ovules within the same cone. Facilitated selfing is thought to be much less advantageous than autonomous selfing because it incurs the costs of having a pollen vector—for example, pollen or ovules lost to herbivory, or pollen loss transfer (Lloyd 1992; Harder and Barrett 1996; Holsinger 1996). Facilitated selfing is also dependent on pollinator availability, thus providing less reproductive assurance (Lloyd 1992).

FACULTATIVE OUTCROSSING (MIXED MATING SYSTEM)

The occurrence of insect visitors in cones of *Cycadeoidea* indicates that these plants may have also been capable of both selfing (either autonomous or facilitated selfing, as described above) and outcrossing, a condition referred to as a mixed mating system or facultative outcrossing (see Goodwillie et al. 2005). This is a strategy common in extant gymnosperms (Williams 2007) and can be considered a "best of both worlds" situation (Becerra and Lloyd 1992). Occasional cross-pollination maintains heterozygosity and preserves genetic diversity, while self-pollination provides reproductive assurance in the absence of pollinators (Lloyd 1992; Goodwillie et al. 2005). The balance between self- and cross-pollination in gymnosperms has been best studied in extant conifers, which often exhibit moderate to high rates of self-pollination (Franklin 1969; Williams 2007). However, pollination alone does not determine the proportion of self-fertilization, but instead determines the frequency and diversity of mating opportunities that post-pollination processes then filter (Harder and Barrett 1996).

Most gymnosperms exhibit mechanisms to promote cross-pollination, including the separation of male and female function spatially into different cones (moneocy) or different plants (dioeocy), and developmentally (i.e., protandry). The bisporangiate cone in *Cycadeoidea* does not provide spatial separation of male and female function, but the delay in ovule development until after disintegration of the synangia may have increased the probability that cross-pollination by insects could have occurred, especially if insects were drawn to cones with mature pollen, as suggested by Crepet (1979).

After pollination, there may be additional processes that promote cross-fertilization despite high rates of self-pollination. In gymnosperms, these are thought to primarily act postzygotically and to include embryo competition and low viability of selfed embryos due to lethal alleles (see Fernando et al. 2005). Reduced seed set and fewer self-progeny than expected are well-documented phenomena in conifers (Franklin 1969; Owens et al. 1991, 2005; Williams 2007). If cycadeoids did experience both self- and cross-pollination, it is possible that these plants also had mechanisms that promoted outcrossing, such as those in extant gymnosperms, increasing the ratio of cross-to-self-fertilization when cross-pollination did occur. It is also possible that *Cycadeoidea* exhibited pre-zygotic mechanisms to promote cross-fertilization, which are common in angiosperms, but are thought to occur much less frequently in gymnosperms (Willson and Burley 1983; Mulcahy and Mulcahy 1987; but see Runions and Owens 1996).

The evolutionary stability of mixed mating systems is highly controversial (see Goodwillie et al. 2005). Lande and Shemske (1985) provided compelling theoretical evidence that selection will move mating systems toward one of two extremes, either outcrossing with high inbreeding depression or selfing with low inbreeding depression. This is because the level of inbreeding depression, the major cost to selfing, should evolve with the selfing rate. As populations self, recessive deleterious alleles are exposed to selection and purged. Thus, over time, the cost of selfing due to inbreeding is reduced and the rate of selfing should increase (Lande and Shemske 1985). However, recent analyses indicate that the selective pressures on mating systems are dependent on many factors and that mixed mating systems are more common than previously thought (Goodwillie et al. 2005). Therefore, the lack of evolutionary stability is far from certain, and it is possible that *Cycadeoidea* was able to maintain a mixed mating system. If *Cycadeoidea* was dependent on a mixed mating strategy, the changing environment of the Late Cretaceous may have contributed to the disruption of this system and to the subsequent decline of the lineage.

POLLINIVORY AND INSECT POLLINATION

It is often suggested that insect pollination evolved from plant–insect interactions involving herbivory on pollen (pollinivory) or ovules. Subsequent specialization of pollinator rewards and flower morphology may have then led to increasingly consistent and specialized relationships between plants and pollinators (e.g., Faegri and van der Pijl 1979; Crepet 1979, 1983).

Crepet (1979) suggested that pollinivory, which was widespread by the Permian (e.g., Krassilov et al. 1998; Krassilov and Rasnitsyn 1999), occurred in *Cycadeoidea* on the basis of his observations that cones with mature pollen exhibited the most insect damage. However, our study indicates that pollen was not present within coprolites or frass pellets isolated from *Cycadeoidea* cones, and there is no other direct evidence that insect visitors were foraging on *Cycadeoidea* pollen. The lack of pollinivory would have made pollen transfer much less efficient because beetles would have had to fortuitously come into contact with pollen (Crepet 1979). This situation, in which insects feed on ovules but not pollen, is thought to be less likely to result in successful pollination than the reverse (Crepet 1979).

It is possible that even in the absence of pollinivory, the beetle–plant interaction in *Cycadeoidea* represents an intermediate step in the evolution of insect pollination. Fertilization could occur successfully without pollen vectors, but the presence of beetles in the cone allowed for fortuitous, incidental insect pollination that was the precursor to more specialized relationships between plants and pollinators. It certainly shows that insect predation on cones was prevalent. If we assume Crepet's (1974) hypothesis that closed cones evolved in response to beetle predation, then it is likely that some insect pollination was occurring before the evolution of the closed cone, such as in the bennettitalean genus *Williamsoniella* (Williamsoniaceae), a bisexual, chasmogamous taxon. It is possible that *Cycadeoidea* evolved a closed cone and a high degree of selfing, whereas members of the Williamsoniaceae maintained an open cone and evolved more efficient insect pollination. However, there is also indirect evidence that selective pressure from insect predation was pushing williamsonians to evolve in the direction of sporophyll closure (Taylor et al. 2009). Thus, plant–beetle interactions in Mesozoic cycadeoids demonstrate the complexity of early plant–insect associations and their potential ramifications on morphology, reproductive biology, and ultimately evolutionary success.

Acknowledgments

We thank Ted Delevoryas for his seminal research on *Cycadeoidea* and other bennettitaleans and for serving as an inspiration to us. We are also grateful to Leo J. Hickey and Linda Klise (Yale University) for providing access to Yale's extensive collection of cycadeoids and for a specimen loan, as well as to Thomas N. Taylor (University of Kansas) for the opportunity to reexamine the *Cycadeoidea* material described in Taylor (1973).

References Cited

Becerra, J. X., and D. G. Lloyd. 1992. Competition-dependent abscission of self-pollinated flowers of *Phormium tenax* (Agavaceae)—a 2nd action of self-incompatibility at the whole flower level. *Evolution* 46: 458–469.

Bose, M. N. 1968. A new species of *Williamsonia* from the Rajmahal Hills, India. *Journal of the Linnean Society, Botany* 61: 121–127.

Crepet, W. L. 1972. Investigations of North American cycadeoids: pollination mechanisms in *Cycadeoidea*. *American Journal of Botany* 59: 1048–1056.

———. 1974. Investigations of North American cycadeoids: the reproductive biology of *Cycadeoidea*. *Palaeontographica*, Abt. B, 148: 144–169.

———. 1979. Insect pollination: a paleontological perspective. *BioScience* 29: 102–108.

———. 1983. The role of insect pollination in the evolution of the angiosperms. In L. Real (ed.), *Pollination Biology*, pp. 29–50. Orlando: Academic Press.

Crepet, W. L., and D. W. Stevenson. 2010. The Bennettitales (Cycadeoidales): a preliminary perspective on this arguably enigmatic group. In C. T. Gee (ed.), *Plants in Mesozoic Time: Morphological Innovations, Phylogeny, Ecosystems*, pp. 215–244. Bloomington: Indiana University Press.

Delevoryas, T. 1963. Investigations of North American cycadeoids: cones of *Cycadeoidea*. *American Journal of Botany* 50: 45–52.

———. 1965. Investigations of North American cycadeoids: microsporangiate structures and phylogenetic implications. *The Palaeobotanist* 14: 89–93.

————. 1968. Investigations of North American cycadeoids: structure, ontogeny, and phylogenetic considerations of cones of *Cycadeoidea*. *Palaeontographica*, Abt. B, 121: 122–133.

Doores, A. S., G. El-Ghazaly, and J. M. Osborn. 2007. Pollen ontogeny in *Ephedra americana* (Gnetales). *International Journal of Plant Sciences* 168: 985–997.

Faegri, K., and L. van der Pijl. 1979. *The Principles of Pollination Ecology*, 3rd ed. New York: Pergamon.

Fernando, D. D., M. D. Lazzaro, and J. N. Owens. 2005. Growth and development of conifer pollen tubes. *Sexual Plant Reproduction* 18: 149–162.

Franklin, E. C. 1969. Inbreeding depression in metrical traits of loblolly pine (*Pinus taeda* L.) as a result of self-pollination. *North Carolina Sate University, School of Forest Resources, Technical Report* 40: 1–19.

Goodwillie, C., S. Kaliz, and C. G. Eckert. 2005. The evolutionary enigma of mixed mating systems in plants: occurrence, theoretical explanations, and empirical evidence. *Annual Review of Ecology Evolution and Systematics* 36: 47–79.

Harder, L. D., and S. C. H. Barrett. 1996. Pollen dispersal and mating patterns in animal-Pollinated plants. In D. G. Lloyd and S. C. H. Barrett (eds.), *Floral Biology: Studies on Floral Evolution in Animal-pollinated Plants*, pp. 140–190. New York: Chapman and Hall.

Holsinger, K. E. 1996. Pollination biology and the evolution of mating systems in flowering plants. *Evolutionary Biology* 29: 107–149.

Klavins, S. D., D. W. Kellogg, M. Krings, E. L. Taylor, and T. N. Taylor. 2005. Coprolites in a Middle Triassic cycad pollen cone: evidence for insect pollination in early cycads? *Evolutionary Ecology Research* 7: 479–488.

Krassilov, V. A., and A. P. Rasnitsyn. 1999. Plant remains from the gut of fossil insects: evolutionary and paleoecological inferences. In *AMBA Projects AM/PFICM98/1.99: Proceedings of the First International Palaeoentomological Conference, Moscow (1998)*, pp. 65–72.

Krassilov, V. A., A. P. Rasnitsyn, and S. A. Afonin. 1998. Pollen morphotypes from the intestine of a Permian booklouse. *Review of Palaeobotany and Palynology* 106: 89–96.

Labandeira, C. C., J. Kvacek, and M. B. Mostovski. 2007. Pollination drops, pollen, and insect pollination of Mesozoic gymnosperms. *Taxon* 56: 663–695.

Lande, R., and D. W. Schemske. 1985. The evolution of self-fertilization and inbreeding depression in plants. I. Genetic models. *Evolution* 39: 24–40.

Lloyd, D. G. 1992. Self- and cross-fertilization in plants. II. The selection of self-fertilization. *International Journal of Plant Sciences* 153: 370–380.

Lupia, R., P. S. Herendeen, and J. A. Keller. 2002. A new fossil flower and associated coprolites: evidence for angiosperm-insect interactions in the Santonian (Late Cretaceous) of Georgia, U.S.A. *International Journal of Plant Sciences* 163: 675–686.

Mulcahy, D. L., and G. B. Mulcahy. 1987. The effects of pollen competition. *American Scientist* 75: 44–50.

Osborn, J. M. 2000. Pollen morphology and ultrastructure of gymnospermous anthophytes. In M. M. Harley, C. M. Morton, and S. Blackmore (eds.), *Pollen and Spores: Morphology and Biology*, pp. 163–185. London: Royal Botanic Gardens, Kew.

Osborn, J. M., and T. N. Taylor. 1994. Comparative ultrastructure of fossil gymnosperm pollen and its phylogenetic implications. In M. H. Kurmann and J. A. Doyle (eds.), *Ultrastructure of Fossil Spores and Pollen*, 99–121. London: Royal Botanic Gardens, Kew.

————. 1995. Pollen morphology and ultrastructure of the Bennettitales: in situ pollen of *Cycadeoidea*. *American Journal of Botany* 82: 1074–1081.

Osborn, J. M., T. N. Taylor, and M. R. de Lima. 1993. The ultrastructure of fossil ephedroid pollen with gnetalean affinities from the Lower Cretaceous of Brazil. *Review of Palaeobotany and Palynology* 77: 171–184.

Owens, J. N., A. M. Colangeli, and S. J. Simpson. 1991. Factors affecting seed set in Douglas-fir (*Pseudotsuga menziesii* (Mirb.) Franco). *Canadian Journal of Botany* 69: 229–238.

Owens, J. N., G. L. Catalano, S. J. Morris, and J. Aitken-Christie. 1995. The reproductive biology of kauri (*Agathis australis*). II. Male gametes, fertilization, and cytoplasmic inheritance. *International Journal of Plant Sciences* 156: 404–416.

Owens, J. N., J. Bennett, and S. L'Hirondelle. 2005. Pollination and cone morphology affect cone and seed production in lodgepole pine seed orchards. *Canadian Journal of Forest Research* 35: 383–400.

Reymanówna, M. 1960. A cycadoidean stem from the western Carpathians. *Acta Palaeobotanica* 1: 3–28.

Rothwell, G. W., and R. A. Stockey. 2010. Independent evolution of seed enclosure in the Bennettitales: evidence from the anatomically preserved cone *Foxeoidea connatum* gen. et sp. nov. In C. T. Gee (ed.), *Plants in Mesozoic Time: Morphological Innovations, Phylogeny, Ecosystems*, pp. 51–64. Bloomington: Indiana University Press.

Rothwell, G. W., W. L. Crepet, and R. A. Stockey. 2009. Is the anthophyte hypothesis alive and well? New evidence from the reproductive structures of Bennettitales. *American Journal of Botany* 96: 296–322.

Runions, J. C., and J. N. Owens. 1996. Evidence of pre-zygotic self-incompatibility in a conifer. In S. J. Owens and P. J. Rudall (eds.), *Reproductive Biology*, pp. 255–264. London: Royal Botanic Gardens, Kew.

Saiki, K., and Y. Yoshida. 1999. A new bennettitalean trunk with unilacunar five-track nodal structure from the Upper Cretaceous of Hokkaido, Japan. *American Journal of Botany* 86: 326–332.

Stebbins, G. L. 1970. Adaptive radiation of reproductive characteristics in angiosperms. I. Pollination mechanisms. *Annual Review of Ecology Evolution and Systematics* 361: 307–326.

Stevenson, D. W., K. J. Norstog, and P. K. S. Fawcett. 1998. Pollination biology of cycads. In S. J. Owens and P. J. Rudall (eds.), *Reproductive Biology in Systematics, Conservation and Economic Botany*, pp. 277–294. London: Royal Botanic Gardens, Kew.

Stockey, R. A., and G. W. Rothwell. 2003. Anatomically preserved *Williamsonia* (Williamsoniaceae): evidence for bennettitalean reproduction in the Late Cretaceous of western North America. *International Journal of Plant Science* 164: 251–262.

Taylor, T. N. 1973. A consideration of the morphology, ultrastructure, and multicellular microgametophyte of *Cycadeoidea dacotensis* pollen. *Review of Palaeobotany and Palynology* 16: 157–164.

Taylor, M. L., B. L. Gutman, N. A. Melrose, A. M. Ingraham, J. A. Schwartz, and J. M. Osborn. 2008. Pollen and anther ontogeny in *Cabomba caroliniana* (Cabombaceae, Nymphaeales). *American Journal of Botany* 95: 399–413.

Taylor, T. N., E. L. Taylor, and M. Krings. 2009. *Paleobotany: The Biology and Evolution of Fossil Plants*, 2nd ed. San Diego: Academic Press.

Watson, J., and C. A. Sincock. 1992. Bennettitales of the English Wealden. *Monograph of the Palaeontographical Society, London* 145: 1–228. (Publ. No. 588).

Wieland, G. R. 1906. *American Fossil Cycads*. Publication No. 34. Washington, D.C.: Carnegie Institution of Washington.

———. 1916. *American Fossil Cycads*, vol. 2, *Taxonomy*. Publication No. 34. Washington, D.C.: Carnegie Institution of Washington.

Williams, C. G. 2007. Re-thinking the embryo lethal system within the Pinaceae. *Canadian Journal of Botany* 85: 667–677.

Willson, M. F., and N. Burley. 1983. *Mate Choice in Plants: Tactics, Mechanisms, and Consequences*. Princeton, N.J.: Princeton University Press.

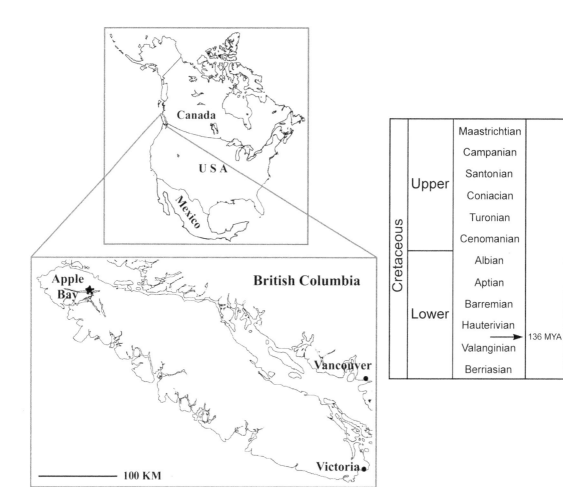

INDEPENDENT EVOLUTION OF SEED ENCLOSURE IN THE BENNETTITALES: EVIDENCE FROM THE ANATOMICALLY PRESERVED CONE *FOXEOIDEA CONNATUM* GEN. ET SP. NOV.

4

Gar W. Rothwell and Ruth A. Stockey

A new anatomically preserved ovulate cone from the Lower Cretaceous of western Canada reveals an independent evolution of ovule enclosure in the Bennettitales. *Foxeoidea connatum* gen. et sp. nov. consists of an ovulate receptacle bearing tightly packed and highly fused seeds and interseminal scales. The cone axis has a eustele from which collateral bundles diverge to vascularize individual seeds and interseminal scales. Interseminal scales are polygonal in cross section and consist of a central bundle surrounded by ground tissue and an epidermis. Seeds are erect with an exposed micropyle and have a multilayered integument with a distinctive inner epidermis. The nucellus separates from the integument at the chalaza, surrounds the megagametophyte, and terminates in a parenchymatous apical plug that is tightly inserted into the base of the micropylar canal. The nucellus is vascularized by a chalazal bundle. There is also a ring of small bundles at the base of the integument. Megagametophytes are represented by a noncellular region at the base of the nucellus. No megaspore membrane is evident. *Foxeoidea* demonstrates that enclosure of seeds in the Bennettitales has been achieved independently of the analogous condition in flowering plants and the Gnetales, adding additional evidence for calling into question the anthophyte hypothesis for the origin of the angiosperm flower.

Fig. 4.1. Geographic and stratigraphic occurrence of the Apple Bay fossil plant locality from which *Foxeoidea* was recovered. *MYA = million years ago.*

Introduction

The Bennettitales is an extinct order of Mesozoic seed plants commonly implicated in the origin of angiosperms. Several vegetative and reproductive characters found in the bennettitaleans are suggestive of a close relationship between the two clades, including syndetocheilic (≈paracytic) stomata, scalariform pitting of the secondary xylem tracheids, and flowerlike bisporangiate cones with seeds at the apex of a receptacle. In several species, the compact cones produce microsporophylls below the seeds, and the terminal fertile zone is subtended and surrounded by bracts (i.e., a putative perianth analog; Arber and Parkin 1907; Stewart

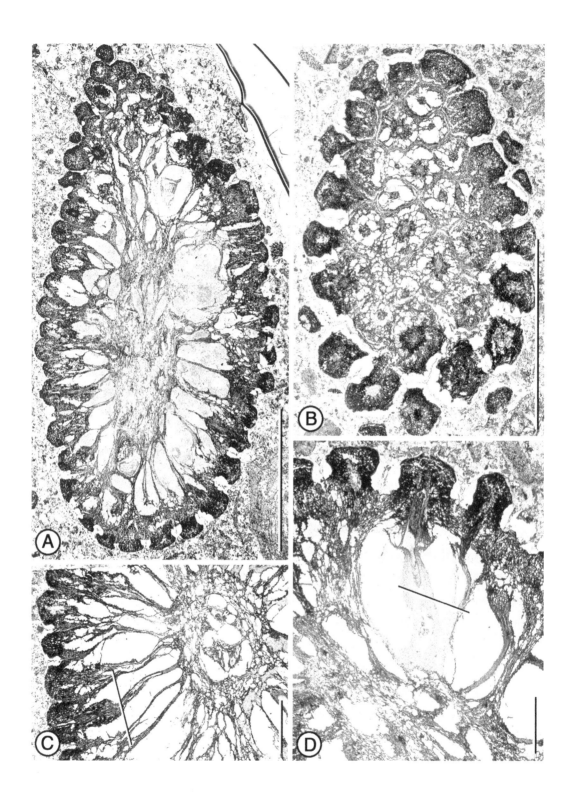

and Rothwell 1993; Taylor et al. 2009). An additional angiosperm-like feature is the partial enclosure of bennettitalean seeds. In bennettitalean cones, enclosure is accomplished by interseminal scales that surround individual seeds on the ovulate region of the receptacle (Wieland 1906; Delevoryas 1959, 1963, 1965, 1968a; Crane 1985).

Hypotheses for a close relationship between the Bennettitales and flowering plants are routinely supported by the results of numerical cladistic studies that incorporate extinct taxa in the analysis of morphological characters and that consistently resolve the Bennettitales, Gnetales, and flowering plants as a clade on the seed plant tree (Crane 1985; Doyle and Donoghue 1986, 1992; Nixon et al. 1994; Rothwell and Serbet 1994; Doyle 1996, 2006; Hilton and Bateman 2006; Rothwell et al. 2009). However, evolutionary relationships among the three groups and the possible role of the Bennettitales in the origin of angiosperms remain incompletely understood and controversial (cf. Friis et al. 2007; Rothwell et al. 2009; Crepet and Stevenson 2010; Cúneo et al. 2010). Whereas some authors hypothesize that the bisporangiate flowerlike cones of the Bennettitales reflect the ancestral morphology from which angiosperm floral organs may have arisen (e.g., Arber and Parkin 1907; Doyle and Donoghue 1987), other authors interpret the similarities to represent parallel evolution of aggregated fertile structures (i.e., cones) and the parallel evolution of seed enclosure; Rothwell et al. 2009).

Although a wealth of compressed bennettitalean remains from around the world has markedly enhanced our understanding of morphological variation and constitutes solid data for inferring the paleobiogeography of the group (e.g., Harris 1969; Watson and Sincock 1992), anatomically preserved specimens provide the strongest evidence for structure and homologies of the reproductive structures (e.g., Solms-Laubach 1891; Lignier 1894; Seward 1912; Stopes 1918; Sharma 1970; Crepet 1974). This is particularly demonstrated by the insightful and detailed investigations of permineralized specimens assignable to both the Cycadeoidaceae and Williamsoniaceae conducted by Theodore Delevoryas from the late 1950s through the 1970s (e.g., Delevoryas 1959, 1960, 1963, 1965, 1968a, 1968b; Crepet and Delevoryas 1972; Delevoryas and Gould 1973). In addition to dramatically increasing our knowledge of the vegetative structure and diversity among species of *Cycadeoidea* Buckland and *Monanthesia* Wieland ex Delevoryas, Delevoryas developed important new insights about the morphology, anatomy, and development of bennettitalean cones and about the homologies of bennettitalean reproductive organs. Together with similar specimens that have been described from around the world for more than 100 years (e.g., Lignier 1894, 1901, 1911; Wieland 1906; Stopes 1918; Sharma 1970, 1973; Crepet 1974; Nishida 1994; Ohana et al. 1998; Rothwell and Stockey 2002; Stockey and Rothwell 2003; Rothwell et al. 2009), the anatomically preserved cones characterized by Delevoryas provide convincing evidence for the structure, diversity, and homologies of bennettitalean reproductive structures.

Fig. 4.2. *Foxeoidea connatum* gen. et sp. nov. Holotype. (A) Oblique section showing general features of cone. P13158D Bot 190, magnification 7.5×. Scale bar = 5 mm. (B) Tangential section showing overall features. Note connate seeds and interseminal scales except toward periphery of cone. P13158 D Top 53, magnification 10×. Scale bar = 5 mm. (C) Cross section of cone showing appendages radiating from central axis. Note appendages connate to near periphery of cone. Line indicates approximate boundary between receptacle and bases of seeds and interseminal scales. P13158 D_1 Side 27, magnification 14×. Scale bar = 1 mm. (D) Oblique section of cone showing histological features of cone axis (lower left), with connate ovules and interseminal scales. Line indicates approximate boundary between receptacle and bases of seeds and interseminal scales. P13158 D Bot 158, magnification 15×. Scale bar = 1 mm.

Among the exquisitely preserved permineralized specimens that make up the diverse Lower Cretaceous flora from Apple Bay near the north end of Vancouver Island in British Columbia, Canada (Stockey and Rothwell 2006), there are several cones assignable to the Bennettitales that consist of a receptacle bearing seeds and interseminal scales. One of the specimens has been described earlier as an immature cone of the genus *Williamsonia* (Rothwell et al. 2009). The specimen upon which the current investigation is focused is also an ovulate receptacle bearing both seeds and interseminal scales, but it differs from all previously known Bennettitales by having the seeds and interseminal scales highly fused both to the receptacle and to each other. The resulting fructification superficially resembles a multiovulate angiosperm carpel, but the seed micropyles are exposed at the exterior. This new seed cone, *Foxeoidea connatum* gen. et sp. nov., extends the range of structural variation known to occur among bennettitalean reproductive structures. It also provides direct evidence for the parallel evolution of seed enclosure in this important clade of Mesozoic gymnosperms and in flowering plants, thereby casting further doubt on the validity of the anthophyte hypothesis sensu Arber and Parkin (1907; Doyle and Donoghue 1987) for the origin of the angiosperm flower.

Materials and Methods

The current study is based on an anatomically preserved seed cone found in Lower Cretaceous (Hauterivian–Valanginian) sediments on the shore of Apple Bay, Vancouver Island, British Columbia, Canada (Fig. 4.1; Stockey et al. 2006). The rhizome is preserved in a calcareous concretion embedded in a graywacke matrix. It is oriented in an oblique longitudinal plane as originally exposed on the upper and lower surfaces of slab P13158 D. The cone was first revealed near its outer surfaces and then serially sectioned toward the center from both sides by the cellulose acetate peel technique (Joy et al. 1956). This included 220 peel sections from the upper surface and 160 sections from the lower surface of the slab. The specimen was subsequently reoriented and peeled in cross section (i.e., 60 sections) from the base toward the apex to expose features in transverse view. Peels for microscopic examination and image capture were mounted on microscope slides with Eukitt mounting medium (O. Kindler GmbH, Freiburg, Germany). Images were captured with a Photophase (Phase One A/S, Frederiksberg, Denmark) digital scanning camera and processed with Adobe Photoshop (San Jose, Calif.). The specimen, peels, and microscope slides are housed in the University of Alberta Paleobotanical Collections, Edmonton, Alberta, Canada (UAPC-ALTA).

Order Bennettitales
Family unknown

FOXEOIDEA ROTHWELL ET STOCKEY GEN. NOV.

Diagnosis. Bennettitalean cones consisting of a dispersed ovulate receptacle covered by closely spaced erect seeds and interseminal scales. Seeds and interseminal scales fused to receptacle and each other up to level of seed micropyles. Vascular traces to seeds and interseminal scales diverging from eustele of receptacle, with single terete bundle entering base of each interseminal scale and seed. Bundle dividing at seed chalaza to vascularize base of nucellus and forming ring of terete bundles in base of integument.

Etymology. The generic name *Foxeoidea* is proposed in honor of Theodore Delevoryas, "Foxy Grandpa," for his numerous enlightening and insightful contributions to our understanding of the Cycadeoidales (=Bennettitales).

FOXEOIDEA CONNATUM ROTHWELL ET STOCKEY SP. NOV.
FIGS. 4.2 AND 4.3; PLATES 4 AND 5

Diagnosis. Ellipsoidal cone consisting of receptacle with attached erect seeds and interseminal scales measuring approximately 2 cm long and 0.8 cm in maximum diameter. Seeds and interseminal scales 2 mm long, constructed of common tissue basally, separating as distinct organs in region of seed micropyle. Receptacle with parenchymatous pith surrounded by eustele of cauline bundles and thin wedges of secondary xylem tracheids. Terete traces of tracheids each diverging from cauline bundle and extending into base of either interseminal scale or seed. Common parenchymatous ground tissue of apparently immature interseminal scales and seed integuments with prominent intercellular spaces; cell lumina with light-colored contents proximally; cell contents becoming darker distally. Nucellus separating from integument near seed base, extending distally to form cylindrical plug of parenchymatous tissue in base of micropylar canal. Megagametophyte represented by noncellular area at base of nucellus, surrounded by degraded-appearing cells with dark contents; megaspore membrane not evident.

Holotype hic designatus. Permineralized slab, peels, and slides of specimen P13158 D housed in the University of Alberta Paleobotanical Collection (UAPC-ALTA), Edmonton, Alberta, Canada; Figures 4.2 and 4.3 and Plates 4 and 5.

Collecting locality. Apple Bay, northern Vancouver Island, British Columbia, Canada (49° 54' 42" N, 125° 10' 40" W; UTM 10U CA 5531083N 343646E).

Stratigraphic position and age. Longarm Formation equivalent. Valanginian–Hautervian boundary, Early Cretaceous.

Etymology. The specific epithet *connatum* refers to the cellular continuity among the receptacle, seeds, and interseminal scales of the cone.

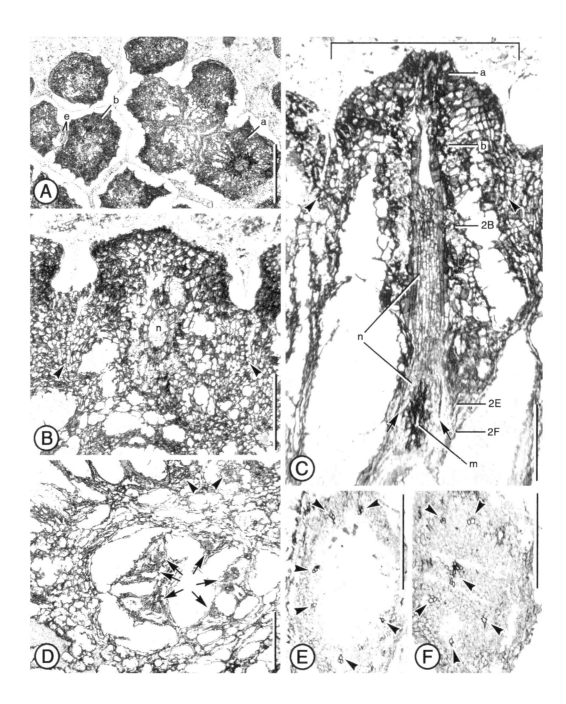

Description
The *Foxeoidea connatum* seed cone is represented by a eustelic receptacle (=cone axis) that bears tightly packed ovules and interseminal scales along the sides and over the apex (Fig. 4.2A–B; Plate 4A, B). The receptacle truncates at the base, as if having been abscised. More proximal organs of the plant are unknown. As is characteristic of immature bennettitalean cones (e.g., Seward 1912; Sharma 1970; Rothwell and Stockey 2002; Stockey and Rothwell 2003), the holotype of *F. connatum* consists largely of a cone axis (=receptacle), with the small seeds and interseminal

scales making up a comparatively narrow zone at the periphery (Fig. 4.2A, C to left of line, D above line; respectively, Plate 4A, C, D.). The cone is ellipsoidal, measuring approximately 2 cm long and up to 0.8 cm wide in the central region. The outer surface of the specimen is covered by a large number of roughly hemispherical protrusions (Fig. 4.2A–D; Plate 4A–D) that represent the apices of seeds and interseminal scales (Figs. 4.2D, 4.3A–C; Plates 4D, 5A–C), which give it a distinctly lumpy appearance.

The cone is nearly circular in cross section at the base, where it shows a small parenchymatous pith with large intercellular spaces (Figs. 4.2C, 4.3D; Plates 4C, 5D). At the basalmost level preserved, the stele is surrounded by a wide zone of cortex, but more distally, seeds and interseminal scales radiate from the cone axis (Fig. 4.2C; Plate 4C). Near the cone base, the eustele has five to seven cauline xylem bundles (Fig. 4.3D, at arrows; Plate 5D), each with a wedge of secondary xylem tracheids radially aligned to the outside. Phloem is incompletely preserved. Oblique sections of the midregion show no secondary xylem (Fig. 4.2A, Plate 4A). To the periphery of the stele, in the parenchymous ground tissue, there are terete bundles (Fig. 4.3D, at arrowheads; Plate 5D) that represent traces to seeds and interseminal scales.

Throughout the pith and cortex of the axis, and toward the bases of the ovules and interseminal scales, parenchymatous cells of the ground tissues are interrupted by large and smaller intercellular spaces (Figs. 4.2A–D, 4.3A–D; Plates 4A–D, 5A–D). The size of such spaces within the appendages diminishes distally, with the parenchyma becoming increasingly compact toward the apices of seeds and interseminal scales (e.g., Figs. 4.2A–C, 4.3C; Plates 4A–C, 5C). We are uncertain whether the development of such spaces has an ontogenetic origin, whether the spaces result from taphonomic alteration (i.e., postmortem tissue shrinkage), or whether it is a combination of the two.

One of the most distinctive features of *Foxeoidea* is the high degree of fusion among seeds and interseminal scales. These organs display a common ground tissue for four-fifths of their length (Fig. 4.2A, C, D; Plate 4A, C, D), with only the tips of interseminal scales and seeds separating from each other (Figs. 4.2D, 4.3A; Plates 4D, 5A) to form the lumpy surface of the cone. At the periphery of the cone, adjacent seeds and interseminal scales are clearly discernible from each other both because they are separated by a space, and because each has a prominent epidermis that extends downward to somewhat below the level where they are separated by the obvious space (i.e., to the arrowheads in Fig. 4.3B, C; Plate 5B, C). At more proximal levels, however, adjacent appendages share a common ground tissue (Figs. 4.2B at center, 4.3B, C; Plates 4B, 5B, C). At the more proximal levels, adjacent seeds and/or interseminal scales are recognizable as distinct organs only because each has its own vascular bundle.

The ratio of seeds to interseminal scales is much higher in *F. connatum* than in other bennettitalean species that have been described previously, ranging between 1:1 and 2:3 in different regions of the cone

Fig. 4.3. *Foxeoidea connatum* gen. et sp. nov. Holotype. (A) Tangential section of cone through apical region of interseminal scales and seeds. Most are free from each other at this level, except those at right center sectioned at level where they become interconnected. Seed marked "a" is sectioned at level "a" in (C); seed marked "b" is sectioned at level "b" in (C); e = epidermis. P13158 D Top 24, magnification 15×. Scale bar = 1 mm. (B) Oblique cross section of seed apex and adjacent seeds and/or interseminal scales at level "6" of (C). Note inner epidermis of integument and incompletely preserved plug of nucellar tissue (n) occluding micropylar canal. Note also that seed and adjacent appendages separated by epidermal cells only in apical region (above arrowheads). P13158 D Bot 220, magnification 40×. Scale bar = 0.5 mm. (C) Longitudinal section of seed showing features of ovule and adjacent seeds and/or interseminal scales. Arrowheads indicate depth to which adjacent organs separated by epidermal layers. Narrow micropylar canal of seed extends within broad multicellular micropyle and displays characteristic inner epidermis of bennettitalean seeds. Bracket indicates width of micropylar end of seed. Nucellus free from integument above chalaza (at arrows), small megagametophyte (m) at base of nucellus (dark area near bottom); nucellus (n) consists of solid parenchymatous tissue distally. Note tip of nucellus plugs basal region of micropylar canal. Lines identified by numbers and letters correspond to levels of cross section in those figures with the corresponding

letters and numbers (i.e.,
B, E, and F). P13158 D Bot
19,0 magnification 40×.
Scale bar = 0.5 mm. (D)
Cross section of cone axis
showing cauline bundles of
eustele (at arrows) in basal
receptacle, and bases of
seeds and interseminal scales
(at top). Traces to seeds or
interseminal scales identified
by arrowheads. P13158 D,
Side 27, magnification 30×.
Scale bar = 1 mm. (E) Cross
section above separation of
nucellus from integument
(i.e., at level "9" of C) showing ring of small bundles
of tracheids in undifferenti-
ated tissue of integument.
P13158 D Bot 95, magnification 50×. Scale bar = 1 mm.
(F) Cross section of seed at
level below fusion of nucellus with integument (i.e., at
level "10" of C), showing
tracheids of nucellus (at central arrowhead) surrounded
by tiny bundles of integumentary tracheids (peripheral arrowheads). P13158 D
Bot. 168, magnification 50×.
Scale bar = 1 mm.

surface. Therefore, each seed is not surrounded by only interseminal scales, and it is common for seeds to lie adjacent to two or even three other seeds (Figs. 4.2B, 4.3A; Plates 4B, 5A). Distal to the level where they separate from each other, interseminal scales and seeds are polygonal in cross section (Figs. 4.2B, 4.3A; Plates 4B, 5A). As is characteristic of other bennettitalean cones, the number of sides is variable and correlated with the number of other organs by which each is surrounded. In *F. connatum*, the number of sides is typically five or six.

Interseminal scales consist of a terete trace surrounded by several layers of parenchymatous ground tissue and a uniseriate epidermis (Fig. 4.3A; Plate 5A). At the most distal levels, they are difficult to distinguish from seeds in cross section (Figs. 4.2B, 4.3A; Plates 4B, 5A). This is because the center of each organ is often incompletely preserved, the seed integument is histologically similar to the cortex of the interseminal scales, and the seed integument is as thick as the cortex of the interseminal scales. However, at slightly more proximal levels, the two organs are easily identified because interseminal scales have a central vascular bundle, while the seeds show a micropylar canal that is lined by a prominent inner integumentary epidermis (Figs. 4.2B, 4.3A–C; Plates 4B, 5A–C).

Seeds of the holotype are relatively immature and are actually ovules in a developmental sense. The megagametophyte is confined to the base of the nucellus (Fig. 4.3C at M; Plate 5C), where it is represented by a noncellular area surrounded by degraded-appearing cells of the mega-gametophyte. As in other bennettitalean seeds, there is no evidence of a megaspore membrane. In longitudinal sections, the seeds show a distinct epidermis and cuticle on adjacent surfaces of the nucellus and integument that extends down to nearly the base of the megagametophyte (Fig. 4.3C, at arrows; Plate 5C). Therefore, the nucellus and integument separate from each other slightly more distally than in other bennettitalean cones (e.g., Rothwell et al. 2009). Distal to the megagametophyte, the nucellus consists of closely spaced parenchyma cells with an epidermis that is most obvious at the apex (Fig. 4.3C; Plate 5C). As is characteristic of bennettitalean seeds, the tip of the nucellus produces a plug of tissue that is inserted into the base of the micropylar canal (Fig. 4.3B, C; Plate 5B, C), rather than forming a pollen chamber like those produced by many other gymnospermous seeds. In *F. connatum*, the nucellar plug extends up to about the level where the seeds and interseminal scales separate from each other (Fig. 4.3C; Plate 5C).

The distinct epidermis on the inner surface of the integument consists of rectangular isodiameteric cells up to the level where the nucellar plug occludes the micropyle (Fig. 4.3A–C; Plate 5A–C). At more distal levels, the cells elongate radially, forming the distinctive radiating pattern (Fig. 4.3A, lower right; Plate 5A) that characterizes most other bennettitalean seeds at the same level (Rothwell et al. 2009). As described above, the integument consists of several layers of parenchyma cells that are histologically indistinguishable from the ground tissue of the interseminal scales (Figs. 4.2A–D, 4.3A–C; Plates 4A–D, 5A–C). Below the apical

Table 4.1. Morphological Complexity and Potential Homologies among the Cones of the Bennettitales, Gnetales, and the Flowers of Angiosperms

Character	Bennettitales	Gnetales	Flowering Plants
1. Bisporangiate fructification	Present	Present	Present
2. Megasporangiate organs distalmost (at apex of receptacle)	Present	Present	Present
3. Microsporangiate organs subtending megasporangiate organs	Present	Present	Present
4. Perianth (= vegetative leaves) subtending fertile organs	Present	Present	Present
5. Morphological complexity of fertile shoot (= cone or flower)	Simple	Compound	Simple, or simple with compound megasporangiate apex[a]
6. Perianth differentiated as sepals and showy petals	Absent	Absent	Present/absent
7. Microsporophyll morphology	Pinnate or pinnate derived[b]	Simple	Simple
8. Synangiate microsporangia	Present	Present (in some)	Present
9. Seeds surrounded/enclosed by interseminal scales	Pair(s) of subtending bracts	Leaf lamina, or single subtending bracts	Subtending bract[a]
10. Anatropous ovules	Absent	Absent	Present/absent
11. Micropyle exposed to environment	Present	Present	Absent
12. Double integument	Absent	Absent	Absent/present
13. Pollen chamber	Absent	Present	Absent
14. Branched haustorial pollen tubes	Present	Absent	Absent
15. Prezygotic selection mediated by tissue?[c]	Ovule	Absent	Carpel[c]

[a] Depending on the ancestral morphology and evolutionary origin of the angiosperm carpel, this organ may represent either a sporophyll or a condensed shoot (Taylor and Kirchner 1996). See text for explanation.

[b] See Rothwell et al. (2009) for detailed discussion.

[c] Prezygotic selection (i.e., regulation of pollen tube growth; for explanation, see Dilcher 2010), when present in angiosperms, is mediated by tissue of the carpel. It is not known whether the Bennettitales had a prezygotic selection mechanism, but if they did, that mechanism had to be mediated by tissues of the ovule that lie between the position of pollination (i.e., micropyle distal to nucellar plug; Stockey and Rothwell 2003). The distinctive radially elongated inner epidermal cells that line the micropyles of bennettitalean seeds (e.g., Fig. 4.2A; Rothwell et al. 2009) and also characterize dispersed fossil seeds with apparent gnetalean affinities (Friis et al. 2007) appear to be the site of pollination in representatives of both groups and may reflect the parallel evolution this aspect of reproductive biology in the two clades.

region of each seed, the integument is split by the large intercellular space that extends peripherally from the cortex of the cone axis (Figs. 4.2D, 4.3C; Plates 4D, 5C). Therefore, that space does not represent the outside of the integument, which is adnate to the surrounding organs except at the apex.

Discussion

As a result of the highly fused nature of the seeds and interseminal scales, affinities of *Foxeoidea* are not obvious at first glance. However, closer examination reveals a combination of characters that is diagnostic for bennettitalean ovulate cones (Delevoryas 1963, 1968a; Sharma 1970, 1973; Crepet 1974; Rothwell et al. 2009). These characters include a eustelic receptacle bearing tightly packed seeds and interseminal scales, each of which is vascularized by a single terete trace. Seeds are erect with an exposed micropyle. The nucellus separates from integument near the chalaza and does not produce a pollen chamber. Rather, it forms a parenchymatous nucellar plug that is tightly inserted into the base of the micropylar canal. The integument consists of several cell layers and has a distinctive inner epidermis of cells that are radially elongated distal to the nucellar plug. No obvious megaspore membrane is preserved at the periphery of the megagametophyte.

Whether *Foxeoidea* produced bisporangiate cones like those of *Cycadeoidea* (e.g., Wieland 1906; Delevoryas 1968a; Crepet 1974), *Cycadeoidella* (e.g., Nishida 1994), *Monanthesia* (Delevoryas 1959), and similar genera, or monosporangiate cones like those of *Williamsonia* (e.g., Harris 1969; Watson and Sincock 1992; Delevoryas and Gould 1973) is unknown because the nature of organs proximal to the ovulate receptacle of *F. connatum* is unknown. However, the combination of highly fused seeds and interseminal scales and a vascularized seed integument clearly distinguishes *Foxeoidea* from all previously described bennettitalean fructifications. There is a possibility that the highly fused nature of *Foxeoidea* seeds and interseminal scales could be related to the immaturity of the specimen, but that is unlikely because tracheids of the xylem and some sclerenchyma have already differentiated in the specimen. Also, specimens of *Cycadeoidea* and *Williamsonia* cones that are preserved at about the same stage of maturity show ovulate sporophylls and interseminal scales that are not fused at all (Rothwell et al. 2009).

Reproductive Structures of the Bennettitales, Gnetales, and Angiosperms

The reproductive biology of most seed plant groups involves the location of seed-producing and pollen-producing organs on either different parts of the same plant or on different individuals. This includes nearly all seed fern clades, cycads, ginkgophytes, and conifers (Stewart and Rothwell 1993; Hilton and Bateman 2006; Taylor et al. 2009; Dilcher 2010). In contrast, the megasporangiate and microsporangiate organs of many Bennettitales, Gnetales, and flowering plants are aggregated into either functional or potentially functional bisporangiate fructifications (i.e., cones or flowers; Stewart and Rothwell 1993; Hilton and Bateman 2006; Taylor et al. 2009). In addition, all three groups produce fructifications in which megasporangiate structures are distal to the microsporangiate structures, and vegetative leaflike structures surround the fertile zone (Table 4.1).

The most commonly discussed and seriously considered proposal for evolution of the angiosperm flower stems from the recognition of that common basic fertile organ structure (i.e., characters 1–3 of Table 4.1; Arber and Parkin 1907). In what was much later named the "anthophyte hypothesis" (Doyle and Donoghue 1987), Arber and Parkin (1907, 1908) interpreted the angiosperm flower as being equivalent to a morphologically simple bisporangiate cone like those produced by many species of the Bennettitales. These authors hypothesized that bisporangiate angiosperm flowers, bennettitalean cones, and gnetalean cones all evolved through a hypothetical intermediate (a progenitor?) by the transformation of mega- and microsporophylls that were surrounded by bracts forming a bisporangiate conelike fructification (Fig. 4.4; Arber and Parkin 1907, 1908; Doyle and Donoghue 1987). Evolution of gnetalean cones also required reduction from a more complex inflorescence (see Doyle and Donoghue 1987 for a summary and explanation).

Numerous reiterations and variations of the Arber and Parkin (1907) proposal have appeared over the intervening century (e.g., Friis et al. 2007), and these are bolstered by the occurrence of an "anthophyte topology" in the results of nearly all numerical cladistic studies that include extinct taxa and use morphological characters (Rothwell and Stockey 2002; Rothwell et al. 2009). However, the results of phylogenetic analyses that use nucleotide sequence characters of living plants typically do not include an anthophyte clade (e.g., Burleigh and Mathews 2004). Although the results of the latter analyses yield several discordant topologies for seed plant phylogeny, almost all of these studies produce trees in which flowering plants and gnetophytes are not closely related sister groups (Burleigh and Mathews 2004). Those results and other data have prompted several recent authors to seriously question the validity of homologies upon which the anthophyte hypothesis (sensu Arber and Parkin 1907) is based (e.g., Doyle 1996, 1998, 2006; Donoghue and Doyle 2000).

A comparison of features that characterize fructifications of the Bennettitales, Gnetales, and flowering plants (Table 4.1) reveals that all of the common characters are associated with the aggregation of fertile parts within a terminal fructification (i.e., characters 1–3 of Table 4.1), whereas the numerous details of the fructifications differ among the three (i.e., characters 4–14 of Table 4.1). Questions about the accuracy of the anthophyte hypothesis are probably best addressed by detailed examinations of fertile structures in the Bennettitales, Gnetales, and flowering plants to determine which may be synapomorphies and which are more likely the result of parallel evolution. Such detailed studies by Delevoryas (1963, 1965, 1968a, 1968b) clarified the structure of bennettitalean microsporophylls and documented their dissimilarity to comparable structures in

Fig. 4.4. Diagram depicting longitudinal view of the "proanthostrobilus" of hypothetical ancestor of the Bennettitales, Gnetales, and flowering plants, upon which the anthophyte hypothesis (sensu Arber and Parkin 1907) is based. Note that fructification consists of a simple cone-bearing basal bracts, pinnate microsporophylls with synangiate microsporangia, and apically positioned, pinnately lobed megasporophylls with marginal seeds. *Reproduced from Arber and Parkin (1907: fig. 4).*

the Gnetales and flowering plants. Fusion among seeds and interseminal scales in *Foxeoidea* provides comparable evidence that seed enclosure has been achieved in different ways among the Bennettitales, Gnetales, and angiosperms (i.e., character 9 of Table 4.1). Whether future investigations will verify or falsify the hypothesis that the Bennettitales, Gnetales, and flowering plants form a clade on the spermatophyte tree, it is becoming increasingly clear that many features of the fertile structures in the three clades do not represent synapomorphies (Table 4.1). Therefore, it is becoming decreasingly probable that the anthophyte hypothesis for the evolution of the angiosperm flower (sensu Arber and Parkin 1907) is accurate.

Acknowledgments

Those of us who have had the opportunity to participate in the development of our scientific discipline find ourselves as part of the flow of a continuously expanding intellectual heritage. Our ability to contribute to this flow has been tremendously enhanced by the personal guidance and the example set for us by Ted Delevoryas. To paraphrase a well-worn but appropriate saying, if we are able to touch the sky, it is because we are standing on the shoulders of giants. Ted, we hope our weight is not too heavy a burden for your shoulders.

We thank Joe Morin, Mike Trask, Gerald Cranham, Sharon Hubbard, Thor Henrich, Graham Beard, Dan Bowen, and Pat Trask of the Vancouver Island Paleontological Society for assistance in specimen collecting at the Apple Bay locality. Dr. William L. Crepet, Cornell University, provided the digitized copy of Figure 4.4. This work was supported in part by the National Science Foundation (grant EF-0629819 to G.W.R. and R.A.S.) and NSERC (grant A-6908 to R.A.S.).

References Cited

Arber, E. A. N., and J. Parkin. 1907. On the origin of angiosperms. *Journal of the Linnean Society of London, Botany* 38: 29–80.

———. 1908. Studies on the evolution of the angiosperms. The relationship of the angiosperms to the Gnetales. *Annals of Botany* 22: 489–515.

Burleigh, J. G., and S. Matthews. 2004. Phylogenetic signal in nucleotide data from seed plants: implications for resolving the seed plant tree of life. *American Journal of Botany* 91: 1599–1613.

Crane, P. R. 1985. Phylogenetic analysis of seed plants and the origin of angiosperms. *Annals of the Missouri Botanical Garden* 72: 716–793.

Crepet, W. L. 1974. Investigations of North American cycadeoids: the reproductive biology of *Cycadeoidea*. *Palaeontographica*, Abt. B, 148: 144–169.

Crepet, W. L., and T. Delevoryas. 1972. Investigations of North American cycadeoids: early ovule ontogeny. *American Journal of Botany* 59: 209–215.

Crepet, W. L., and D. W. Stevenson. 2010. The Bennettitales (Cycadoidales): a preliminary perspective on this arguably enigmatic group. In C. T. Gee (ed.), *Plants in Mesozoic Time: Morphological Innovations, Phylogeny, Ecosystems*, pp. 215–244. Bloomington: Indiana University Press.

Cúneo, N. R., I. Escapa, L. Villar de Seoane, A. Artabe, and S. Gnaedinger. 2010. Review of the cycads and bennettitaleans from the Mesozoic of Argentina.

In C. T. Gee (ed.), *Plants in Mesozoic Time: Morphological Innovations, Phylogeny, Ecosystems*, pp. 187–212. Bloomington: Indiana University Press.

Delevoryas, T. 1959. Investigations of North American cycadeoids: *Monanthesia*. *American Journal of Botany* 46: 657–666.

———. 1960. Investigations of North American cycadeoids: trunks from Wyoming. *American Journal of Botany* 47: 778–786.

———. 1963. Investigations of North American cycadeoids: cones of *Cycadeoidea*. *American Journal of Botany* 50: 45–52.

———. 1965. Investigations of North American cycadeoids: microsporangiate structures and phylogenetic implications. *The Palaeobotanist* 14: 89–93.

———. 1968a. Investigations of North American cycadeoids: structure, ontogeny and phylogenetic considerations of cones of *Cycadeoidea*. *Palaeontographica*, Abt. B, 121: 122–133.

———. 1968b. Some aspects of cycadeoid evolution. *Botanical Journal of the Linnean Society* 61: 137–146.

Delevoryas, T., and R. E. Gould. 1973. Investigations of North American cycadeoids: *Williamsonia* cones from the Jurassic of Oaxaca, Mexico. *Review of Palaeobotany and Palynology* 15: 27–42.

Dilcher, D. L. 2010. Major innovations in angiosperm evolution. In C. T. Gee (ed.), *Plants in Mesozoic Time: Morphological Innovations, Phylogeny, Ecosystems*, pp. 97–116. Bloomington: Indiana University Press.

Donoghue, M. J., and J. A. Doyle. 2000. Seed plant phylogeny: demise of the anthophyte hypothesis? *Current Biology* 10: R106–R109.

Doyle, J. A. 1996. Seed plant phylogeny and the relationships of Gnetales. *International Journal of Plant Sciences* 157: S3–S39.

———. 1998. Molecules, morphology, fossils and the relationships of angiosperms and Gnetales. *Molecular Phylogeny and Evolution* 9: 448–462.

———. 2006. Seed ferns and the origin of angiosperms. *Journal of the Torrey Botanical Society* 133: 169–209.

Doyle, J. A., and M. J. Donoghue. 1986. Seed plant phylogeny and the origin of angiosperms: an experimental cladistic approach. *Botanical Review* 52: 321–431.

———. 1987. The origin of angiosperms: a cladistic approach. In E. M. Friis, W. G. Chaloner, and P. R. Crane (eds.), *The Origins of Angiosperms and Their Biological Consequences*, pp. 17–50. Cambridge: Cambridge University Press.

———. 1992. Fossils and seed plant phylogeny reanalyzed. *Brittonia* 44: 89–106.

Friis, E. M., P. R. Crane, K. R. Pedersen, S. Bengston, P. C. J. Donoghue, G. W. Grimm, and M. Stampanoni. 2007. Phase contrast X-ray microtomography links Cretaceous seeds with Gnetales and Bennettitales. *Nature* 450: 549–552.

Harris, T. M. 1969. *The Yorkshire Jurassic Flora III. Bennettitales*. London: British Museum (Natural History).

Hilton, J., and R. M. Bateman. 2006. Pteridosperms are the backbone of seed-plant phylogeny. *Journal of the Torrey Botanical Society* 133: 119–168.

Joy, K. W., A. J. Willis, and W. S. Lacey. 1956. A rapid cellulose peel technique in palaeobotany. *Annals of Botany*, n.s., 20: 635–637.

Lignier, O. 1894. Végétative fossiles de Normandie. Structure et affinities du *Bennettitites Morieri* (Sap. & Mar.). *Mémoirs de la Societié linnéenne de Normandie, Caen* 18: 5–78.

———. 1901. Végétative fossiles de Normandie. III. Étude anatomique du *Cycadeoidea micromyela* Mor. *Mémoirs de la Sociétié linnéenne de Normandie, Caen* 20: 329–370.

———. 1911. Le *Bennettites Morieri* (Sap. et Mar.) Lignier se reproduisait probablement par parthénogénèse. *Bulletin Societié Botanique de France* 59: 425–428.

Nishida, H. 1994. Morphology and the evolution of Cycadeoidales. *Journal of Plant Research* 107: 479–492.

Nixon, K. C., W. L. Crepet, D. W. Stevenson, and E. M. Friis. 1994. A reevaluation of seed plant phylogeny. *Annals of the Missouri Botanical Garden* 81: 484–533.

Ohana, T., T. Kimura, and S. Chitaley. 1998. *Bennetticarpus yezoites* sp. nov. (Bennettitales) from the Upper Cretaceous of Hokkaido, Japan. *Paleontological Research* 2: 108–119.

Rothwell, G. W., and R. Serbet. 1994. Lignophyte phylogeny and the evolution of the spermatophytes: a numerical cladistic analysis. *Systematic Botany* 19: 443–482.

Rothwell, G. W., and R. A. Stockey. 2002. Anatomically preserved *Cycadeoidea* (Cycadeoidaceae), with a reevaluation of the systematic characters for the seed cones of Bennettitales. *American Journal of Botany* 89: 1447–1458.

Rothwell, G. W., W. L. Crepet, and R. A. Stockey. 2009. Is the anthophyte hypothesis alive and well? New evidence from the reproductive structures of Bennettitales. *American Journal of Botany* 96: 296–322.

Seward, A. C. 1912. A petrified *Williamsonia* from Scotland. *Philosophical Transactions of the Royal Society of London B* 203: 101–126.

Sharma, B. D. 1970. On the structure of *Williamsonia* cf. *W. scotica* from the Middle Jurassic Rocks of Rajmahal Hills, India. *Annals of Botany* 34: 289–296.

———. 1973. Anatomy of the peduncle of *Williamsonia* collected from the Jurassic of Amarjola in the Rajmahal Hills, India. *Botanique* 4: 93–100.

Solms-Laubach, H. 1891. On the fructification of *Bennettites gibbsonianus* Carr. *Annals of Botany* 5: 419–454.

Stewart, W. N., and G. W. Rothwell. 1993. *Paleobotany and the Evolution of Plants.* New York: Cambridge University Press.

Stockey, R. A., and G. W. Rothwell. 2003. Anatomically preserved *Williamsonia* (Williamsoniaceae): evidence for bennettitalean reproduction in the Late Cretaceous of western North America. *International Journal of Plant Sciences* 164: 251–262.

———. 2006. The last of the pre-angiospermous vegetation: a Lower Cretaceous flora from Apple Bay, Vancouver Island. In *Abstracts of Advances in Paleobotany. Recognizing the Contributions of David L. Dilcher and Jack A. Wolfe on the Occasion of Their 70th Birthday.* Gainesville: Florida Museum of Natural History, University of Florida. http://www.flmnh.ufl.edu/ paleobotany/meeting/abstract.htm#Stockey.

Stockey, R. A., G. W. Rothwell, and S. A. Little. 2006. Relationships among fossil and living Dipteridaceae: anatomically preserved *Hausmannia* from the Lower Cretaceous of Vancouver Island. *International Journal of Plant Sciences* 167: 649–663.

Stopes, M. C. 1918. New bennettitalean cones from the British Cretaceous. *Philosophical Transactions of the Royal Society of London B* 208: 389–440.

Taylor, D. W., and G. Kirchner. 1996. The origin and evolution of the angiosperm carpel. In D. W. Taylor and L. J. Hickey (eds.), *Flowering Plant Origin, Evolution, and Phylogeny,* pp. 116–140. New York: Chapman & Hall.

Taylor, T. N., E. L. Taylor, and M. Krings. 2009. *Paleobotany: The Biology and Evolution of Fossil Plants,* 2nd ed. San Diego: Academic Press.

Watson, J., and C. A. Sincock. 1992. *Bennettitales of the English Wealden.* London: Palaeontographical Society Monographs.

Wieland, G. R. 1906. *American Fossil Cycads.* Publication No. 34. Washington, D.C.: Carnegie Institution of Washington.

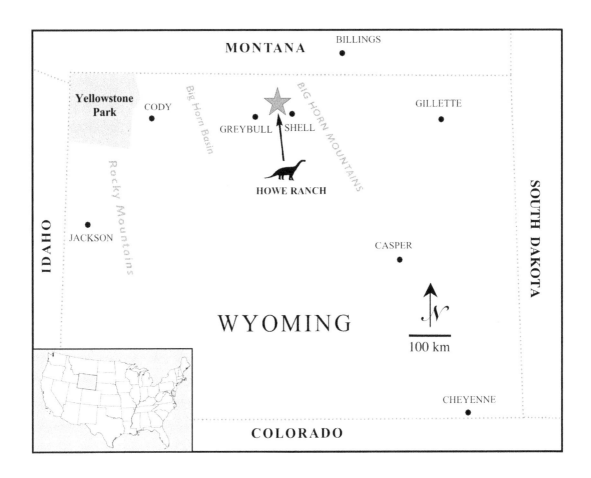

A MOSAIC OF CHARACTERS IN A NEW WHOLE-PLANT *ARAUCARIA, A. DELEVORYASII* GEE SP. NOV., FROM THE LATE JURASSIC MORRISON FORMATION OF WYOMING, U.S.A.

5

Carole T. Gee and William D. Tidwell

A new fossil flora consisting of many hundreds of specimens was recovered from the Late Jurassic Morrison Formation (Brushy Basin Member) from a new quarry, the Howe-Stephens Quarry, on the Howe Ranch in north-central Wyoming. These coalified plant compressions occur alongside the remains of a diverse dinosaur fauna and represent the plant parts (woody shoots with leaves, ovulate cones, cone scales, seeds, and pollen cones) pertaining to a single whole-plant species of *Araucaria*, a new species described here as A. *delevoryasii* sp. nov. Gee. This species is distinguished by the large size of the seed cones, the spatulate shape of the cone scales, the nonembedded position of the seed on the surface of the cone scale, the large size of the pollen cones, and the long, awnlike, tapered form of the microsporophylls. The occurrence of pollen cones of this size and morphology sets a minimum geological age of Late Jurassic for the morphological innovation of large sized pollen cones and awnlike microsporophylls in the genus. A. *delevoryasii* does not fit into any existing section of *Araucaria* but displays a mosaic of araucarian characters. A global survey of araucarian reproductive remains in the Mesozoic was compiled to facilitate comparisons between fossil taxa. The survey underlines the prevalence of *Brachyphyllum* type foliage and *Araucariacites* pollen among Mesozoic *Araucaria* spp., as well as points out the existence of several whole-plant *Araucaria* spp. in the fossil record.

Fig. 5.1. Location of Howe Ranch in north-central Wyoming. (Inset) Location of the state of Wyoming in the Western Interior of the United States. *Map modified from Ayer (2000).*

Introduction

Interest in the Araucariaceae has been revitalized in the past 15 years, despite a long-standing acquaintance exceeding two centuries between botanists and this family of Southern Hemisphere conifers. The resurgence in attention in the systematics of this group is due in great part to the discovery in 1994 of *Wollemia nobilis* Jones, Hill et Allen (1995), a new member of the family, which formerly consisted of only two extant genera, *Agathis* Salisbury and *Araucaria* de Jussieu. In the meantime, *Wollemia*-like fossils have also been recognized from the Cretaceous

(Chambers et al. 1998; Cantrill and Falcon-Lang 2001; Cantrill and Raine 2006).

In addition to the discovery of *Wollemia* and *Wollemia*-like fossils, the recent application of phylogenetic analysis of the molecular data to the problem of phylogenetic relationships within the family has yielded some expected—and unexpected—results (Gilmore and Hill 1997; Setoguchi et al. 1998; Codrington et al. 2002). The monophyly of *Araucaria* and *Agathis* is, for example, well supported by an analysis based on the *rbcL* gene. This *rbcL* phylogeny corresponds well to the current recognition of four sections with the genus *Araucaria* (Setoguchi et al. 1998), although relationships between the sections are still unclear. In this phylogeny, for example, section *Bunya* occupies a derived position (Setoguchi et al. 1998), although fossil remains pertaining to this section are among the first to appear in the rock record, as far back as the Early Jurassic (Tidwell and Ash 2006). The recognition of a new section of *Araucaria* known only from fossil remains, section *Yezonia* (Ohsawa et al. 1995), has only added more fuel to the fire of new interest in the family.

That the evolutionary history of *Araucaria* is not as straightforward as was once thought has already been recognized by Stockey (1994) because its fossil record is based to a large extent on detached plant parts, such as ovulate cones, cone scales, or shoots with leaves, that often occur as solitary and isolated plant organs. Thus, the discovery, study, and reconstruction of "whole plants" of *Araucaria* from deep time are essential to identifying the mosaic of characters exhibited by fossil species, which might sort out differently in the living species. Moreover, the appearance of novel characters in fossil araucarians can serve as time calibration points for molecular clock studies.

This chapter describes a new species of a whole-plant *Araucaria*, namely *Araucaria delevoryasii* Gee sp. nov., from the Late Jurassic Morrison Formation of north-central Wyoming, U.S.A. Included here is a global survey of araucarian remains in the Mesozoic, with a focus on reproductive parts.

Geological Setting

The new fossil plant remains originate from deposits of the Late Jurassic Morrison Formation at the Howe Ranch located near Greybull in north-central Wyoming, U.S.A. (Fig. 5.1), on the western slopes of the Big Horn Mountains. There, the Morrison Formation crops out in a southwest–northeast trend, with the Big Horn Basin to the west and the Big Horn Mountains to the east (Ayer 2000). In general, the Morrison Formation is a laterally extensive but relatively thin unit of clastic sediments (see map by Hotton and Baghai-Riding 2010), which is renowned for its high concentration of well-preserved dinosaur skeletons (Foster 2007). Although the age of this formation was under dispute for decades, recent integrated efforts by geologists and paleontologists have led to the consensus that the Morrison Formation was deposited in the Late Jurassic and for the most part during the Kimmeridgian, although sedimentation

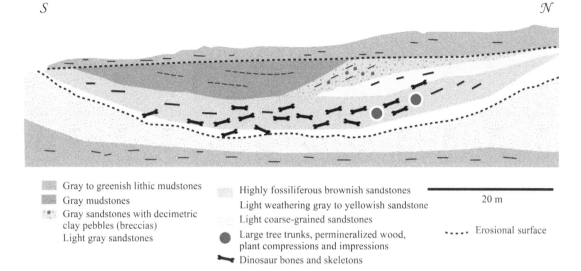

S   N

Gray to greenish lithic mudstones	Highly fossiliferous brownish sandstones
Gray mudstones	Light weathering gray to yellowish sandstone
Gray sandstones with decimetric clay pebbles (breccias)	Light coarse-grained sandstones
Light gray sandstones	Large tree trunks, permineralized wood, plant compressions and impressions
	Dinosaur bones and skeletons

20 m

····· Erosional surface

may have started in the late Oxfordian and continued into the Tithonian in some places (Litwin et al. 1998; Kowallis et al. 1998; Steiner 1998; Turner and Peterson 2004).

At the Howe Ranch, the Morrison Formation consists primarily of sandstones and mudstones with rare layers of limestone and reaches a thickness of 55 m. Carbonaceous plant material is usually concentrated in lenses in cross-bedded sandstones, which range from a few centimeters to several decimeters in thickness (Ayer 2000).

The fossil plant remains presented in this chapter occurred in the matrix of a dinosaur bone bed, the Howe-Stephens Quarry (henceforth called HSQ), on the Howe Ranch, where they were found among bones and skeletons (Fig. 5.2). In most cases, the plants were collected during the systematic excavation of the bone bed, although their locations within the pit were not noted on the quarry maps on which the individual dinosaur bones were plotted (Esther Premru, personal communication to C.T.G., 2004). Nevertheless, the plant and dinosaur fossils were found in the same stratum in immediate proximity to one another (Esther Premru, personal communication to C.T.G., 2004), an unusual occurrence that is limited to a very few sites in the Morrison (cf. Tidwell et al. 1998) or elsewhere in the world. The bone-bed matrix is made up of a cross-bedded beige sandstone, which is interpreted as part of a point bar sequence (Jacques Ayer, personal communication to C.T.G., 2006). Sedimentological and taphonomical work on the HSQ bone bed is currently being carried out by Jacques Ayer at the University of Neuchâtel, Switzerland.

The plant remains are preserved as coalified compressions and permineralized wood, as well as pollen and spores. In a few cases, leaf cuticle is also present. We will describe here the vast majority of the compression flora—that is, various plant organs of the Araucariaceae. The rest of the compression flora, the wood flora, and the palynoflora will be described separately.

Fig. 5.2. Schematic cross section through the Howe-Stephens Quarry. The dinosaur bones and fossil plants (coalified compressions, impressions, and permineralized wood) are intimately associated in the same bone bed. *Redrawn from a stratigraphic section mapped by Jacques Ayer.*

Brief History of Collecting on the Howe Ranch

The fossiliferous nature of the sediments on the Howe Ranch was first discovered by Barnum Brown of the American Natural History Museum in New York in the 1930s (Bird 1985). In a well-publicized, 6-month-long field campaign in 1934, Brown recovered over 2500 scattered bones representing at least 25 individuals of the sauropods *Diplodocus*, *Barosaurus*, and *Apatosaurus* from a now-famous site known as the Howe Quarry (Siber 2004); most of these specimens were subsequently lost through fire or neglect (Siber 2004).

Excavation in the Howe Quarry was restarted by Hans-Jakob Siber of the Sauriermuseum (SMA) in Aathal in Switzerland in 1990. After the bone bed was played out, prospecting for dinosaurs continued elsewhere on the Howe Ranch, resulting in the discovery of several other dinosaur-bearing sites on the property. In 1992, Siber and his crew started digging at a new site roughly 450 m southwest of the original Howe Quarry and struck a productive deposit of bones and plant fossils. This new pit, initially called the G Quarry, became the HSQ and was excavated each field season for 14 years, from 1992 to 2003. The HSQ has turned out to be a truly productive bone bed, yielding 12 dinosaur individuals, ranging from completely articulated to incomplete skeletons, to which pertain 98% of the approximately 2500 bones (Ayer 2000). These represent the remains of eight individuals of sauropods (including one juvenile), two ornithopods, one stegosaur, and one theropod. The material is on display or is deposited at the Sauriermuseum Aathal near Zurich, Switzerland (Ayer 2000).

Materials and Methods

The fossil flora was assembled by two sets of collectors at different times, albeit from the same quarry. The larger part of the flora was collected continuously over a period of 14 years during the Siber excavation of the HSQ. These specimens are deposited in the Siber paleobotany collection of the Sauriermuseum in Aathal, near Zurich, Switzerland, organized by pit and year of excavation. The Siber collection consists of both coalified compression material and permineralized wood. The other part of the flora was collected by W.D.T., Sidney R. Ash, and Leith S. Tidwell from one corner of the quarry during a brief field campaign in 1994. These specimens, composed of only coalified compressions but no permineralizations, are housed at the Museum of Paleontology at Brigham Young University in Provo, Utah.

The compression fossils occur as dark, coalified remains. Their preservation ranges from moderately good to poor. The matrix consists of a well-sorted, very fine to fine-grained sandstone, which ranges in color from pale yellowish orange or dark yellow orange to grayish orange to pale yellowish brown (Rock-Color Chart Committee 1980), depending on the percentage of fine, dark organic particles mixed in with the sand grains.

The compression flora is made up of more than several hundred specimens, although this is only a rough estimate, as several to numerous individual plant parts, such as detached cone scales or seeds, can often be observed on the same piece of rock. Extensive study of the compression fossils was made and measurements were undertaken at the University of Bonn.

Order Coniferales
Family Araucariaceae
Genus *Araucaria* de Jussieu

ARAUCARIA DELEVORYASII GEE SP. NOV.
FIGS. 5.3A–H; 5.4A, B, D–F, I–L; PLATES 6 AND 7

Diagnosis. Evergreen tree with unisexual cones. Ovulate cone large, virtually spherical, up to 8.5 cm in diameter, composed of spatulate ovuliferous scale–bract complexes, each bearing a single unwinged seed at the base of the ovuliferous scale. Ovuliferous scale fused to bract, free in distal half. Bract woody, expanded laterally into symmetrical wings only in distal half of bract. Seed resting on adaxial surface of ovuliferous scale, not embedded or sunken into tissue. Upon maturity, ovulate cone drops to the ground, and scale–bract complexes are abscised. Seed separates from scale–bract complex.

Pollen cone large, ovate to narrowly ovate, to about 60 mm long and 28 mm wide. Microsporophyll numerous, long, narrow, awnlike; each gradually tapered and slightly curved toward cone apex.

Branches and twigs bearing helically arranged, simple, entire, and apetiolate leaves. Leaf shape trullate (trowel shaped) to rhomboid. Entire leaf, including apex, closely appressed to stem.

Holotype. SMA HRFLO 308 (a cone scale, Fig. 5.3H; Plate 6H)

Paratypes. SMA HRFLO 306, 307 (cone scales, Fig. 5.3F, G; Plate 6F, G).

Locality. Howe-Stephens Quarry, Howe Ranch, near Greybull in north-central Wyoming.

Stratigraphy. Morrison Formation (Brushy Basin Member).

Age. Late Jurassic.

Etymology. This species honors Dr. Theodore Delevoryas, professor emeritus at the University of Texas at Austin, where Ted ably and kindly guided C.T.G.'s first steps into the field of Mesozoic paleobotany.

Repositories. Sauriermuseum Aathal in Aathal, Switzerland, and Brigham Young University Museum of Paleontology in Provo, Utah.

Ovulate Cones

The seed-bearing woody cones occur as partially flattened, carbonaceous compressions and protrude at times up to 2 cm from the matrix surface (Fig. 5.3A; Plate 6A). The cones reach a diameter up to 8.5 cm and are assumed to be virtually spherical, as cones are found in nearly equal amounts oriented either perpendicular (in polar view; Fig. 5.3A, B; Plate

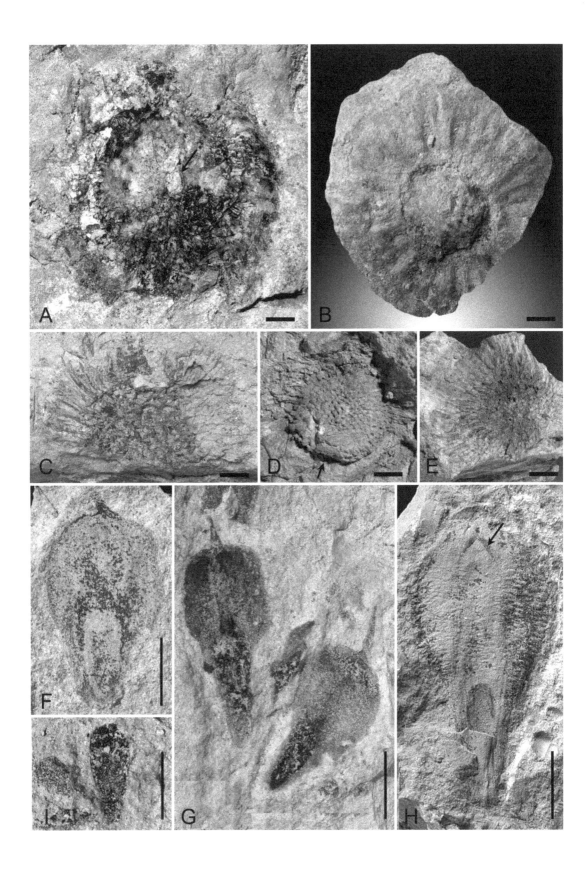

6A, B) or parallel (in lateral view; Fig. 5.3C–E; Plate 6C–E) to the bedding plane. Cones in lateral view have a carbonaceous center composed of a spiraling pattern of rhomboids representing cone scales that have broken off near their base. This may grade into irregular rhomboid sections cut through the distal ends of intact cone scales and ultimately to an outer ring of splayed cone scales that appear thin edged in lateral view (Fig. 5.3C–E; Plate 6C–E). Cones in proximal view have a centrally situated cone peduncle up to 9 mm in diameter (Fig. 5.3A; Plate 6A), which is surrounded at its distal end by a radiating splay of cone scales. These cone scales are generally indistinct in outline as a result of overlapping and usually measure 3 cm, but sometimes nearly 4 cm, in length (Fig. 5.3A; Plate 6A). One specimen is broken transversely, showing a section through the cone axis (Fig. 5.3B; Plate 6B). This internal axis is broad and massive, up to 3 cm in diameter; the ratio of cone axis diameter to mature cone is 1:2.3.

Ovuliferous Scale–Bract Complex

The ovuliferous scale–bract complexes (which are referred to as "cone scales" in this chapter) commonly occur detached from the ovulate cones and are spatulate to narrowly spatulate in shape with a pointed distal apophysis and tapered base (Fig. 5.3F–H; Plate 6F–H). They usually measure 3.1 cm in total length from tip to base and 1.4 cm in width at their widest point; the largest is 4.4 cm long. Each ovuliferous scale is fused to its woody bract for the better part of its length and bears a single unwinged seed on its adaxial side, although it is common to find cone scales without the actual seed attached (Fig. 5.3F, G; Plate 6F, G). The ovuliferous scale is strap shaped and free distally for roughly one-half of its length (Fig. 5.3H; Plate 6H). This free distal tip measures 6–7 mm in length and 6 mm in width at its widest point.

The woody bract is laterally expanded over the distal two-thirds of its length, forming the characteristic spatulate shape of the cone scale (Fig. 5.3F–H; Plate 6F–H). The seed on the cone scale, represented by the seed itself (preserved in part in Fig. 5.3H; Plate 6H) or clearly marked by a seed scar (Fig. 5.3F, G; Plate 6F, G), is obovate in outline and medially situated at the base of the cone scale (Fig. 5.3F; Plate 6F). Seeds measure up to 16 mm long and 5.5 mm wide—roughly a 3:1 length-to-width ratio (Figs. 5.3I, 5.4E; Plates 6I, 7E).

The cone scales occasionally occur in close association with branches, as well as with pollen cones and seeds. Detached seeds commonly occur in masses in the matrix (Fig. 5.4E; Plate 7E), and these masses are frequently encountered in the sediment.

Fig. 5.3. *Araucaria delevoryasii* Gee sp. nov. (A) Proximal view of an ovulate cone. Note splayed cone scales and peduncle (arrow). SMA HRFLO 301. (B) Natural transverse section through an ovulate cone, in which cone scales radiate from a wide, internal cone axis. SMA HRFLO 302. (C) Lateral view of one-half of an ovulate cone. Note splayed cone scales, as well as rhomboid sections through distal ends of the cone scales on right. SMA HRFLO 303. (D) Lateral view of an immature ovulate cone. Note cone scales projecting from the edge of the cone, as well as the rhomboid scars in the center of the cone where they have broken off. Arrow indicates the attachment site of the peduncle, which has not been preserved. SMA HRFLO 304. (E) Lateral view of an ovulate cone, with a carbonaceous center and splayed cone scales along the edge of the cone. SMA HRFLO 305. (F) Cone scale with the spatulate shape, pointed distal apophysis, and obovate seed scar diagnostic of this species. Paratype, SMA HRFLO 306. (G) Two cone scales, each displaying a long distal apophysis. Paratypes, SMA HRFLO 307. (H) A large cone scale in which the tip (arrow) of the long, strap-shaped tip of the ovuliferous scale is clearly visible. At the base of the cone scale, part of a seed is still attached; the darker area distal to the broken edge of the seed is the seed scar. Holotype, SMA HRFLO 308. Note the natural size variation in size among cone scales in (F–H), all depicted here at the same size. (I) Detached seed. SMA HRFLO 309. Note the congruency between seed size and shape to those of the seed scars on the cone scales. Scale bars = 1 cm.

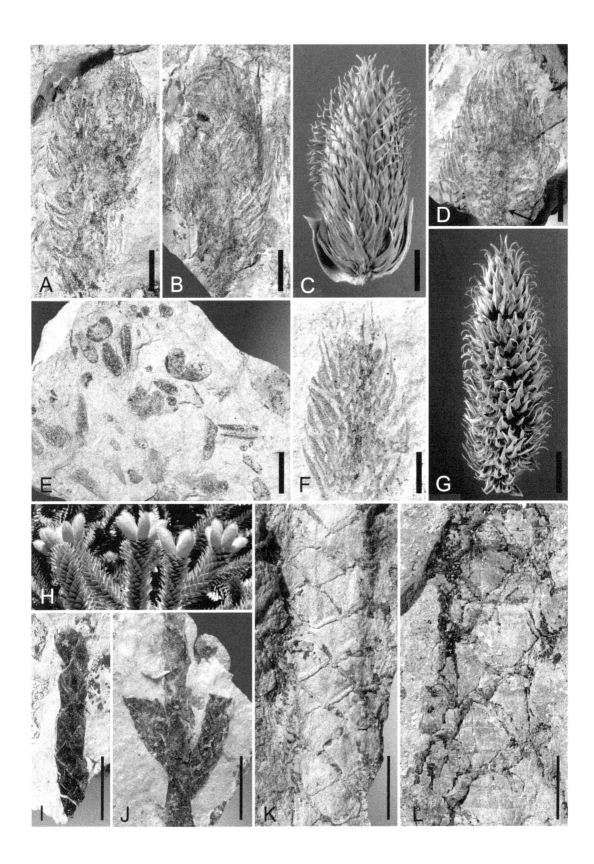

Pollen Cone

The pollen-bearing cones usually occur on the sediment surface in lateral view (Fig. 5.4A, B, D, F; Plate 7A, B, D, F) but are occasionally found in transverse section. The cones are ovate to narrowly ovate in outline and bear numerous, spirally arranged microsporophylls. Small cones measure 14–20 mm in diameter, to about 35 mm long, while larger cones are generally 60 mm long to 28 mm wide. Microsporophylls are long and narrow and measure roughly 2 mm in width (Fig. 5.4F; Plate 7F). They originate from the cone axis at nearly perpendicular angles; after about one-third of their length, they abruptly change direction to 90 degrees (parallel to the cone axis) and project in a slightly curved manner toward the distal tip of the cone. The individual microsporophylls taper very gradually throughout their length, forming a slightly curved, long awn shape with a pointed tip, giving the cone a distinctive hirsute appearance (Fig. 5.4A, B, D, F; Plate 7A, B, D, F). The cone axis in the center of the cone is 2 mm wide (Fig. 5.4D; Plate 7D).

The pollen cones are borne singly by a relatively stout branch, up to 13 mm in width, although it is common to find several of them occurring next to one another in the rock. Unlike the seed cones, the pollen cones do not shatter upon impact with the substrate. The smaller cones most likely represent immature cones, while the larger ones represent a later, more mature stage. One fossil specimen (Fig. 5.4F; Plate 7F) displays widely spaced microsporophylls and represents a cone from which the pollen has been shed.

Woody Shoot and Leaf

Woody shoots with leaves occur mostly as carbonaceous compressions, or occasionally as three-dimensional structures with a carbonaceous surface and a sandstone infilling up to 3.4 cm in diameter, and are rarely branched. The longest unbranched shoot is 21 cm in length, and the single example of branching has subopposite branches (Fig. 5.4J; Plate 7J).

Leaves are simple, entire, apetiolate, scalelike, and helically arranged around the shoot (Fig. 5.4I–L; Plate 7I–L). Leaves of slender shoots (Fig. 5.4I, J; Plate 7I, J) are imbricate, appearing trullate (trowel shaped) when entirely exposed, otherwise rhomboid in shape, with faint striations that may mirror the venation running from base to apex. Leaf apices are cuspidate, and in nearly all cases not spreading from shoot axis, but closely appressed to the shoot like the rest of the leaf. The one example of a leaf with an exposed base on a slender shoot exhibits a length-to-width ratio of about 2.

Leaves of thicker shoots (Fig. 5.4K, L; Plate 7K, L) are not imbricate but clearly separated from one another by a gap of 1–3 mm, rhomboid or subrhomboid in shape, slightly elevated in the center of rhomboid, reminiscent of a cushion, and otherwise closely appressed to the shoot.

Fig. 5.4. Late Jurassic *Araucaria delevoryasii* Gee sp. nov., with living *Araucaria araucana* (Mol.) K. Koch for comparison. (A) Lateral view of a pollen cone of *A. delevoryasii*. SMA HRFLO 310. Note long, awnlike microsporophylls. (B) Lateral view of another *A. delevoryasii* pollen cone, again showing the hirsute appearance typical of these pollen cones. SMA HRFLO 311. (C) Immature pollen cone of *A. araucana*, with long, awnlike microsporophylls. (D) Proximal end of an *A. delevoryasii* pollen cone lacking some of its microsporophylls, showing its central cone axis (arrow). SMA HRFLO 312. Compare slenderness of cone axis here to the thick axis in ovulate cone in Figure 5.3B. (E) Numerous detached seeds of *A. delevoryasii*, showing the typical abundance of fossils in the matrix. SMA HRFLO 313. (F) Part of a mature pollen cone of *A. delevoryasii*, showing the loose arrangement of microsporophylls typical of a spent cone. SMA HRFLO 314. (G) Mature pollen cone of *A. araucana* with loosely arranged microsporophylls. (H) Immature pollen cones borne by *A. araucana* at the Botanical Garden of the University of Bonn, Germany. An immature cone from this tree is figured in (C), and a mature cone from the previous growing season is shown in (G). (I) Twig of *A. delevoryasii* with *Brachyphyllum* type leaves, SMA HRFLO 315. (J) A rare branched twig of *A. delevoryasii*. SMA HRFLO 316. (K) Branch of *A. delevoryasii*. SMA HRFLO 317. Note that the leaves retain their general rhomboid shape and relation to one another on the woody shoot, regardless of shoot girth. (L) Larger branch of

A. delevoryasii. SMA HRFLO 318. The thicker the branch, the larger the individual leaves and the farther away the leaves are pulled apart from one another, increasing the gap between them. The twigs and branches in (I–L) are all depicted here at the same size. Scale bars = 1 cm.

Leaf bases and apices are clearly exposed, at obtuse angles. The largest leaf is 11 cm long and 13 cm wide. Leaf length-to-width ratio is variable, ranging from 1 to 0.5.

As a result of their persistent (evergreen) nature on the shoots, the leaves alter slightly in morphology and relative placement to one another as the shoot gains in girth. Specifically, the leaves change in shape from trowel shaped to rhomboid as the shoot increases in diameter. Relative placement on the shoot axis varies from densely imbricate (Fig. 5.4I, J; Plate 7I, J) to distinctly separate from one another (Fig. 5.4K, L; Plate 7K, L). However, these are merely growth-related changes that do not significantly affect the general appearance of the leaves and woody shoots, as they are readily recognizable as pertaining to the same plant.

Remarks

Of the many hundreds of compression fossils recovered from the HSQ, virtually all of them represent plant organs pertaining to *Araucaria*. The few other compression fossils in the Siber collection that obviously do not pertain to the Araucariaceae are limited to a single shoot of *Equisetum*, several cycad seeds, and an enigmatic, unidentified lamina. Thus, in view of their frequent and constant association with one another, their morphological affinity to extant araucarian plant organs, and the paucity of other plant taxa at this limited collecting site, the fossil remains described in this chapter are considered as the plant parts of a single species of *Araucaria*, although they were not found in organic connection with one another. A similar approach was adopted by Gandolfo et al. (1997) in regard to a Late Cretaceous gleicheniaceous fern, in which numerous biologically and phylogenetically allied plant parts that were found in a similar, species-poor deposit were assigned to the same species.

In the case of *Araucaria* fossils from the HSQ, assignment to the family Araucariaceae is clearly evident in the morphology of each plant part. When compared to the extant genera *Araucaria*, *Agathis*, and *Wollemia*, shared characters include the globose shape of the fossil ovulate cones, the dehiscence of cone scales at maturity, and the single seed on each cone scale. Furthermore, there is greater similarity of the fossil material with extant *Araucaria* and *Wollemia* than with *Agathis*. This shows up in the extremely broad nature of the cone axis relative to the cone's width, the long, pointed distal apophysis of the cone scale, and the bilateral symmetry of the cone scale and seed complex.

However, in the case of the HSQ fossils, the plant remains pertain to *Araucaria* and not to *Wollemia* or *Agathis*. This can clearly be observed in the unfused distal end of the ovuliferous scale on the fossil cone–scale complex (Fig. 5.3H; Plate 6H), a character that occurs only in *Araucaria*. The HSQ seeds are also unwinged, a condition found in some species of living *Araucaria* (e.g., Wilde and Eames 1952; Krüssmann 1983), while the seeds of *Agathis* and *Wollemia* are winged (Whitmore 1980; Jones et al. 1995). The closely appressed, scalelike form of the fossil leaves is also a character found in some species of *Araucaria* (e.g., Wilde and Eames

1952; Krüssmann 1983), while *Agathis* and *Wollemia* have strap-shaped leaves that spread freely from the shoot axis (Whitmore 1980; Jones et al. 1995).

Furthermore, the unusual morphology of the fossil pollen cones from the HSQ is known among living *Araucaria*, but not *Wollemia* or *Agathis*. The pollen cones of extant species of *Araucaria* have either scale-like or elongated microsporophylls (e.g., Schütt et al. 2004; Del Fueyo et al. 2008). In the case of the fossil pollen cones, the microsporophylls are elongated and awnlike in their form, resembling the pollen cones borne by living *Araucaria araucana* (Molina) K. Koch (Fig. 5.4C, G, H; Plate 7C, G, H; cf. Del Fueyo et al. 2008). When compared side by side to one another (Fig. 5.4A–G; Plate 7A–G), the fossil cones and extant *A. araucana* cones also exhibit similar morphologies at different stages of maturity. The immature fossil pollen cones (Fig. 5.4A, B; Plate 7A, B) and the immature pollen cones of *A. araucana* (Fig. 5.4C; Plate 7C) both have closely packed microsporophylls, while mature fossil cones (Fig. 5F; Plate 7F) and the mature cones of *A. araucana* (Fig. 5.4G; Plate 7G), from which pollen has been shed, have a looser, laxer arrangement of microsporophylls.

Ovulate Cones

Comparison of the large, globose ovulate cones from the HSQ with other Mesozoic species of araucarian ovulate cones clearly shows that the new fossil cones represent a previously undescribed species in regard to its large size and virtually spherical shape. Although other araucarian cones have been reported from the Morrison Formation, which were collected in Utah (cf. Tidwell 1990; Stockey 1994; Dayvault and Hatch 2007) and probably represent more than one taxon, they are generally elongate in form and half the size in diameter of *Araucaria delevoryasii*.

To gain an overview of araucarian plant remains in the fossil record, a global survey was made of Mesozoic occurrences of *Araucaria* that concentrated on reproductive parts (Table 5.1). Judging from the more than 30 different taxa that have been reported, araucarian ovulate cones are not uncommon in the fossil record. To facilitate comparison between ovulate cones, 14 taxa were looked at in greater detail, focusing on their general shape, length, width, size, relative size, and type of preservation (Table 5.2). Because these cones are quite variable in shape, cone length was multiplied by cone width to get a rough approximation of size and to make size comparison possible independent of shape. The product of length and width is a rectangle into which the cone can be inscribed. These taxa were then sorted into size classes of 10 cm^2 intervals in regard to their size, and the distribution of the size classes were plotted as a bar graph (Fig. 5.5).

As a result of this sorting, the 14 fossil ovulate cones fall into two distinct size classes: normal and exceptionally large (Fig. 5.5). The

Comparisons

Table 5.1. Global survey of Mesozoic Species of *Araucaria* on the Basis of the Literature, Organized Stratigraphically, with an Emphasis on Reproductive Plant Parts

Taxon, Author Citation	Ovulate Cone	Detached Cone Scale	Pollen Cone	Pollen Type	Leaf Type	Geological Age	Locality	Reference
Araucarites rudicula Axsmith et Ash[a]	X	X				Late Triassic (Carnian)	Chinle Fm., Petrified Forest, Arizona, U.S.A.	Axsmith and Ash 2006
Araucarites charcottii Harris[a]	X	X				Late Triassic/Early Jurassic	Scoresby Sound, East Greenland	Harris 1935; McElwain et al. 1999
Araucarites bindrabunensis Vishnu-Mittre[a]	X					Jurassic	Rajmahal Hills, India	Vishnu-Mittre 1954
Unnamed ovulate cone	X					Jurassic	Rajmahal Hills, India	Bose and Jain 1974
Araucarites nipaniensis Singh		X				Jurassic	Rajmahal Hills, India	Singh 1956
Araucarites stockeyi Tidwell et Ash[a]	X	X				Early Jurassic (Hettangian)	Moenave Fm., Utah, U.S.A.	Tidwell and Ash 2006
Araucaria sp.		X				Early Jurassic (Hettangian)	South Hadley Mbr., Portland Fm., Massachusetts, U.S.A.	Axsmith et al. 2008
Araucaria mirabilis (Spegazzini) Calder[a]	X				BR	Middle Jurassic	Cerro Cuadrado Petrified Forest, Argentina	Calder 1953; Stockey 1975
Araucarites phillipsii Carruthers[a] & *Brachyphyllum mamillare* Lindley et Hutton ex Brongniart	X	X	X	A	BR	Middle Jurassic (Aalenian–Bathonian)	Ravenscar Group, Yorkshire, England	Harris 1979; van Konijnenburg-van Cittert and Morgans 1999
Araucaria sphaerocarpa Carruthers emend. Stockey[a]	X					Middle Jurassic (Bajocian)	Inferior Oolite Fm., southern England	Carruthers 1866; Stockey 1980a
Unnamed cones[a]	X				BR	Middle–Late Jurassic	Chubut Province, Argentina	Escapa and Cúneo, personal communication to C.T.G., 2008

Table 5.1 (continued).

Taxon, Author Citation	Ovulate Cone	Detached Cone Scale	Pollen Cone	Pollen Type	Leaf Type	Geological Age	Locality	Reference
Araucaria antarctica Gee		X			BR	Late Jurassic	Mt. Flora Fm., Hope Bay, Antarctic Peninsula	Gee 1989
Araucarites falsanii (Saporta) Barale & *Masculostrobus graiterensis* Allenbach et van Konijnenburg-van Cittert	X	X		A, CA	PO	Late Jurassic (Oxfordian)	Vellerat Fm., Jura Mts., Switzerland	Allenbach and van Konijnenburg-van Cittert 1997
Araucaria brownii Stockey[a]	X					Late Jurassic (Oxfordian–Kimmeridgian)	Osmington Oolite Series, England	Stockey 1980b
Unnamed cones[a]	X					Late Jurassic (Kimmeridgian–Tithonian)	Morrison Fm., Utah, U.S.A.	Tidwell 1990; Stockey 1994; Dayvault and Hatch 2007
Araucaria delevoryasii Gee sp. nov.[a]	X	X	X		BR	Late Jurassic (Kimmeridgian–Tithonian)	Morrison Fm., Wyoming, U.S.A.	Gee and Tidwell, this chapter
Araucarites macropterus Feistmantel		X				Late Jurassic	Rajmahal Stage, Rajmahal Hills, India	Feistmantel 1877; Sukh-Dev and Bose 1964
Araucaria indica (Sahni) Sukh-Dev et Zeba-Bano (syn. *A. cutchensis*)	X	X			PO	Late Jurassic–Early Cretaceous	Jabalpur Stage, Jabalpur Formation, India	Pant and Srivastava 1968; Sukh-Dev and Zeba-Bano 1978
Araucarites pedreranus Barale		X			PA	Early Cretaceous	lithographic limestones, Montsech, Spain	Barale 1989
Araucarites rogersii Seward emend. Brown	X	X				Early Cretaceous	Kirkwood Fm., Cape Province, South Africa	Seward 1903; Brown 1977
Araucarites mesozoica Walkom	X				PO	Early Cretaceous	Maryborough (Marine Series), Queensland, Australia	Walkom 1918

Table 5.1 (continued).

Taxon, Author Citation	Ovulate Cone	Detached Cone Scale	Pollen Cone	Pollen Type	Leaf Type	Geological Age	Locality	Reference
Araucarites chilensis Baldoni & Brachyphyllum feistmantelii (Halle) Sahni	X		X	A	BR	Early Cretaceous (pre-Aptian)	Springhill Fm., Santa Cruz Prov., Argentina	Baldoni 1979
Araucarites baqueroensis Archangelsky	X					Early Cretaceous (Aptian)	Baqueró Fm., Santa Cruz Prov., Argentina	Archangelsky 1966
Araucarites minimus Archangelsky	X					Early Cretaceous (Aptian)	Baqueró Fm., Santa Cruz Prov., Argentina	Archangelsky 1966
Alkastrobus peltatus Del Fueyo et Archangelsky			X	CY		Early Cretaceous (Aptian)	Ticó Fm., Santa Cruz Prov., Argentina	Del Fueyo and Archangelsky 2005
Brachyphyllum irregulare Archangelsky et Gamerro			X	B	BR	Early Cretaceous (Aptian)	Ticó Fm., Santa Cruz Prov., Argentina	Archangelsky and Gamerro 1967
Nothopehuen brevis Del Fueyo & Brachyphyllum mirandai Archangelsky			X	A	BR	Early Cretaceous (Aptian)	Ticó Fm., Santa Cruz Prov., Argentina	Del Fueyo 1991
Cf. Araucaria sp.[a]	X	X			BR	Early Cretaceous (late Aptian)	Crato Fm., Araripe Basin, Brazil	Kunzmann et al. 2004
Araucarites citadelbastionensis Cantrill et Falcon-Lang		X				Early Cretaceous (late Albian)	Neptune Glacier Fm., Alexander Island, Antarctica	Cantrill and Falcon-Lang 2001
Araucaria nipponensis Stockey, Nishida et Nishida[a]	X	X				Late Cretaceous (Turonian)	Upper Yezo and Miho Groups, Japan and Russia	Stockey et al. 1994

Table 5.1 (continued).

Taxon, Author Citation	Ovulate Cone	Detached Cone Scale	Pollen Cone	Pollen Type	Leaf Type	Geological Age	Locality	Reference
Araucaria vulgaris (Stopes et Fujii) Ohsawa, Nishida et Nishida[a] & *A. nihongii* Stockey, Nishida et Nishida	X	X			BR	Late K (Turonian–Santonian)	Upper Yezo Group, Hokkaido, Japan	Stockey et al. 1992; Ohsawa et al. 1995
Pertaining to the Araucariaceae, but not necessarily to *Araucaria*								
Unnamed, detached coniferous cones			X	OPT	BR	Jurassic	Rajmahal Hills, India	Bose and Hsü 1953
Agathis jurassica White & *Araucarites grandis* Walkom	X	X	X		PO	Early–Middle Jurassic	Talbragar Fish Beds, SW Australia	Walkom 1921; White 1981; see also Stockey 1982, 1994; Cantrill 1992
Araucarites polycarpa Tension-Woods	X					Early Cretaceous	Queensland, Australia	Tension-Woods 1883; Walkom 1918, 1919
Araucarites vulcanoi Duarte	X					Early Cretaceous (Aptian)	Santana Fm., Araripe Basin, NE Brazil	Duarte 1993
Araucaria cone scale		X				Early Cretaceous (Aptian)	Koonwarra Fossil Bed, Victoria, Australia	Drinnan and Chambers 1986
Immature ?araucacarian cones	X					Early Cretaceous (late Albian)	Neptune Glacier Fm., Alexander Island, Antarctica	Cantrill and Falcon-Lang 2001
Araucarites wollemiaformis Cantrill et Falcon-Lang		X				Early Cretaceous (late Albian)	Neptune Glacier Fm., Alexander Island, Antarctica	Cantrill and Falcon-Lang 2001
Wairarapaia mildenhallii Cantrill et Raine (syn. *Araucarites* sp. Mildenhall et Johnston)	X			A	X	Mid-Cretaceous (late Albian–Cenomanian)	Split Rock Fm. and Springhill Fm., New Zealand	Cantrill and Raine 2006

Table 5.1 (continued).

Taxon, Author Citation	Ovulate Cone	Detached Cone Scale	Pollen Cone	Pollen Type	Leaf Type	Geological Age	Locality	Reference
Araucaria scale type A		X				Late Cretaceous (Maastrichtian)	Matakaea Group, eastern Otago, New Zealand	Pole 1995
Araucaria scale type B		X				Late Cretaceous (Cenomanian)	Taratu Fm., eastern Otago, New Zealand	Pole 1995
Araucaria scale type C		X				Late Cretaceous (Cenomanian)	Taratu Fm., eastern Otago, New Zealand	Pole 1995

Note. Taxa consisting of only foliage, leafy shoots, or wood were excluded from this overview. X = present; A = Araucariacites pollen; B = Balmeiopsis pollen; CA = Calliala-sporites pollen; CY = Cyclusphaera pollen; OPT = other pollen type; BR = Brachyphyllum type leaves; PA = Pagiophyllum type leaves; PO = Podozamites type leaves.
[a] Ovulate cone species examined in greater detail in Table 5.2.

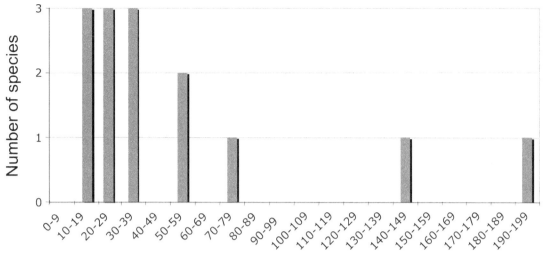

Size classes of ovulate cones (cm²)

exceptionally large species pertain to the large (approximately 14 cm diameter; size class 140–149 cm²) spherical cone of *Araucaria sphaerocarpa* Carruthers emend. Stockey and to the very long (approximately 18 cm in length; size class 190–199 cm²), narrowly obovate, unnamed cones from Argentina (Ignacio Escapa and Rubén Cúneo, personal communication to C.T.G., 2008). Cones of the normal size class vary continuously from small (4 cm diameter; size class 10–19 cm²) to large (8.5 diameter; size class 70–79 cm²), with a size-class average of 34 (Table 5.2, Fig. 5.5). The cones of *A. delevoryasii* are distinctive in that they sort out at the top of this range. Clearly, no other fossil ovulate cones of the normal size class reported thus far come close in size to the ovulate cones of *Araucaria delevoryasii*, as there is a considerable gap in size from *A. delevoryasii* to the next smallest cones. There is an even larger size gap between *A. delevoryasii* and the two exceptionally large species of cones (Fig. 5.5).

In terms of shape (Table 5.2), the fossil ovulate cones are quite variable, ranging in degrees from globose to elongate. Of the globose cones, most are actually elliptical in shape; few are virtually spherical like *Araucaria delevoryasii*. Of the elongate cones, most are narrower at their apex than at their base (narrowly ovate), but one unusual species is broader at its apex than at its base (narrowly obovate). Concerning type of preservation (Table 5.2), most ovulate cones occur as compressions rather than as permineralizations.

Cone Scales

Unusual, too, in *Araucaria delevoryasii* is the rounded yet clearly spatulate shape of its cone scale. No other fossil cone scale, whether attached to a seed cone or found detached, has such a distinctive shape. Other detached cone scales, usually assigned to *Araucaria* or *Araucarites*, are

Fig. 5.5. Size classes of Mesozoic seed cones of *Araucaria*. For this comparison, cone length (mm) was multiplied by cone width (mm) to get a rough approximation of size. The seed cones of *A. delevoryasii* are by far the largest cones of the normal size class, occurring in the 70–79 cm² class size.

Table 5.2. A Detailed Look at the 14 Species of Ovulate Cones of *Araucaria* from the Mesozoic in Regard to Cone Size, Size Relative to Other Cones, Cone Shape, and Type of Preservation

Species of Fossil Ovulate Cone	Length × Width (cm)	Product of Length × Width (cm²)	Relative Size	General Shape of Cone	Preservation	Reference
Araucarites rudicula	6.35 × 1.94	12	Much smaller	Elongate	Compression	Axsmith and Ash 2006
Araucarites charcottii	6 × 2.5	15	Much smaller	Elongate	Compression	Harris 1935
Araucarites bindrabunensis	7.5 × 4.4	33	Near average	Elongate	Permineralization	Vishnu-Mittre 1954
Cf. *Araucarites stockeyi*	At least 9.3 × 6	Minimum 56	Larger	Generally elongate	Compression	Tidwell and Ash 2006
Araucaria mirabilis	8 × 6.5	52	Larger	Elliptical	Permineralization	Calder 1953; Stockey 1975
Araucarites phillipsii	6 × 4	24	Smaller	Elliptical	Compression	Harris 1979; van Konijnenburg-van Cittert and Morgans 1999
Araucaria sphaerocarpa	13 × 15	195	Extremely large	Globose/elliptical	Permineralization	Stockey 1980a
Unnamed cones (from Argentina)	18 × 8	144	Very large	Narrowly obovate	Permineralization and compression	Escapa and Cúneo, personal communication to C.T.G., 2008
Araucaria brownii	Minimum 5 (diameter)	Minimum 25	Smaller	Globose	Permineralization	Stockey 1980b
Unnamed cone (from Morrison Fm.)	6.5 × 3.6	23	Smaller	Narrowly elliptical	Compression	Stockey 1994
Araucaria delevoryasii	8.5 × 8.5	72	Much larger	Nearly spherical	Compression	Gee and Tidwell, this chapter
Araucaria sp.	8.3 × 4.7	39	Near average	Elongate	Compression	Kunzmann et al. 2004
Araucaria nipponensis	6 (diameter)	36	Near average	Widely elliptical	Permineralization and compression	Stockey et al. 1994
Araucaria vulgaris/A. nihongii	4 × 4	16	Much smaller	Globose	Compression	Stockey et al. 1992; Ohsawa et al. 1995

almost always broadly cuneate (wedge shaped), rarely narrowly cuneate (e.g., new species to be described by Escapa and Cúneo, personal communication to C.T.G., 2008), and sometimes rounded rhombic (e.g., *Araucaria indica* [Sahni] Sukh-Dev et Zeba-Bano 1978). Infrequent cone scale shapes are subcordate (e.g., *Araucarites falsanii* [Saporta] Barale; cf. Allenbach and van Konijnenburg-van Cittert 1997) or broadly cuneate on a narrow basal stalk (*Araucaria rudicula* Axsmith et Ash 2006).

The unique spatulate shape of the cone scale of *Araucaria delevoryasii* is the major reason why a cone scale and not an entire cone was selected as the holotype of the new species. Furthermore, while the ovulate cones of A. *delevoryasii* are distinctive in their moderately large size, the shape, size, and other details of the individual cone scales are difficult to make out in the entire cones.

Seeds

In regard to general size and shape, the seeds of *Araucaria delevoryasii* resemble the *Araucaria* seeds known from the fossil record. There is nonetheless one major difference between them. In Mesozoic seeds of *Araucaria*, particularly in those in which the internal anatomy is apparent (i.e., in permineralized material), it has been observed that the seeds are deeply embedded into the tissue of the cone scale. This includes, for example, A. *brownii* Stockey, A. *mirabilis* (Spegazzini) Calder, *Araucaria nipponensis* Stockey, Nishida et Nishida, A. *sphaerocarpa* Carruthers emend. Stockey, and A. *vulgaris* (Stopes et Fujii) Ohsawa, Nishida et Nishida/A. *nihongii* Stockey, Nishida et Nishida. However, this is not the case in A. *delevoryasii*, in which the seeds rest on the surface of the ovuliferous scale and are not deeply sunken into or surrounded by it. In this single character, A. *delevoryasii* more closely resembles *Wollemia* and *Agathis* than any species of *Araucaria*.

Pollen Cones

Pollen cones pertaining to *Araucaria* are relatively rare in the fossil record. However, compared with other dispersed araucarian plant parts, the pollen cones have been placed in the widest range of taxa: *Brachyphyllum mamillare* Lindley et Hutton emend. Harris (e.g., Harris 1979), *Araucariostrobus* Krasser (e.g., Kunzmann et al. 2004), *Alkastrobus* Del Fueyo (e.g., Del Fueyo and Archangelsky 2005), *Maculostrobus* Seward (e.g., Allenbach and van Konijnenburg-van Cittert 1997), and *Nothopehuen* Del Fueyo (e.g., Del Fueyo 1991). All of these fossil pollen cones consist of microsporophylls that are relatively small, imbricate, and scalelike. Pollen cones of *Araucaria delevoryasii*, on the other hand, bear awnlike microsporophylls with a long, thin, slightly curved distal tip. Although this sort of morphology has not been previously described in the fossil araucarian pollen cones, as mentioned earlier, it is found among living species of *Araucaria*, such as in *A. araucana* (e.g., Schütt et al. 2004; Del Fueyo et al. 2008), indicating that it is clearly within the range of morphologies exhibited by the genus. The awnlike microsporophylls of *A. delevoryasii* also sets a minimum geological age of Late Jurassic for the first appearance of this character in the genus *Araucaria*.

Another striking difference between *Araucaria delevoryasii* and other fossil pollen cones is size. As pointed out by others (Del Fueyo and Archangelsky 2005; Archangelsky and Del Fueyo 2010), all araucarian pollen cones described from the fossil record so far are extremely small (approximately 2 mm long) to small (13 mm long), and these small sizes are at odds with pollen cones borne by living *Araucaria*, which are relatively large (Earle 1999). The 6-cm-long pollen cones of *A. delevoryasii*, however, illustrate that large-sized pollen cones did exist among the Mesozoic araucarians.

Foliage

On the basis of leaf proportions, the leaf-bearing shoots in the HSQ flora would be assigned to the form genus *Brachyphyllum* Lindley et Hutton ex Brongniart if they had not been found in close physical, frequent, and virtually exclusive association with the reproductive remains described above. Leaves of *Brachyphyllum* are artificially differentiated from *Pagiophyllum* Heer in having a free part of the leaf that is shorter than the width of its basal cushion (Harris 1979). Species of both *Brachyphyllum* and *Pagiophyllum*, as well as of *Podozamites* Braun and *Geinitzia* Endlicher, have been found attached to unequivocal reproductive remains of *Araucaria* or exhibit leaf cuticle features that pertain only to the Araucariaceae (Stockey and Ko 1986: Stockey and Atkinson 1993), indicating their biological association to araucarias. However, they have also been found to pertain to other, completely unrelated groups of conifers, such as the Cheirolepidiaceae (van Konijnenburg-van Cittert 1987; van der Ham et al. 2003; Archangelsky and Del Fueyo 2010), Hirmerellaceae (Harris 1979), and Podocarpaceae (Archangelsky and Del Fueyo 1989, 2010), which is why they are considered form genera.

A Mosaic of Characters

The plant organs of *Araucaria delevoryasii* exhibit a suite of characters not encountered in any other section or species of *Araucaria*, living or fossil. In regard to size of the ovulate cone, *A. delevoryasii* is moderately large among fossil taxa (Table 5.2, Fig. 5.5), but relatively small in the face of the huge cones of living *A. bidwillii* Hook or *A. araucana*, for example. The spatulate shape of the cone scales is unique to *A. delevoryasii*. The position of the seed on the surface of the cone scale, and not deeply embedded in the tissue, is not characteristic of any fossil or modern species of *Araucaria*, but is typical of *Agathis* and *Wollemia*.

There are five sections of the genus *Araucaria*, four of which (*Eutacta, Bunya, Intermedia*, and *Araucaria*) are based on living species (cf. Wilde and Eames 1952; Setoguchi et al. 1998) and a fifth (*Yezonia*) is known only from the Late Cretaceous (Ohsawa et al. 1995; see discussion below). In its scalelike foliage, *Araucaria delevoryasii* resembles the section *Eutacta*, which has reduced leaves and in this way differs from all other sections, which have large, spreading leaves. However, the cone scales of *A. delevoryasii* are not light and samara-like, nor are they thinly winged, as those in the section *Eutacta*.

Araucaria delevoryasii is similar to the section *Bunya* because both *A. delevoryasii* and *A. bidwillii* have seeds that are shed from the cone scale. In the other extant sections, the seed is retained on the cone scale, even after the cone scales detach from the ovulate cone (Wilde and Eames 1952). In section *Bunya*, represented today by only one species, *A. bidwillii*, the cone scales are characterized by thick, lateral woody wings, which are formed by the extension of the bract on both sides of the ovuliferous scale. In *A. delevoryasii*, the bract is only expanded in the distal half of the cone scale, resulting in its diagnostic spatulate shape. In the other sections, the cone scale has either thin wings (*Eutacta, Intermedia*) or are nutlike and unwinged (*Araucaria*, syn. *Columbea*). *Araucaria delevoryasii* does not share any characters with *Intermedia* that are diagnostic for the section.

There is one major similarity between *Araucaria delevoryasii* and section *Araucaria*, which consists today of only two species, both of which are native to South America, *A. araucana* and *A. angustifolia* (Bertol.) Kuntze. As discussed earlier, the fossil species of *A. delevoryasii* and the living species of *A. araucana* both have pollen cones with long, awnlike microsporophylls, a character that has not yet been reported from other fossil or extant species in the genus. Although these two species have no other diagnostic characters in common, it should be noted that pollen cones have provided the best taxonomic characters elsewhere in the family Araucariaceae. Microsporophyll characteristics were used with good effect, for example, to differentiate the 13 species and two subspecies of *Agathis* from one another (Whitmore 1980).

In summary, *Araucaria delevoryasii* possesses a mosaic of characters that are displayed by various fossil and living species of *Araucaria* but are not limited to any one section. The novel combination of characters in this species provides evidence that there was greater taxonomic diversity in the genus in the past. Greater diversity in the Mesozoic Araucariaceae has also recently been pointed out by other paleobotanical studies (Cantrill 1992; Cantrill and Raine 2006 and references therein).

The survey of *Araucaria* species in the Mesozoic record in Table 5.1 shows that the reproductive remains of this genus occur throughout most of the Mesozoic, starting from the Late Triassic and continuing beyond the Late Cretaceous. To date, the first appearance of the genus in form of

Global Survey of Mesozoic *Araucaria*

ovulate cones and cone scales occurs in the early Late Triassic (Carnian; Axsmith and Ash 2006). By the Middle and Late Jurassic, *Araucaria* has spread worldwide, occurring in both the Northern Hemisphere (United States and Europe) and Southern Hemisphere (India, Argentina, and the Antarctic Peninsula) (Table 5.1). In the Cretaceous, *Araucaria* continues to spread and maintain its global geographical range, and it is found in other parts of the world, such as in Australia, Japan, and Russia, although it is especially abundant in Argentina (Table 5.1). The cosmopolitan distribution of the genus in the mid and late Mesozoic contrasts with its restriction today to virtually only the Southern Hemisphere. The paleo-biogeography of the family Araucariaceae has recently been a topic of renewed discussion (e.g., Kershaw and Wagstaff 2001; Kunzmann 2007).

The survey of Mesozoic araucarians carried out here (Table 5.1) also clearly supports the close relationship between fossil *Araucaria* and *Brachyphyllum* type leaves. The frequent attachment of araucarian pollen cones to woody shoots bearing *Brachyphyllum* type foliage, as well as the co-occurrence of araucarian seed cones or cone scales in the same deposits, strongly suggest that *Brachyphyllum* was a major type of foliage among the Mesozoic species of *Araucaria*. Other types of foliage associated with araucarian seed cones or cone scales, albeit to a much lesser extent, are *Pagiophyllum* and *Podozamites*, both of which have leaves with lamina larger than those of *Brachyphyllum*. Similarly, it is evident from this survey (Table 5.1) that an association between fossil *Araucaria* plant parts and *Araucariacites* type pollen is common among Mesozoic species. Fossil *Araucariacites* type pollen is similar in morphology to that of the pollen found in all living species of *Araucaria* today. As Archangelsky and Del Fueyo (2010) point out, however, a number of other pollen types have been found in situ in Mesozoic araucarian pollen cones in Gondwana, such as *Cyclusphaera* and *Balmeiopsis*, while *Callialasporites* has been associated with fossil *Araucaria* elsewhere (van Konijnenburg-van Cittert 1971; Batten and Dutta 1997; Allenbach and van Konijnenburg-van Cittert 1997).

In the last few decades, enough fossil evidence has come together to be able to assemble several whole-plant *Araucaria* species. The first known whole-plant araucaria was described by Harris (1979), based on *Brachyphyllum mamillare* (woody shoots with foliage, cuticle, attached pollen cones, and in situ pollen) and *Araucarites phillipsii* Carruthers (associated cone scales), from the Middle Jurassic of Yorkshire, England. The second whole-plant araucaria was described by Ohsawa et al. (1995) as *Araucaria vulgaris* (Stopes et Fuji) Ohsawa, Nishida et Nishida from the Late Cretaceous of Hokkaido, Japan. This plant is composed of anatomically preserved wood and shoots with attached foliage and an ovulate cone, and is based on *Yezonia vulgaris* Stopes et Fujii, *Brachyphyllum vulgare* (Stopes et Fujii) Seward, and *Araucaria nihongii* Stockey, Nishida et Nishida. The new species of *Araucaria delevoryasii* (woody shoots with leaves, associated pollen cones, and associated seed cones, cone scales, and seeds) described in this chapter is the third whole-plant species of *Araucaria* known from the Mesozoic.

Araucaria delevoryasii Gee sp. nov. is described as a whole-plant fossil species of *Araucaria* from the Late Jurassic Morrison Formation in north-central Wyoming. This new species is represented by entire seed cones, as well as by associated seed-bearing cone scales, detached seeds, pollen cones, and woody shoots with *Brachyphyllum* type leaves. Diagnostic are the relatively large size and virtually spherical shape of the ovulate cones, the characteristic rounded spatulate shape of the cone scale, the separation of the seed from the cone scale at maturity, and the nonembedded nature of the seed on the cone scale, as well as the large size and the long-awned microsporophylls of the pollen cones. The occurrence of these pollen cones indicates a minimum geological age of Late Jurassic for the morphological innovation of large pollen cones with awnlike microsporophylls in the genus. When compared with fossil and living taxa of *Araucaria*, *A. delevoryasii* exhibits a mosaic of characters that do not show up in any one section or species, which thus suggests that there was more diversity in the Mesozoic than is reflected in current systematics. In a global survey of fossil araucarian remains, it is evident that *A. delevoryasii* differs from nearly all other species in the relatively large size of its ovulate cones, but is similar in having *Brachyphyllum* type leaves. The global survey also shows clearly that *Brachyphyllum* type foliage and *Araucariacites* pollen are prevalent among Mesozoic species of *Araucaria*; it also points out the existence of several species of whole-plant araucarias in the Mesozoic record.

Conclusions

Acknowledgments

This chapter is dedicated to Ted Delevoryas for his outstanding contributions to the Mesozoic paleobotany of North America. We thank reviewer Ruth A. Stockey for her suggestions for improvement to an earlier version of this chapter. C.T.G. expresses her gratitude to Hans-Jakob Siber (Sauriermuseum Aathal, Switzerland) for the loan of material, Jacques Ayer (Université de Neuchâtel, Switzerland) for the use of his stratigraphic section, and Georg Oleschinki and Dorothea Kranz (University of Bonn, Germany) for their superb photography and graphic illustration, respectively. W.D.T. thanks Sidney R. Ash (University of New Mexico) for assistance in the field. The support of the German Science Foundation (Deutsche Forschungsgemeinschaft) to C.T.G. is gratefully acknowledged. This is contribution no. 57 of the DFG Research Unit 533 on the Biology of the Sauropod Dinosaurs: The Evolution of Gigantism.

References Cited

Allenbach, R., and J. H. A. van Konijnenburg-van Cittert. 1997. On a small flora with araucariaceous conifers from the Röschenz Beds of Court, Jura Mountains, Switzerland. *Eclogae Geologicae Helvetiae* 90: 571–579.

Archangelsky, S. 1966. New gymnosperms from the Ticó Flora, Santa Cruz Province, Argentina. *Bulletin of the British Museum (Natural History), Geology* 13: 261–295.

Archangelsky, S., and G. M. Del Fueyo. 1989. *Squamastrobus* gen. nov., a fertile podocarp from the early Cretaceous of Patagonia, Argentina. *Review of Palaeobotany and Palynology* 59: 109–126.

———. 2010. Endemism of Early Cretaceous conifers in Western Gondwana. In C. T. Gee (ed.), *Plants in Mesozoic Time: Morphological Innovations, Phylogeny, Ecosystems*, pp. 247–268. Bloomington: Indiana University Press.

Archangelsky, S., and J. C. Gamerro. 1967. Pollen grains found in coniferous cones from the Lower Cretaceous of Patagonia (Argentina). *Review of Palaeobotany and Palynology* 5: 179–182.

Axsmith, B. J., and S. R. Ash. 2006. Two rare fossil cones from the Upper Triassic Chinle Formation in Petrified Forest National Park, Arizona, and New Mexico. In W. G. Parker, S. R. Ash, and R. B. Irmis (eds.), *A Century of Research at Petrified Forest National Park: Geology and Paleontology. Museum of Northern Arizona Bulletin* 62: 82–94.

Axsmith, B. J., I. H. Escapa, and P. Huber. 2008. An araucarian conifer bract–scale complex from the Lower Jurassic of Massachusetts: implications for estimating phylogenetic and stratigraphic congruence in the Araucariaceae. *Palaeontologia Electronica* 11 (3): 11.3.13A.

Ayer, J. 2000. *The Howe Ranch Dinosaurs*. Aathal, Switzerland: Sauriermuseum Aathal.

Baldoni, A. M. 1979. Nuevos elementos paleoflorísticos de la tafoflora de la Formación Spring Hill, limite Jurásico–Cretácico subsuelo de Argentina y Chile austral. *Ameghinana* 16: 103–119.

Barale, G. 1989. Sur trois nouvelles espèces de Coniférales du Crétacé inférieur d'Espagne: intéréts paléoécologiques et stratigraphiques. *Review of Palaeobotany and Palynology* 61: 303–318.

Batten, D. J., and R. J. Dutta. 1997. Ultrastructure of exine of gymnospermous pollen grains from Jurassic and basal Cretaceous deposits in Northwest Europe and implications for botanical relationships. *Review of Palaeobotany and Palynology* 99: 25–54.

Bird, R. T. 1985. *Bones for Barnum Brown: Adventures of a Dinosaur Hunter*. Edited by V. T. Schreiber. Fort Worth: Texas Christian University Press.

Bose, M. N., and K. P. Jain. 1974. A megastrobilus belonging to the Araucariaceae from the Rajmahal Hills, Bihar, India. *The Palaeobotanist* 12: 229–231.

Bose, M. N., and J. Hsü. 1953. On some coniferous cones, probably of *Brachyphyllum*, from the Jurassic of the Rajmahal Hills, Bihar, India. *Proceedings of the National Institute of Sciences of India* 19: 203–209.

Brown, J. T. 1977. On *Araucarites rogersii* Seward from the Lower Cretaceous Kirkwood Formation of the Algoa Basin, Cape Province, South Africa. *Palaeontologia Africana* 20: 47–51.

Calder, M. G. 1953. A coniferous petrified forest in Patagonia. *Bulletin of the British Museum (Natural History), Geology* 2 (2): 99–138.

Cantrill, D. J. 1992. Araucarian foliage from the Lower Cretaceous of southern Victoria, Australia. *International Journal of Plant Sciences* 153: 622–645.

Cantrill, D. J., and H. J. Falcon-Lang. 2001. Cretaceous (Late Albian) coniferales of Alexander Island, Antarctica. 2. Leaves, reproductive structures and roots. *Review of Palaeobotany and Palynology* 115: 119–145.

Cantrill, D. J., and J. I. Raine. 2006. *Wairarapaia mildenhallii* gen. et sp. nov., a new araucarian cone related to *Wollemia* from the Cretaceous (Albian–Cenomanian) of New Zealand. *International Journal of Plant Sciences* 167: 1259–1269.

Carruthers, W. 1866. On araucarian cones from the Secondary Beds of Britain. *Geological Magazine* 3: 249–252.

Chambers, T. C., A. N. Drinnan, amd S. McLoughlin. 1998. Some morphological features of Wollemi pine (*Wollemia nobilis*: Araucariaceae) and their comparison to Cretaceous plant fossils. *International Journal of Plant Sciences* 159: 160–171.

Codrington, T. A., L. J. Scott, K. Scott, G. C. Graham, M. Rossetto, M. Ryan, T. Whiffin, R. Henry, and K. G. Hill. 2002. Unresolved phylogenetic position of *Wollemia*, *Araucaria* and *Agathis*. In M. Wilcox and R. Bieleski (eds.), *International Araucariaceae Symposium*, p. 107. Auckland: International Dendrology Society.

Dayvault, R. D., and H. S. Hatch. 2007. Conifer cones from the Jurassic and Cretaceous rocks of eastern Utah. *Rocks & Minerals* 82: 382–396.

Del Fueyo, G. M. 1991. Una nueva Araucariaceae cretacica de Patagonnia, Argentina. *Ameghiniana* 28: 149–161.

Del Fueyo, G. M., and S. Archangelsky. 2005. A new araucarian pollen cone with in situ *Cyclusphaera* Elsik from the Aptian of Patagonia, Argentina. *Cretaceous Research* 26: 757–768.

Del Fueyo, G. M., M. A. Caccavari, and E. A. Dome. 2008. Morphology and structure of the pollen cone and pollen grain of the *Araucaria* species from Argentina. *Biocell* 32: 49–60.

Drinnan, A. N., and T. C. Chambers. 1986. Flora of the Lower Cretaceous Koonwarra Fossil Bed (Korumburra Group), South Gippsland, Victoria. In P. A. Jell and J. Roberts (eds.), *Plants and Invertebrates from the Lower Cretaceous Koonwarra Fossil Bed, South Gippsland, Victoria*, pp. 1–77. *Association of the Australasian Palaeontologists*, Memoir 3: 1–77.

Duarte, L. 1993. Restos de Araucariáceas da Formação Santana–Membro Crato (Aptiano), NE do Brasil. *Anais da Academia Brasileira de Ciências* 65: 357–362.

Earle, C. J. 1999. Araucariaceae Henkel & W. Horst (1865). *Gymnosperm Database*. http://www.biologie.uni-hamburg.de/b-online/earle/ar/index.htm.

Feistmantel, O. 1877. Jurassic (Liassic) flora of the Rajmahal Group from Golapil, near Ellore, South Godavari. *Memoirs of the Geological Survey of India, Palaeontologia Indica*, ser. 2, vol. 1, pt. 3: 163–190.

Foster, J. 2007. *Jurassic West: The Dinosaurs of the Morrison Formation and Their World*. Bloomington: Indiana University Press.

Gandolfo, M. A., K. C. Nixon, W. L. Crepet, and G. E. Ratcliffe. 1997. A new fossil fern assignable to Gleicheniaceae from Late Cretaceous sediments of New Jersey. *American Journal of Botany* 84: 483–493.

Gee, C. T. 1989. Revision of the Late Jurassic/Early Cretaceous flora from Hope Bay, Antarctica. *Palaeontographica*, Abt. B, 213: 149–214.

Gilmore, S., and K. D. Hill. 1997. Relationships of the Wollemi Pine (*Wollemia nobilis*) and a molecular phylogeny of the Araucariaceae. *Telopea* 7: 275–291.

Harris, T. M. 1935. The fossil flora of Scoresby Sound, East Greenland. Part 4, Ginkgoales, Coniferales, Lycopodiales and isolated fructifications. *Meddelelser om Grønland* 112: 1–176.

———. 1979. *The Yorkshire Jurassic Flora. V. Coniferales*. London: British Museum (Natural History).

Hotton, C. L., and N. L. Baghai-Riding. 2010. Palynological evidence for conifer dominance within a heterogeneous landscape in the Late Jurassic Morrison Formation, U.S.A. In C. T. Gee (ed.), *Plants in Mesozoic Time: Morphological Innovations, Phylogeny, Ecosystems*, pp. 295–328. Bloomington: Indiana University Press.

Jones, W. G., K. D. Hill, and J. M. Allen. 1995. *Wollemia nobilis*, a new living Australian genus and species in the Araucariaceae. *Telopea* 6: 173–176.

Kershaw, P., and B. Wagstaff. 2001. The southern conifer family Araucariaceae: history, status, and value for paleoenvironmental reconstruction. *Annual Review of Ecology and Systematics* 32: 397–414.

Kowallis, B. J., E. H. Christiansen, A. L. Deino, F. Peterson, C. E. Turner, M. J. Kunk, and J. D. Obradovich. 1998. The age of the Morrison Formation. *Modern Geology* 22: 235–260.

Krüssmann, G. 1983. *Handbuch der Nadelhölzer*, 2nd ed. Berlin: Verlag Paul Parey.

Kunzmann, L. 2007. Araucariaceae (Pinopsida): aspects in palaeobiogeography and palaeobiodiversity in the Mesozoic. *Zoologischer Anzeiger* 246: 257–277.

Kunzmann, L., B. A. R. Mohr, and M. E. C. Bernardes-de-Oliveira. 2004. Gymnosperms from the Lower Cretaceous Crato Formation (Brazil). I. Araucariaceae and *Lindleycladus* (incertae sedis). *Mitteilungen aus dem Museum für Naturkunde in Berlin, Geowissenschaftliche Reihe* 7: 155–174.

Litwin, R. J., C. E. Turner, and F. Peterson. 1998. Palynological evidence on the age of the Morrison Formation, Western Interior U.S. *Modern Geology* 22: 297–319.

McElwain, J. C., D. J. Beerling, and F. I. Woodward. 1999. Fossil plants and global warming at the Triassic–Jurassic boundary. *Science* 285: 1386–1390.

Ohsawa, T., H. Nishida, and M. Nishida. 1995. *Yezonia*, a new section of *Araucaria* (Araucariaceae) based on permineralized vegetative and reproductive organs of *A. vulgaris* comb. nov. from the Upper Cretaceous of Hokkaido, Japan. *Journal of Plant Research* 108: 25–39.

Pant, D. D., and G. K. Srivastava. 1968. On the cuticular structure of *Araucaria* (*Araucarites*) *cutchensis* (Feistmantel) comb. nov. from the Jabalpur Series, India. *Journal of the Linnean Society (Botany)* 61: 201–206.

Pole, M. 1995. Late Cretaceous macrofloras of eastern Otago, New Zealand: gymnosperms. *Australian Systematic Botany* 8: 1067–1106.

Rock-Color Chart Committee. 1980. *Rock Color Chart.* Boulder, Colo.: Geological Society of America.

Schütt, P., H. Weisgerber, H.-J. Schuck, K. J. Lang, B. Stimm, and A. Roloff (eds.). 2004. *Lexikon der Nadelbäume.* Hamburg: Nikol-Verlag.

Setoguchi, H., T. A. Osawa, J.-C. Pintaud, T. Jaffré, and J.-M. Veillon. 1998. Phylogenetic relationships within Araucariaceae based on *rbcL* gene sequences. *American Journal of Botany* 85: 1507–1516.

Seward, A. C. 1903. Flora of the Uitenhage Series. Fossil floras of the Cape Colony. *Annals of the South African Museum* 4: 1–22.

Siber, H.-J. 2004. Important elements of information to understand the conflict. A short history of the Howe Ranch and the early dinosaur discoveries. http://www.sauriermuseum.ch/bilder/grabung04/Conflict_Information.pdf.

Singh, G. 1956. *Araucarites nipaniensis* sp. nov., a female araucarian cone-scale from the Rajmahal Series. *The Palaeobotanist* 5: 54–65.

Steiner, M. B. 1998. Age, correlation, and tectonic implications of Morrison Formation paleomagnetic data, including rotation of the Colorado Plateau. *Modern Geology* 22: 261–281.

Stockey, R. A. 1975. Seeds and embryos of *Araucaria mirabilis. American Journal of Botany* 62: 856–868.

———. 1980a. Anatomy and morphology of *Araucaria sphaerocarpa* Carruthers from the Jurassic Inferior Oolite of Bruton, Somerset. *Botanical Gazette* 141: 116–124.

———. 1980b. Jurassic araucarian cone from southern England. *Palaeontology* 23: 657–666.

———. 1982. The Araucariaceae: an evolutionary perspective. *Review of Palaeobotany and Palynology* 37: 133–154.

———. 1994. Mesozoic Araucariaceae: morphology and systematic relationships. *Journal of Plant Research* 107: 493–502.

Stockey, R. A., and I. J. Atkinson. 1993. Cuticle micromorphology of *Agathis* Salisbury. *International Journal of Plant Sciences* 154: 187–225.

Stockey, R. A., and H. Ko. 1986. Cuticle micromorphology of *Araucaria* de Jussieu. *Botanical Gazette* 147: 508–548.

Stockey R. A., H. Nishida, and M. Nishida. 1992. Upper Cretaceous araucarian cones from Hokkaido: *Araucaria nihongii* sp. nov. *Review of Palaeobotany and Palynology* 72: 27–40.

———. 1994. Upper Cretaceous araucarian cones from Hokkaido and Saghalien: *N. nipponensis* sp. nov. *International Journal of Plant Sciences* 155: 800–809.

Sukh-Dev and M. N. Bose. 1964. On some conifer remains from Bansa, South Rewa Gondwana Basin. *The Palaeobotanist* 12: 59–69.

Sukh-Dev and Zeba-Bano. 1978. *Araucaria indica* and two other conifers from the Jurassic-Cretaceous rocks of Madhya Pradesh, India. *The Palaeobotanist* 25: 496–508.

Tension-Woods, J. E. 1883. On the fossil flora of the coal deposits of Australia. *Proceedings of the Linnean Society of New South Wales* 8: 37–167.

Tidwell, W. D. 1990. Preliminary report on the megafossil flora of the Upper Jurassic Morrison Formation. *Hunteria* 2 (8): 1–8.

Tidwell, W. D., and S. R. Ash. 2006. Preliminary report on the Early Jurassic flora from the St. George Dinosaur Discovery Site, Utah. In J. D. Harris, S. G. Lucas, J. A. Spielmann, M. G. Lockley, A. R. C. Milner, and J. I. Kirkland (eds.), *The Triassic–Jurassic Terrestrial Transition: New Mexico Museum of Natural History and Science Bulletin* 37: 414–420.

Tidwell, W. D., B. B. Britt, and S. R. Ash. 1998. Preliminary floral analysis of the Mygatt-Moore Quarry in the Jurassic Morrison Formation, west-central Colorado. *Modern Geology* 22: 341–378.

Turner, C. E., and F. Peterson. 2004. Reconstruction of the Upper Jurassic Morrison Formation extinct ecosystem—a synthesis. *Sedimentary Geology* 167: 309–355.

van der Ham, R. W., J. M., J. H. A. van Konijnenburg-van Cittert, R. W. Dortangs, G. F. W. Herngreen, and J. van der Burgh. 2003. *Brachyphyllum patens* (Miquel) comb. nov. (Cheirolepidiaceae?): remarkable conifer foliage from the Maastrichtian type area (Late Cretaceous, NE Belgium, SE Netherlands). *Review of Palaeobotany and Palynology* 127: 77–97.

van Konijnenburg-van Cittert, J. H. A. 1971. In situ gymnosperm pollen from the Middle Jurassic of Yorkshire. *Acta Botanica Neerlandica* 20: 1–97.

———. 1987. New data on *Pagiophyllum maculosum* Kendall and its male cone from the Jurassic of North Yorkshire. *Review of Palaeobotany and Palynology* 51: 95–105.

van Konijnenburg-van Cittert, J. H. A., and H. S. Morgans. 1999. *The Jurassic Flora of Yorkshire*. Palaeontological Association Field Guide to Fossils No. 8. London: Palaeontological Association.

Vishnu-Mittre. 1954. *Araucarites bindrabunensis* sp. nov., a petrified megastrobilus from the Jurassic of Rajmahla Hills, Bihar. *The Palaeobotanist* 3: 103–108.

Walkom, A. B. 1918. Mesozoic floras of Queensland. Part II. The flora of the Maryborough (Marine) Series. *Queensland Geological Survey* 262: 1–20.

———. 1919. Mesozoic floras of Queensland. Part III and IV. The floras of the Burrum and Styx River Series. *Queensland Geological Survey* 263: 1–76.

———. 1921. Mesozoic floras of New South Wales. Part I. Fossil plants from the Cockabuta Mountain and Talbragar. *Memoirs of the Geological Survey of New South Wales, Palaeontology* 12: 1–21.

White, M. E. 1981. Revision of the Talbragar Fish Bed Flora (Jurassic) of New South Wales. *Records of the Australian Museum* 3: 695–721.

Whitmore, T. C. 1980. A monograph of *Agathis*. *Plant Systematics and Evolution* 135: 41–69.

Wilde, M. H., and A. J. Eames. 1952. The ovule and "seed" of *Araucaria bidwilli* with discussion of the taxonomy of the genus. II. Taxonomy. *Annals of Botany*, n.s., 16: 27–47.

Fig. 6.1. The seed fern *Medullosa*. (A) *Medullosa* had a treelike habit and produced seeds and pollen organs on separate leaves or even on separate plants. (B) Prepollen-bearing organ. (C–F) Seed-bearing organs. (G, H) Prepollen *Monoletes*. Fossil insects, such as *Palaeodictyoptera,* may have transported pollen because the pollen grains were too large for effective wind transport. *Source: Retallack and Dilcher (1988).*

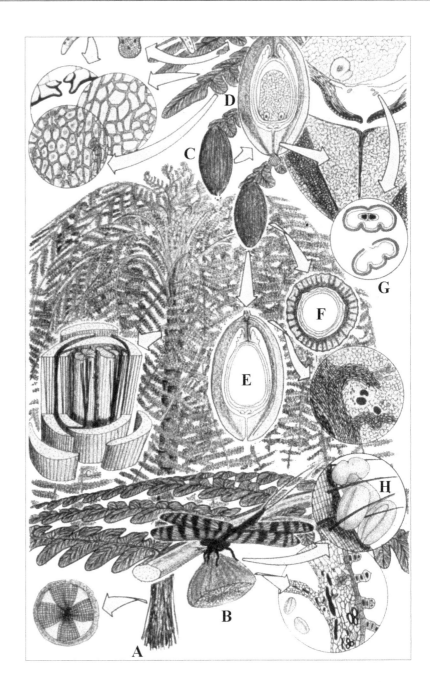

MAJOR INNOVATIONS IN ANGIOSPERM EVOLUTION

6

David L. Dilcher

Both the evolution of flowering plants and the evolution of flowers involved many innovations which extended over 75 million years. These innovations are presented here in six steps and center upon the reproductive biology of the flowering plants, often involving coevolution with animals to increase the potential for outcrossing. These innovations are as follows: (1) positioning of pollen and seed organs in close proximity to one another, (2) closure or closures of the carpel in two stages, loose closure and complete closure; (3) origin or origins of ornate, radially symmetric (polysymmetric) flowers; (4) development of elongate styles, (5) origin or origins of ornate bilaterally symmetric (monosymmetric) flowers; and (6) production of nutritious seeds and fleshy fruits. Angiosperm origins and early evolution are a Mesozoic event that continued to evolve into the Cenozoic. Angiosperms are products of their biotic and abiotic environments. Their potential to use the animals around them to carry their genetic material and disperse the fruits and seeds has allowed them to occupy and dominate much of the world's vegetation.

Introduction

As a major contributor to the science of plant evolution, paleobotany is concerned with the innovations that have appeared in plants throughout geological time. Plant evolution is most readily documented through the comparative morphology of plants, whether extant or fossil. This is as true as for the study of lower vascular plants, e.g., lycopods, sphenopsids, ferns, and major extinct groups of early land plants (Stewart and Rothwell 1993; Pryer et al. 2001), as it is for the angiosperms (Manos et al. 2007). Comparative anatomical studies of fossil taxa have led, for example, to the recognition of extinct angiosperm families (Retallack and Dilcher 1981; Sun et al. 2002), as well as major extinct groups of seed plants, e.g., the Pteridospermophyta (seed ferns) and the Bennettitales (Stewart and Rothwell 1993).

Great strides have been made in understanding the relationships of extant angiosperms to one another with the help of molecular systematics (Chase et al. 2003; Soltis et al. 2005). The results of molecular systematics should be viewed as botanical end points after millions of years of evolution. They are important in understanding the relative relationships of living angiosperm taxa to one another. However, molecular systematics organizes angiosperm taxa according to their molecular similarities rather than providing insights into the relevant steps or innovations that were pivotal in the origin and early evolution of the angiosperms.

Comparative analysis of angiosperm morphology and anatomy must be character based, and their relationships and evolutionary trends should be firmly based on the characters of living and fossil angiosperms. The morphological databases for angiosperms are still limited in their availability, number of characters analyzed, and diversity of taxa sampled, but the number of taxa examined is increasing (some that are available include Nixon et al. 1994; Doyle and Endress 2000; Doyle 2006, 2008). Still, they lag far behind the molecular data available. As these morphological databases grow, so will our ability to place the fossil records of angiosperms in the proper context of angiosperm evolution. We will continue to recognize extinct families (e.g., Retallack and Dilcher 1981; Sun et al. 2002) and even orders of angiosperms. And as the detailed, character-based fossil record of angiosperms grow, character-based evolutionary studies combined with those that are molecular based will become integrated into the study of angiosperms (Crepet et al. 2004). This research must be grounded in the extinct angiosperms that we are only now coming to recognize and value. To understand the evolutionary history of the flowering plants, it is of utmost importance that extinct families and genera of angiosperms be presented in textbooks alongside those of the extant angiosperms.

The most conspicuous character of the flowering plants as a whole is the flower, and flowers are all about sex and successful reproduction. Thus, this chapter will focus on the evolutionary innovations involved in the reproductive biology of the angiosperms that have been documented in the paleontological record. These innovations did not all happen at the same time, and perhaps not even in any single ancestral group of the flowering plants. This study should not be viewed as a compilation of secondary material, but rather a new interpretation of evidence from the fossil record. Our starting point is the ancestral seed ferns that bear reproductive organs, which may present homologies with those of the flowering plants (Doyle 2006).

Evolutionary Innovations

There are six major evolutionary innovations considered in this chapter, all concerned with angiosperm reproduction and the evolution of flowers and fruits. These are positioning of pollen and seed organs in close proximity to one another; closure of the carpel in two stages—loose closure and complete closure; origin or origins of ornate, radially symmetric (polysymmetric) flowers; development of elongate styles; origin or origins of ornate bilaterally symmetric (monosymmetric) flowers; and production of nutritious seeds and fleshy fruits.

Paleozoic seed plants, as typified by the seed fern *Medullosa* (Fig. 6.1), generally produced both seeds and pollen-bearing organs on leaves. However, the pollen organs of *Medullosa* (known as *Dolerotheca*, Fig. 6.1B) were not positioned in close proximity to the seed-bearing organs (*Pachytesta*, Fig. 6.1C–F) but were probably borne on separate leaves or even on separate plants. The prepollen of *Medullosa*, known as *Monoletes* (Fig. 6.1G, H), is quite large, up to 0.5 mm in length, and it has been suggested that it was most likely carried by an insect from the pollen organ to the micropyle (and pollination droplet) of the ovule (Dilcher 1979).

Ovules and pollen-bearing organs of seed ferns of the later Paleozoic and Mesozoic are always found separately from one another and were most likely borne on separate leaves or different individual plants. Pollination may often have been mediated by insects and perhaps also by wind.

There is some indication that insects were associated with some Mesozoic plants such as the seed ferns and the Bennettitales (Crepet 1974; Labandeira 2002; Osborn and Taylor 2010). Many of these plants had separate pollen and ovule-bearing organs, although the pollen organs were positioned in close proximity with the ovules as in *Williamsoniella* and *Cycadeoidea*. In both of these genera, the pollen organs are attached to the same axis that terminated in the seed-bearing receptacles. These pollen organs appear to have been derived from pollen-bearing leaves that were organized in whorls and subtended the seed-bearing receptacle. This is the same general plan that the angiosperms followed. However, although the bennettitaleans may not be on the path of angiosperm evolution (Doyle 2008; Crepet and Stevenson 2010; Rothwell and Stockey 2010), they demonstrate the strong coevolutionary influence that insect pollinators played on mid-Mesozoic plant reproduction. The visitation of insects to the reproductive organs of the bennettitaleans illustrates that the potential to bring the pollen-bearing and ovule-bearing organs together occurred in another group of seed plants totally independent of the angiosperms (Crepet 1974; Doyle 2008). Seed ferns also experienced the same biotic pressures, and it is most likely from this line that the angiosperms evolved as hinted at by the organization of the paleobotany textbook by Andrews (1961) and seriously discussed by Doyle (2006, 2008) and Retallack and Dilcher (1981).

In the angiosperms, seed-bearing leaves are the plant part requisite for the origin of the seed-bearing carpels (see Doyle 2006). Seed-bearing carpels are borne terminally on one of the earliest known examples of a complete reproductive flowering plant axis, *Archaefructus*. In *Archaefructus*, the pollen organs are located immediately below the carpels (Fig. 6.2), and thus, the pollen-bearing and ovule-bearing organs are both brought into close proximity to one another. Through this proximity, it would be possible for insect visitors to both deposit pollen on the carpels containing the young ovules and to pick up pollen from adjacent stamens. This efficient system would also be a benefit in wind pollination: the branches of one plant, growing in proximity of other individuals,

Fig. 6.2. *Archaefructus sinensis* fertile axis. The seed-bearing carpels are located on the same axis, with the carpels terminal (arrow) and the stamen below them. *Source: Sun et al. (2002).*

might produce pollen in sufficient quantities that a gentle wind could carry the pollen to receptive stigmatic surfaces of the carpels of nearby plants. For several reasons outlined by Sun et al. (2002), it appears most probable that *Archaefructus* extended its reproductive shoots above the surface of the water, contrary to the submerged flowering presented by Friis et al. (2003) and Endress (2008). Bringing ovule-bearing and pollen-bearing organs close together appears to have been a feature of the earliest angiosperms, if not a basic character established early on in the history of flowering plants, that continues to be expressed in many flowers today.

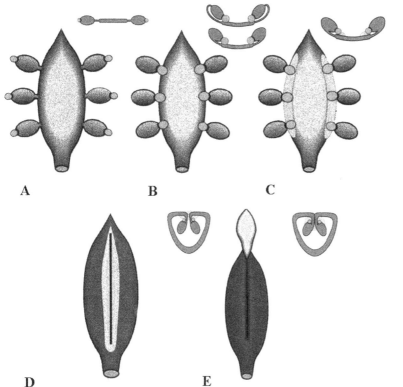

Fig. 6.3. Schematic diagrams of hypothetical leaf-bearing ovules with pollination droplets and early stages of carpel evolution. (A) Ovules along the leaf margin. (B) Ovules folded in toward the leaf center. (C) Pollination droplet fluid coalescing on the leaf surface over and between the ovules. (D) Leaf margins rolling up, with the pollination droplet fluid filling the area along the slightly open leaf or carpel margins and covering the micropyles of the ovules. (E) Morphologically closed leaf forming a closed carpel containing ovules. *Modified by T. A. Lott from Stützel and Röwekamp (1997).*

A   B   C

D   E

Closure of the Carpel in Two Stages

The closed carpel is the defining feature of the angiosperms. This feature is used to define the earliest angiosperm in the fossil record. Characters in pollen, leaves, and wood, while informative, cannot be used unequivocally to document the occurrence of the first angiosperms (Sun et al. 1998). Thus, although the earlier records of angiosperm or angiosperm-like pollen extend back to the Triassic and are well documented in the Hauterivian of the Lower Cretaceous (Cornet 1989; Hughes 1994; Brenner 1996; Hochuli and Feist-Burkhardt 2004), they are not indisputable evidence of the earliest angiosperms but interesting and important data of angiosperm-like characters early in the fossil record. Geochemical data suggests a pre-Mesozoic angiosperm ancestry (Taylor et al. 2006). The earliest closed carpel, however, can be documented in sediments 125 million years old (Sun et al. 1998, 2002).

Here I hypothesize that the earliest carpels of flowering plants and perhaps their immediate ancestors did not have tightly closed carpels. But when male and female reproductive organs came in close proximity and were aggregated together on a single shoot, then angiosperms developed functionally closed carpels. Paleozoic seed ferns clearly produced leaf-borne seeds with micropyles that produced pollination droplets (fig. 1 of Retallack and Dilcher 1988). The ovules of some seed ferns were borne along the leaf margins or embedded into the tissue of the leaf margin (Meitang et al. 1992). If these (Fig. 6.3A) or similar seed ferns folded their leaf margins inwards, the micropyles would all be pointing inward,

Fig. 6.4. Reconstruction of *Archaefructus sinensis*, showing the vegetative portions below the water surface and the reproductive organs just at or slightly above the water surface. *Modified by D. L. Dilcher from an illustration by K. Simons and D. Dilcher.*

and the pollination droplets might coalesce (Fig. 6.3B) as a fluid on the leaf surface (Stützel and Röwekamp 1997). This would have enhanced pollination, especially if the motile gametes of prepollen (Poort et al. 1996) were involved. The continued folding of the leaf and inrolling of the co-joined leaf margins would produce a partially closed carpellike structure filled with pollination droplets or a pollination fluid (Fig. 6.3C) that may have been biochemically active and functional in a pollen-incompatibility system inhibiting the pollen tube growth of unsuitable pollen (Dilcher 1995, 1996).

The partial closure of the carpel must have been the innovation that occurred to avoid self-pollination when the pollen organs and ovule-producing organs came into close proximity to one another. Although Zavada (1984: 63) considered that "there is no fossil evidence suggesting the early occurrence of self-incompatibility in angiosperms," the very presence of ovule and pollen organs positioned in close proximity to one another, as demonstrated in *Archaefructus* (Fig. 6.2), shows now that there is evidence for the existence of self-incompatibility at least 125 million years ago. Kubitzki and Gottlieb (1984: 457) wrote that "the chemical potential of the angiosperms paved the way for intense evolutionary interactions between plants and animals," which certainly was essential for promoting outcrossing in the development of self incompatibility (Zavada 1984 and references cited therein).

Continued closure of the carpel (Fig. 6.3D, E) probably likely followed partial closure of the carpel in order to prevent any interference with the incompatibility nature of the pollination fluid in the slightly open carpels. The most obvious interference with the pollination fluid, which is so important in ensuring outcrossing, would have been water that could have washed away, diluted, or otherwise compromised the incompatible nature of the pollination fluid. This would have been especially true for *Archaefructus* (Fig. 6.4; Plate 8), an aquatic plant with reproductive organs held just above the water surface (Sun et al. 2002), or for early flowering plants growing in open riparian environments open to heavy dew, rain, or occasional flooding. Certainly the closing of the carpel was not a protective response to the appetites of insects hungry for young ovules, as they had learned to drill through the carpel wall to attack the seeds very early in angiosperm evolution (Dilcher et al. 2007). Also, flies may have fed on the nutrient-rich fluid that filled the partially closed carpels and provided the incompatibility for pollen that guarded against self-pollination. The closed carpel ensures that the pollen incompatibility system of angiosperms is secure against external environmental factors. Endress and Igersheim (2000) group angiosperms into four categories on the basis of the nature of their carpel closure. They recognize the importance of carpel closure as the active site of pollen tube growth, and male gametophyte interactions and incompatibility: "the carpel site where carpel closure takes place is a hot spot zone of events in reproductive biology" (Endress and Igersheim 2000: S213). The types of carpels recognized are open ovules surrounded by secretions (type 1), intermediate closure and secretions, to total closure (type 4). The basal angiosperms, as recognized by molecular systematics, "all have angiospermy type 1" (Endress and Igersheim 2000: S213).

The evolution of the reproductive biology of the male gametophyte of terrestrial plants (Fig. 6.5) played a major role in the potential for plant evolution during the Mesozoic (Poort et al. 1996). Paleozoic land plants had freely dispersed spores that produced motile sperm cells. This was followed by seed plants with prepollen that produced haustorial feeding tubes and released motile sperm cells in the proximity of the micropyle.

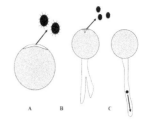

Fig. 6.5. Schematic diagrams representing the male gametophytes of various terrestrial plants. (A) Spore which releases motile sperm into the environment. (B) Prepollen which releases motile sperm cells in the area of the ovule. (C) Pollen which produces a pollen tube for the delivery of a nonmotile sperm nucleus into the ovule. The male gametophyte is essential in land plant reproduction, and each of these examples functions in a different manner. *Modified by T. A. Lott from Poort et al. (1996).*

Fig. 6.6. A selection of Aptian- and Albian-age flowers from the Early Cretaceous. (A) Petallike organs from the Crato Formation of Brazil, Aptian–Albian age, that suggest perhaps seven or more imbricate petals radially organized. Specimen CPCA 174. *Photo by D. L. Dilcher. Scale bar = 1 cm.* (B) Reconstruction of *Archaeanthus* from the upper Albian showing petals and sepals well differentiated. *Source: Dilcher and Crane (1984).* Scale bar = 3 cm. (C) Rose Creek flower with petals and sepals and a well-developed nectiferous ring of tissue at their bases. Scale bar = 1 cm. *Source: Basinger and Dilcher (1984).*

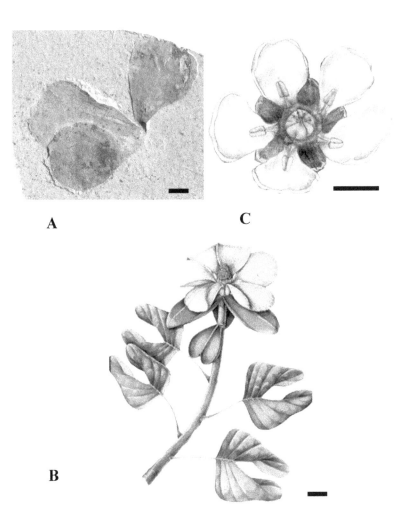

Spores are known from the Silurian to Recent, while prepollen is known from the Carboniferous and continued into the Permian. Prepollen is found still today in *Ginkgo* and the cycads. By the Mesozoic, true pollen with pollen tubes, which deliver nonmotile sperm nuclei, had evolved and are found in most seed-bearing land plants today.

Origin or Origins of Ornate, Radially Symmetrical (Polysymmetric) Flowers

The earliest reproductive axes of flowering plants known, i.e., *Archaefructus* (Sun et al. 1998, 2002) and *Hyrcantha* (Leng and Friis 2003, 2006; Dilcher et al. 2007), lack evidence of accessory organs such as petals or sepals. *Hyrcantha* has some scars below the carpels that could represent pollen organs or bracts, while *Archaefructus* consists of numerous individual terminal carpels that are proximally subtended by numerous stamen-bearing short shoots. No bracts or bract scars are evident (Fig. 6.2) on the reproductive axes of *Archaefructus*. However, the reproductive shoots may have borne vegetative leaves below the fertile portions of the axes. As far as it is known now, it appears that the earliest angiosperms lacked showy or conspicuous flowers (Sun et al. 2002).

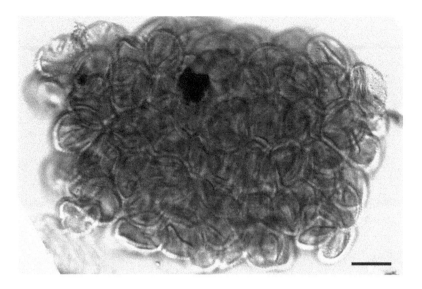

Fig. 6.7. One of the clumps of approximately 200 pollen grains of *Tricolpites labeonis* isolated from Albian sediments of the Dakota Formation. Scale bar = 10 µm. *Photo by S. Hu.*

Later angiosperms probably produced showy flowers, but they did not all follow the same pattern. This means that angiosperms may be considered a monophyletic group defined by the closed carpel (or functionally closed carpel). The aggregation of stamens and carpels on the same axis may or may not have followed a single path of evolution (Dilcher 1979). Stamens may have associated with carpels in various ways (e.g., Euphorbiaceae), and certainly they have been reduced, modified, and lost at various times. Finally, the showy flower, as well as the nonshowy flower, must have had multiple origins during of the radiation of angiosperms from very early times onward.

The flower, as a reproductive unit of the angiosperms, is constructed from the same basic organs available for modification during development of the shoot apex. The flower represents a specially differentiated shoot apex operating under specific genetic control (Krizek and Fletcher 2005). I suggest that the ancestral angiosperm may have been bisexual or unisexual from a very ancient time. Early in angiosperm evolution, the sexual nature of the reproductive shoot may have been modified in various lines of radiation. The organization of the reproductive organs predates the organization of the sterile perianth parts surrounding them. This explains why there is much more homology of the reproductive organs within the flowering plants than there is homology of the perianth parts. As clearly stated by Soltis et al. (2005: 275), "Character-state reconstructions of basal angiosperms . . . indicate that a differentiated perianth of what have been termed sepals and petals has evolved multiple times in these plants."

The basic plan of the reproductive shoot or flower is similar to an axillary shoot in the axis of a leaf, bract, or terminal shoot. Both of these are seen early in the history of flowers, such as in *Archaefructus* and *Hyrcantha* (Sun et al. 2002; Dilcher et al. 2007). The genetics controlling ancient flowering regulated the position and abundance of reproductive organs and only later must also have regulated sterile organs such as petals

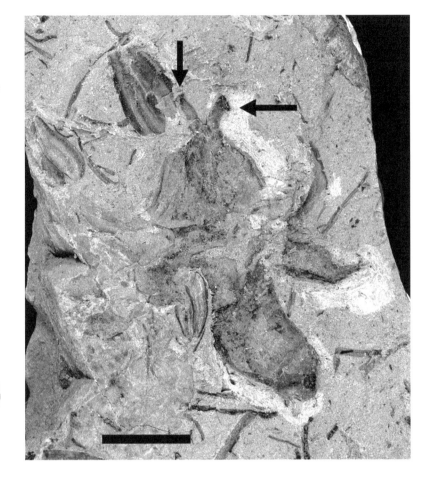

Fig. 6.8. A lateral compression of the Rose Creek flower in an early fruiting stage of development. The styles (arrows) are still attached to the loosely adnate fruits. UF15713-4026, Rose Creek, Nebraska. Scale bar = 5 mm. *Photo by D. L. Dilcher and Hongshan Wang.*

Fig. 6.9. A schematic diagram of the pollination of an ovule to show that the speed of the growth of the pollen tube is crucial in pollination. *Redrawn by T. A. Lott.*

and sepals. Some reproductive axes may have been unisexual in their ancestral state and remained that way as the rest of their lineage radiated. It is entirely possible that not all flowering plants had bisexual ancestors, although most fossils and extant angiosperms are bisexual today and are probably derived from bisexual ancestors. In the angiosperm reproductive shoot, some leaves are fertile, with the terminal ones bearing ovules; some proximal axes or leaves bear pollen organs that are subtended by vegetative leaves modified to attract pollinators and to produce nectar (petals) or are subtended by leaves modified to protect the whole structure or attract pollinators (sepals). All of these organs become aggregated as this shoot is reduced into a foreshortened axis that has been modified into a reproductive shoot whose differentiation is controlled by a set of genes whose activity has been grouped into the ABC and E model of floral organ patterning (Krizek and Fletcher 2005; see explanations by Taylor 2010). It is possible that some flowers never developed the accessory organs we expect for the "perfect" flower. Therefore, it should not be expected that all flowers have homologous organs. I suggest that all flowers had the same potential for the development of similar floral organs, but different developmental genes may have been fixed in populations early in flowering plant evolution, which allowed for the bursts of evolution to respond to changing biotic and abiotic environments.

Petals and sepals are well known from the Albian (Fig. 6.6) (Dilcher and Crane 1984; Balthazar et al. 2005) and are well developed as early as the Aptian–Albian (Mandarim-de-Lacerda et al. 2000; Bernardes-de-Oliveira et al. 2003) (Fig. 6.6A). Tepals are also well known from the Albian (Friis et al. 1988; Drinnan et al. 1991; Pedersen et al. 1994; Balthazar et al. 2007) and well developed by the Aptian–Albian (Mohr and Eklund 2003; Mohr and Bernardes-de-Oliveira 2004). Nectaries are present in fossil flowers by the late Albian (Basinger and Dilcher 1984) (Fig. 6.6C). These are examples of actual preservation of the petals, sepals, and nectaries in the fossil record, rather than the inference of their presence from scars found on the fossil remains of fruits. Until we actually find Barremian-age petals, we can only presume that they must have evolved very soon after the occurrence of *Archaefructus* 125 million years ago or had been present in other flowering plants at that time.

The presence of pollen clumps consisting of from 10, 50, 100, to 200 grains, consisting of a variety of angiosperm pollen types in the Albian (Fig. 6.7), suggests that by this time, the angiosperms were accommodating their insect pollinators by increasing the potential for successful transport of large numbers of pollen as one pollen load (Hu et al. 2008). A fossil flower with elongate styles and elevated stigmas occurs in sediments of the same age (Basinger and Dilcher 1984). The Rose Creek flower (Fig. 6.6C) also has nectaries, showy petals, and stiff sepals that would support insect pollinators. Its styles are attached to immature fruits (Fig. 6.8), which would have provided for elevated stigmas and the potential for large numbers of pollen grains to be delivered at one time. This means that the potential for pollen tube competition was set up as early as the late Albian. Experimental work on pollen tubes in living plants (e.g., Davis et al. 1987; Schlichting et al. 1987) has demonstrated that when several pollen grains arrive at the stigma at the same time, they begin growing their pollen tubes at nearly the same time (Fig. 6.9). The fastest-growing pollen tubes results in the successful fertilization of the ovules in the carpel. Field tests have demonstrated that the fastest-growing pollen tubes result in the most vigorous plants in the next generation. Thus, by the late Albian, some angiosperms were already actively self-selecting their own genetics.

Zavada (1984) has suggested that the earliest known angiosperm pollen has features indicating SSI, the self-incompatibility system in which pollen germination and growth are stopped at the site of the stigma by means of recognition substances in the pollen wall. He also believes that GSI, the self-incompatibility system in which pollen tube growth is stopped in the style by means of recognition substances in the pollen intine, evolved at a later time. This general trend seems to agree with the fossil record, because

Development of the Elongate Style

Fig. 6.10. Fossil flower with bilateral symmetry (monosymmetry) from middle Eocene, approximately 45 million years old, UF15815–5388. Lamkin Clay Pit, Tennessee. Scale bar = 2 mm. *Photo by D. L. Dilcher and Hongshan Wang.*

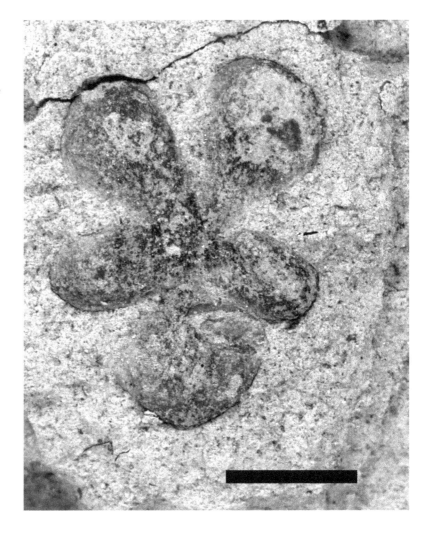

the Rose Creek flower, with its elongate style, shows up 105 million years ago, compared to the 125-million-year-old *Archaefructus.*

Origin or Origins of Ornate Bilaterally Symmetrical (Monosymmetric) Flowers

During most of the Cretaceous, flowering plants produced polysymmetric (also known as actinomorphic or radial) flowers (Fig. 6.6C). Polysymmetry is a common character in many angiosperm families, and some families may have some genera that are polysymmetric and others that are monosymmetric. Other families of flowering plants are entirely monosymmetric. The character of monosymmetry (also known as zygomorphy or bilateral symmetry) should be viewed as a derived feature in flowering plants. This character has been derived independently in many different angiosperm families (Soltis et al. 2005). Although fossil flowers are rare in the fossil record, there are a few that demonstrate this monosymmetry (Fig. 6.10). Gandolfo et al. (1998) described one of the earliest known zygomorphic flowers, *Dressiantha*, from the Upper Cretaceous,

about 90 million years ago. Some families, such as the Leguminosae, with both polysymmetric and monosymmetric genera, were in existence by the middle Eocene (Herendeen et al. 1992). Other families that are monosymmetric, such as the Orchidaceae, probably also evolved in the early Cenozoic (Paleogene) when the biotic environment changed with the evolution of social insects that were faithful pollinators of monosymmetric flowers (Labandeira 2002). Specific monosymmetric flower shapes often accommodate specific pollinators.

Thus, the continued evolution of the flower and various flowering plant families in the Late Cretaceous and Paleogene should be viewed in the context of their biotic environment. The burst of evolution in the legumes that occurred in the Eocene (Herendeen and Dilcher 1992; Herendeen et al. 1992) is another example of the angiosperms' success with insect pollinators. The continued radiation of flowering plant diversity during the Paleogene (Wilf et al. 2003; Wing et al. 2005) is the result of the new cooperative evolutionary strategies that developed between these new pollinators and new taxa of flowering plants.

Fig. 6.11. A bowl full of permineralized fruits and seeds from the middle Eocene, Clarno Formation, western Oregon. The existence of large nuts in the Eocene suggest that angiosperms were attracting animal dispersers for their fruits and seeds as early as 45 million years ago. *Photo by T. Bones.*

Production of Nutritious Seeds and Fleshy Fruits

Most angiosperms known from the Cretaceous have small fruits and seeds (Tiffney 1984; Wing and Tiffney 1987a, 1987b; Dilcher 1996; Eriksson et al. 2000; Eriksson 2008; Pedersen et al. 2007; Friis et al. 1997, 2000). Tiffney (1984) has suggested that the angiosperms had small fruits and seeds until the Late Cretaceous, and that angiosperm seed and fruit size began to increase during the radiation of rodents and fruit and seed eating birds in the Paleocene and Eocene.

Eriksson et al. (2000) have analyzed the proportion of animal-dispersed fruits at the Famelicão locality in western Portugal, which was dated originally as Barremian to Aptian (120 million years ago) but re-dated by Heimhofer et al. (2007) as early Albian (112 million years ago). In their examination of 32 fruit taxa, they found some to be berries and several others to be drupes on the basis of the presence of an epidermal layer surrounding seeds; this included *Anacostia*, which was described as a one-seeded berry by Friis et al. (1997, 2000). Because these fleshy fruits represented 24.5% of the fruits of this flora, Eriksson et al. (2000) concluded that nearly one-quarter of the early angiosperms of this flora were dispersed by animals. Even with the revision in the age of the flora by 8 million years, it is questionable whether such a large proportion of early angiosperms were already dispersed by animals by the early Albian, and in a later paper (Eriksson 2008), he recognized this. Eriksson (2008) also examined hypotheses related to changes in angiosperm seed and fruit size. These are coevolution, recruitment, and life form, and he concluded that none of them is fully supported by the evidence. Of these, I think that the coevolution of bird and rodent radiation in the Paleocene and Eocene must have had a profound effect on angiosperms. This is because of the potential for diaspore dispersal, which tends to reduce inbreeding and promotes outcrossing. Clearly, more work needs to be done to elucidate the interaction between animals and the early flowering plants in the dispersal of fruits and seeds in the Early Cretaceous, the Late Cretaceous, and the Paleogene.

On the basis of the identifications of leaves, pollen, fruits, and seeds throughout the Cenozoic record, it is evident that angiosperms have modified the size, nutritional value, and presentation of their fruits to please animal dispersers during the Paleocene and early Eocene. The numbers of big, edible fruits and seeds from the middle Eocene (Stone 1973; Manchester 1987), for example, in the form of large nuts (Fig. 6.11), are strong evidence for the rapid development of these angiosperm–animal interactions. Again, this coevolution is a means of promoting genetic outcrossing, as it promotes the dispersal of reproductive propagules some distance away from the parent population. This innovation in the evolution of angiosperms resulted in the rapid radiation of many fruit and seed types in angiosperms (Fig. 6.12; Plate 9), which in turn fueled the radiation of bird and mammal groups in the early Cenozoic.

Fig. 6.12. (A) Fruits of *Viburnum* sp. (B) Fruits of *Callicarpa americana* L. (C) Flower of *Anemone* sp. (D) Flowers of *Bougainvillea* sp. (E) Flower/bract complex of *Bougainvillea* sp. (F) Flower of *Nelumbo* sp. (G) Flowers of *Viola* sp. *(A–D, F) Photos by D. L. Dilcher. (E, G) Photos by S. Dilcher.*

Conclusions

All aspects of angiosperm evolution are, and have been over millions of years, controlled by what is most successful for the genetics of the plant. Outcrossing is perhaps the strongest of all the selection pressures in angiosperm evolution. The major innovations involved in angiosperm evolution discussed here are: (1) the close aggregation of stamens and carpels, (2) closure of the carpel in two stages, (3) evolution of ornate radially symmetric flowers, (4) evolution of elevated stigmas on elongated styles, (5) evolution of bilaterally symmetric flowers, and (6) evolution of fleshy fruits and nutritious seeds. Each one of these innovations resulted in some genetic benefits to the plants bearing these characters, and all are directly involved with plant–animal coevolution. The driving force of flowering plant evolution is the reproductive advantages produced as a result of plant–animal coevolution, which have greatly influenced the historical modifications of reproductive organs in plants—that is, flowers. The flowers and fruits of the angiosperms, as we know them today, did not appear suddenly but had a long history of innovations stretching over 140 million years, or perhaps even 240 million years. The numerous unique cooperative ventures established between plants and animals, as demonstrated by the extreme success of both groups, show that cooperation far surpasses competition as an important evolutionary strategy.

Acknowledgments

I thank Carole Gee for her initiative to organize this special Festschrift and her helpful comments, along with those from an anonymous reviewer, to improve the chapter. Thanks also to Terry Lott for his help and to my students and postdocs, who have taught me so much. I wish to add to this volume a personal and heartfelt thanks and a word of gratitude to Ted Delevoryas. He has been a Ph.D. advisor par excellence, a colleague, a counselor, and a good personal friend to me for nearly 50 years. Thanks, Ted. This chapter is dedicated to you.

References Cited

Andrews, H. N. 1961. *Studies in Paleobotany.* New York: Wiley.

Balthazar, M. V., K. R. Pedersen, and E. M. Friis. 2005. *Teixeiraea lusitanica,* a new fossil flower from the Early Cretaceous of Portugal with affinities to Ranunculales. *Plant Systematics and Evolution* 255: 55–75.

Balthazar, M. V., K. R. Pedersen, P. R. Crane, M. Stampanoni, and E. M. Friis. 2007. *Potomacanthus lobatus* gen. et sp. nov., a new flower of probable Lauraceae from the Early Cretaceous (early to middle Albian) of Eastern North America. *American Journal of Botany* 94: 2041–2053.

Basinger, J. F., and D. L. Dilcher. 1984. Ancient bisexual flowers. *Science* 224: 511–513.

Bernardes-de-Oliveira, M. E., D. L. Dilcher, A. M. Franca Barreto, F. Ricardi-Branco, B. Mohr, and M. C. De Castro-Fernandes. 2003. La flora del miembro Crato, Formación Santana, cretácico temprano de la cuenca de Araripe, noreste del Brasil. 10th Congreso Geologico Chileno 2003, Universidad de Concepción, Abstracts.

Brenner, G. J. 1996. Evidence for the earliest stage of angiosperm pollen evolution: a paleoequatorial section from Israel. In D. W. Taylor and L. J. Hickey (eds.), *Flowering Plant Origin, Evolution and Phylogeny,* pp. 91–115. New York: Chapman and Hall.

Chase, M. W., K. Bremer, B. Bremer, M. Thulin, P. F. Stevens, D. Soltis, P. S. Soltis, and J. Reveal. 2003. An update of the Angiosperm Phylogeny Group classification for the orders and families of flowering plants: APG II. *Botanical Journal of the Linnean Society* 141: 399–436.

Cornet, B. 1989. Late Triassic angiosperm-like pollen from the Richmond Rift Basin of Virginia, U.S.A. *Palaeontographica*, Abt. B, 213: 37–87.

Crepet, W. L. 1974. Investigations of North American cycadeoids: the reproductive biology of *Cycadeoidea. Palaeontographica*, Abt. B, 148: 144–169.

Crepet, W. L., and D. W. Stevenson. 2010. The Bennettitales (Cycadeoidales): a preliminary perspective on this arguably enigmatic group. In C. T. Gee (ed.), *Plants in Mesozoic Time: Morphological Innovations, Phylogeny, Ecosystems,* pp. 215–244. Bloomington: Indiana University Press.

Crepet, W. L., K. C. Nixon, and M. A. Gandolfo. 2004. Fossil evidence and phylogeny: the age of major angiosperm clades based on mesofossil and macrofossil evidence from Cretaceous deposits. *American Journal of Botany* 91: 1666–1682.

Davis, L. E., A. G. Stephenson, and J. A. Winsor. 1987. Pollen competition improves performance and reproductive output of the common zucchini squash under field conditions. *Journal of the American Society of Horticultural Science* 112: 712–716.

Dilcher, D. L. 1979. Early angiosperm reproduction: an introductory report. *Review of Palaeobotany and Palynology* 27: 291–328.

———. 1995. Plant reproductive strategies: using the fossil record to unravel current issues in plant reproduction. In P. C. Hoch and A. G. Stephenson (eds.), *Experimental and Molecular Approaches to Plant Biosystematics. Monographs in Systematic Botany* 53: 187–198.

———. 1996. La importancia del origen de las angiospermas y como formaron el mundo alrededor de ellas. *Conferencias VI Congreso Latinoamericano de Botanica, Mar Del Plata—Argentina 1994,* pp. 29–48. London: Royal Botanic Gardens, Kew.

Dilcher, D. L., and P. R. Crane. 1984. *Archaeanthus:* an early angiosperm from the Cenomanian of the Western Interior of North America. *Annals of the Missouri Botanical Garden* 71: 351–383.

Dilcher, D. L., G. Sun, Q. Ji, and H. Li. 2007. An early infructescence *Hyrcantha decussate* (comb. nov.) from the Yixian Formation in northeastern China. *Proceedings of the National Academy of Sciences U.S.A.* 104: 9370–9374.

Doyle, J. A. 2006. Seed ferns and the origin of angiosperms. *Journal of the Torrey Botanical Society* 133: 169–209.

———. 2008. Integrating molecular phylogenetic and paleobotanical evidence on origin of the flower. *International Journal of Plant Sciences* 169: 816–843.

Doyle, J. A., and P. K. Endress. 2000. Morphological phylogenetic analysis of basal angiosperms: comparison and combination with molecular data. *International Journal of Plant Sciences* 161 (6 Suppl.): S121–S153.

Drinnan, A. N., P. R. Crane, E. M. Friis, and K. R. Pedersen. 1991. Angiosperm flowers and tricolpate pollen of buxaceous affinity from the Potomac Group (mid-Cretaceous) of Eastern North America. *American Journal of Botany* 78: 153–176.

Endress, P. K. 2008. Perianth biology in the basal grade of extant angiosperms. *International Journal of Plant Sciences* 169: 844–862.

Endress, P. K., and A. Igersheim. 2000. Gynoecium structure and evolution in basal angiosperms. *International Journal of Plant Sciences* 161 (6 Suppl.): S211–S223.

Eriksson, O. 2008. Evolution of seed size and biotic seed dispersal in angiosperms: paleoecological and neoecological evidence. *International Journal of Plant Sciences* 169: 863–870.

Eriksson, O., E. M. Friis, K. R. Pedersen, and P. R. Crane. 2000. Seed size and dispersal systems of Early Cretaceous angiosperms from Famalico, Portugal. *International Journal of Plant Sciences* 16: 319–329.

Friis, E. M., P. R. Crane, and K. R. Pedersen. 1988. Reproductive structures of Cretaceous Platanaceae. *Biologiske Skrifter* 31: 1–55.

———. 1997. *Anacostia,* a new basal angiosperm from the Early Cretaceous of North America and Portugal with trichotomocolpate/monocolpate pollen. *Grana* 36: 225–244.

Friis, E. M., K. R. Pedersen, and P. R. Crane. 2000. Reproductive structure and organization of basal angiosperms from the Early Cretaceous (Barremian or Aptian) of western Portugal. *International Journal of Plant Sciences* 161 (6 Suppl.): S169–S182.

Friis, E. M., J. A. Doyle, P. K. Endress, and Q. Leng. 2003. *Archaefructus:* angiosperm precursor or specialized early angiosperm? *Trends in Plant Science* 8: 369–373.

Gandolfo, M. A., K. C. Nixon, and W. L. Crepet. 1998. A new fossil flower from the Turonian of New Jersey: *Dressiantha bicarpellata* gen. et sp. nov. (Capparales). *American Journal of Botany* 85: 964–974.

Heimhofer, U., P. A. Hochuli, S. Burla, and H. Weisseert. 2007. New records of Early Cretaceous angiosperm pollen from Portuguese coastal deposits: implications for the timing of the early angiosperm radiation. *Review of Palaeobotany and Palynology* 144: 39–76.

Herendeen, P. S., and D. L. Dilcher, eds. 1992. *Advances in Legume Systematics, Part 4. The Fossil Record*. London: Royal Botanic Gardens, Kew.

Herendeen, P. S., W. L. Crepet, and D. L. Dilcher. 1992. The fossil history of the Leguminosae and biogeographic implications. In P. S. Herendeen and D. L. Dilcher (eds.), *Advances in Legume Systematics, Part 4. The Fossil Record*, pp. 303–316. London: Royal Botanic Gardens, Kew.

Hochuli, P. A., and S. Feist-Burkhardt. 2004. A boreal cradle of angiosperms? Angiosperm-like pollen from the Middle Triassic of the Barents Sea (Norway). *Journal of Micropalaeontology* 23: 97–104.

Hu, S., D. L. Dilcher, D. M. Jarzen, and D. W. Taylor. 2008. Early steps of angiosperm–pollinator coevolution. *Proceedings of the National Academy of Sciences U.S.A.* 105: 240–245.

Hughes, N. F. 1994. *The Enigma of Angiosperm Origins*. Cambridge: Cambridge University Press.

Krizek, B. A., and J. C. Fletcher. 2005. Molecular mechanisms of flower development: an armchair guide. *Nature Reviews Genetics* 6: 688–698.

Kubitzki, K., and O. R. Gottlieb. 1984. Phytochemical aspects of angiosperm origin and evolution. *Acta Botanica Neerlandica* 33: 457–486.

Labandeira, C. C. 2002. The history of associations between plants and animals. In C. M. Herrera, and O. Pellmyr (eds.), *Plant–Animal Interactions: an Evolutionary Approach*, pp. 26–261. London: Blackwell Science.

Leng, Q., and E. M. Friis. 2003. *Sinocarpus decussatus* gen. et sp. nov., a new angiosperm with basally syncarpous fruits from the Yixian Formation of Northeast China. *Plant Systematics and Evolution* 241: 77–88.

———. 2006. Angiosperm leaves associated with *Sinocarpus* Leng et Friis infructescences from the Yixian Formation (mid-Early Cretaceous) of NE China. *Plant Systematics and Evolution* 262: 173–187.

Manchester, S. R. 1987. The fossil history of the Juglandaceae. *Monographs of the Missouri Botanical Garden* 2: 11–137.

Mandarim-de-Lacerda, A., D. L. Dilcher, A. Barreto, M. Bernardes-de-Oliveira, and D. Pons. 2000. Reproductive structures of Magnoliophytes of the Santana Formation, late Aptian/early Albian, Chapada do Araripe, Brazil. 31st Session of the International Geological Congress, 6–17 August 2000, Rio de Janeiro, Brazil, Abstracts.

Manos, P. S., P. S. Soltis, D. E. Soltis, S. R. Manchester, S. H. Oh, C. D. Bell, D. L. Dilcher, and D. E. Stone. 2007. Phylogeny of extant and fossil Juglandaceae inferred from the integration of molecular and morphological data sets. *Systematic Biology* 56: 412–430.

Meitang, M., D. L. Dilcher, and Z. H. Wan. 1992. A new seed-bearing leaf from the Permian of China. In B. S. Venkatachala, D. L. Dilcher, and H. K. Maheshwari (eds.), *Essays in Evolutionary Plant Biology. The Palaeobotanist* 41: 98–109.

Mohr, B. A. R., and M. E. C. Bernardes-de-Oliveira. 2004. *Endressinia brasiliana*, a magnolialean angiosperm from the Lower Cretaceous Crato Formation (Brazil). *International Journal of Plant Science* 165: 1121–1133.

Mohr, B. A. R., and H. Eklund. 2003. *Araripia florifera*, a magnoliid angiosperm from the Lower Cretaceous Crato Formation (Brazil). *Review of Palaeobotany and Palynology* 126: 279–292.

Nixon, D. C., W. L. Crepet, D. W. Stevenson, and E. M. Friis. 1994. A reevaluation of seed plant phylogeny. *Annals of the Missouri Botanical Garden* 81: 484–533.

Osborn, J. M., and M. L. Taylor. 2010. Pollen and coprolite structure in *Cycadeoidea* (Bennettitales): implications for understanding pollination and mating systems in Mesozoic cycadeoids. In C. T. Gee (ed.), *Plants in Mesozoic Time: Morphological Innovations, Phylogeny, Ecosystems*, pp. 35–49. Bloomington: Indiana University Press.

Pedersen, K. R., E. M. Friis, P. R. Crane, and A. N. Drinnan. 1994. Reproductive structures of an extinct platanoid from the Early Cretaceous (latest Albian) of eastern North America. *Review of Palaeobotany and Palynology* 80: 291–303.

Pedersen, K. R., M. Balthazar, P. R. Crane, and E. M. Friis. 2007. Early Cretaceous flora structures and in situ tricolpate-striate pollen: new early eudicots from Portugal. *Grana* 46: 176–196.

Poort, R. J., H. Visscher, and D. L. Dilcher. 1996. Zoidogamy in fossil gymnosperms: the centenary of a concept, with special references to prepollen of late Paleozoic conifers. *Proceedings of the National Academy of Sciences U.S.A.* 93: 11713–11717.

Pryer, K. M., H. Schneider, A. R. Smith, R. Cranfill, P. Wolf, J. S. Hunt, and S. D. Sipes. 2001. Horsetails and ferns are a monophyletic group and the closest living relatives to seed plants. *Nature* 409: 618–622.

Retallack, G., and D. L. Dilcher. 1981. Early angiosperm reproduction: *Prisca reynoldsii* gen. et sp. nov. from mid-Cretaceous coastal deposits in Kansas, U.S.A. *Palaeontographica*, Abt. B, 179: 103–137.

———. 1988. Reconstructions of selected seed ferns. *Annals of the Missouri Botanical Garden* 75: 1010–1057.

Rothwell, G. W., and R. A. Stockey. 2010. Independent evolution of seed enclosure in the Bennettitales: evidence from the anatomically preserved cone *Foxeoidea connatum* gen. et sp. nov. In C. T. Gee (ed.), *Plants in Mesozoic Time: Morphological Innovations, Phylogeny, Ecosystems*, pp. 51–64. Bloomington: Indiana University Press.

Schlichting, C. D., A. G. Stephenson, L. E. Davis, and J. A. Winsor. 1987. Pollen competition and offspring variance. *Evolutionary Trends in Plants* 1: 35–39.

Soltis, D. E., P. E. Soltis, P. K. Endress, and M. W. Chase. 2005. *Phylogeny and Evolution of Angiosperms*. Sunderland: Sinauer Associates.

Stewart, W. N., and G. W. Rothwell. 1993. *Paleobotany and the Evolution of Plants*. New York: Cambridge University Press.

Stone, D. E. 1973. Patterns in the evolution of amentiferous fruits. *Brittonia* 25: 371–384.

Stützel, T., and I. Röwekamp. 1997. Bestäubungsbiologie bei Nacktsamern. *Palmengarten* 61: 100–109.

Sun, G., D. L. Dilcher, S. Zheng, and Z. Zhou. 1998. In search of the first flower: a Jurassic angiosperm, *Archaefructus* from N. E. China. *Science* 282: 1692–1695.

Sun, G., Q. Ji, D. L. Dilcher, S. Zheng, K. C. Nixon, and X. Wang. 2002. Archaefructaceae, a new basal angiosperm family. *Science* 296: 899–904.

Taylor, D. W. 2010. Implications of fossil floral data on understanding the early evolution of molecular developmental controls of flowers. In C. T. Gee (ed.), *Plants in Mesozoic Time: Morphological Innovations, Phylogeny, Ecosystems*, pp. 119–169. Bloomington: Indiana University Press.

Taylor, D. W., H. Li, J. Dahl, F. J. Fago, D. Zinniker, and J. M. Moldowan. 2006. Biogeochemical evidence for the presence of the angiosperm molecular fossil oleanane in Paleozoic and Mesozoic non-angiospermous fossils. *Paleobiology* 32: 179–190.

Tiffney, B. H. 1984. Seed size, dispersal syndromes, and the rise of the angiosperms: evidence and hypothesis. *Annals of the Missouri Botanical Gardens* 71: 551–576.

Wilf, P., N. R. Cúneo, K. R. Johnson, J. F. Hicks, S. L. Wing, and J. D. Obradovich. 2003. High plant diversity in Eocene South America: evidence from Patagonia. *Science* 300: 122–125.

Wing, S. L., and B. H. Tiffney. 1987a. Interactions of angiosperms and herbivorous tetrapods through time. In E. M. Friis, W. G. Chaloner, and P. R. Crane (eds.), *The Origins of Angiosperms and Their Biological Consequences*, pp. 203–224. Cambridge: Cambridge University Press.

———. 1987b. The reciprocal interaction of angiosperm evolution and tetrapod herbivory. *Review of Paleobotany and Palynology* 50: 179–210.

Wing, S. L., G. J. Harrington, F. A. Smith, J. I. Bloch, D. M. Boyer, and K. H. Freeman. 2005. Transient flora changes and rapid global warming at the Paleocene–Eocene boundary. *Science* 310: 993–996.

Zavada, M. S. 1984. The relation between pollen exine sculpturing and self-incompatibility mechanisms. *Plant Systematics and Evolution* 147: 63–78.

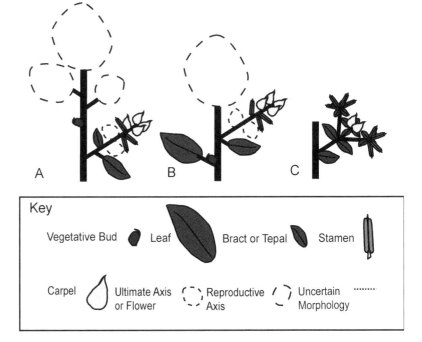

Fig. 7.1. Diagram of the RAs (inflorescences) and URAs (flowers) of *Archaefructus eoflora* (A), *A. liaoningensis* (B), and *Spanomera* (C). The number of RAs and URAs and organs are reduced in number for clarity. See the key for organs and oval patterns designating the additional URAs and RAs. The relative sizes of the ovals show relative maturity: the larger the size, the greater the maturity.

IMPLICATIONS OF FOSSIL FLORAL DATA ON UNDERSTANDING THE EARLY EVOLUTION OF MOLECULAR DEVELOPMENTAL CONTROLS OF FLOWERS

7

David Winship Taylor

A review of Early Cretaceous angiosperm flowers shows that all groups of reproductive developmental control genes were present, but that variation in some characteristics indicates some were not as canalized as they are now in living angiosperms. For example, many reproductive organ characters were uniform, whereas flower and inflorescence structures were variable. Molecular evidence for these genes suggests that most existed in the angiosperm sister groups and fewer were uniquely angiospermous. These new insights have led to the new Organs First, Flowers Later model for the evolution of angiosperm reproductive parts that is proposed in this chapter. This integration will provide a better understanding of the evolution of floral developmental controls and suggestions for new directions in research.

Introduction

There has been a revolution in our understanding of the basis of reproduction in flowering plants. The initial work was able to show a sequence of developmental events that result in the production of flowers (e.g., Coen and Meyerowitz 1991), but since then, our increased knowledge shows that developmental controls actually form a network with enhancers, repressors, and feedback among different levels. The speed at which new data have been collected and integrated into models is seen in the nearly annual reviews of the floral development (e.g., Battey and Tooke 2002; Theissen et al. 2002; Zik and Irish 2003; Buzgo et al. 2005; Krizek and Fletcher 2005; Soltis et al. 2005a; Theissen and Melzer 2007) that show that there are fundamental similarities in angiosperm floral development.

Parallel to the increase in developmental data has been the growth in interest of the evolution of these controls (e.g., Friedman et al. 2004; Frohlich 2006a; Scutt et al. 2007). To best understand morphological evolution, knowledge of the underlying development is needed. The main approach has been examining the distribution of fundamental control genes in a wide variety of angiosperms and preferably other seed plants. These data are then used to construct molecular phylogenies, allowing interpretations of the appearance and function of the controls through time.

Yet no matter how powerful the molecular phylogenetic approach is to understanding the evolution of developmental controls, it neglects the huge and growing data from the fossil record (e.g., Friis et al. 2006). In the contemporaneous 20 years, data on fossil flowers have blossomed. There are now ancient fossils from exquisite, three-dimensionally preserved flowers to reproductive organs attached to whole plants (summarized by Friis et al. 2006). The morphological diversity of flowers from the Early Cretaceous (the earliest period with unequivocal flowering plants) is evident and shows the presence of morphology not found in living plants. These data need to be integrated into our understanding of the evolution of development as the morphology of the fossils can feed back on our understanding of developmental controls.

In this chapter, I will provide a broad review of the evolution of developmental controls of flowers. I will start by summarizing the structure of a select number of well-preserved Early Cretaceous fossil angiosperm flowers. This will be followed by a review of our knowledge of each of the groups of genetic developmental controls based on the literature. These sections include a summary of what we know from *Arabidopsis*, followed by a discussion of the evolution of the genes based on phylogenetic and molecular comparisons, and the implications of fossil morphology. I will follow with a discussion of the overall implications of the fossils and evolutionary data to indicate the extent of expression and timing of major groups of developmental controls. Finally, I will propose a developmental and evolutionary model for the evolution of the angiosperm reproductive structures.

Methodology

Several types of Early Cretaceous fossils were selected for this study. In particular, I concentrated on well-documented fossils with reproductive organs attached to axes. This allowed the variation of floral and inflorescence axes to be examined. In addition, a select number of either early or morphologically interesting dispersed flowers were included to show the variation during this time. In general, the fossils were selected from the paper by Friis et al. (2006), and although their revised ages in this chapter were used, they were supplemented by more recent publications whenever possible.

To aid in describing organs that were not clearly homologous, I avoided the use of the terms *inflorescence* and *flower* unless they were clearly justified. Instead, I use the following two terms. The first is *reproductive axis* (RA), which is defined as a system with reproductive organs including all branches that have leaflike equivalents that are smaller than vegetative leaves at the nodes. In most living angiosperms, these would be considered inflorescences. These may be composed of a single axis, or they may be branched, with the branches opposite (dichasial) or alternate (monochasial). The morphology refers to whether there is more than one axis in the reproductive region (multiple) and whether these axes are found alone (simple) or are in groups (compound). The second term is *ultimate reproductive axis* (URA), which is defined as the last axis to which reproductive organs are attached. Typically, the internodes are shorter than those between the attachments of the URA to the RA. In most living angiosperms, these would be considered flowers. A ratio of URAs of similar size (and presumably age) to those of other sizes was calculated. Several other terms are used as well. *Vegetative continuation* refers to whether the vegetative axes are generally produced from the axillary buds (sympodial growth) or from the continuation of the apical bud and the axillary buds (pseudomonopodial). Finally, *clustered* refers to the condition where two or more organs have a common stalk, not a distinct receptacle.

Three species had sufficient specimens to examine the developmental stages of reproductive development. These data on development were collected by measurements from published works, photographs of types, and published descriptions.

Finally, the evolution of the genes was determined in four ways. First, in some cases, there have been phylogenetic analyses of gene sequences from many species. These studies provided the strongest support for distribution of the genes. It should be noted that even if the gene exists, it may not have the same expression, so caution should be taken. Second, I used GenBank to obtain the *Arabidopsis* FASTA gene sequence and did a protein–protein Basic Local Alignment Search Tool search (BLASTp) to find possible orthologs. I used the results to see what groups had similar sequences and the new distance tree of results to confirm that they were closely related. Third, I used the reports of known orthologs from the literature. Last, I used the fossil record as recorded here, as well as the general morphological fossil record of gymnosperms (Stewart and Rothwell 1993; Taylor et al. 2009). All three were used to determine whether the genes were preangiosperm, found throughout angiosperms and in at least one gymnosperm or other nonangiosperm; all-angiosperm, found in the basal dicot grade, monocots, and eudicots; possible all-angiosperm, found in monocots and eudicots; or restricted to specific clades.

Species with Different Developmental Stages

Review of
Relevant Fossils

Relatively few fossils show different stages in the development of RAs (equivalent to inflorescences in living plants) and URAs (equivalent to flowers in living plants). Several fossils show variation in the sizes of the RAs but have few details on the development of the reproductive structures, and these include *Hyrcantha* (*Sinocarpus*), *Endressinia*, *Araripia*, and *Caspiocarpus*. Only three species of Early Cretaceous fossils have reproductive organs at different stages: *Archaefructus eoflora*, *A. liaoningensis*, and *Spanomera* (Fig. 7.1). The general morphology of all the fossils is summarized in Table 7.1, along with a brief description of the important characteristics.

Table 7.1. Summary of the Structure of Select Early Cretaceous Fossils with Reproductive Organs

Taxon	*Archaefructus eoflora*	*Archaefructus liaoningensis*	*Spanomera*	*Bevhalstia*	*Archaefructus sinensis*	*Hyrcantha (Sinocarpus)*
Geological age	Barremian	Barremian–Aptian	Albian–Cenomanian	Barremian–Aptian	Barremian–Aptian	Barremian–Aptian
Characteristics of Vegetative Structures						
Leaf size	3 cm	1 cm	—	3 cm	4 cm	At least 2 cm, dispersed
Petiole	Long	Long		Long to none	Long	Long
Phyllotaxy	Alternate	Alternate		Alternate, opposite	Alternate	Alternate
Continuation	Axillary, sympodial	Axillary, sympodial		Apical, axillary	Axillary, sympodial	—
Nodes	Straight	—		Deflected	Deflected	Deflected
Characteristics of Reproductive Axis						
Placement	Axillary, terminal	Axillary, terminal	—	Axillary, terminal	Axillary, terminal	Axillary, terminal
Branching arrangement	None, and monochasial (alternate), terminal oldest	None (monochasial? [alternate]), terminal oldest	Bract alternate, probably monochasial (alternate)	None	None and monochasial (alternate) terminal oldest	Monochasial (alternate), and dichasial (opposite), terminal oldest
Morphology	Multiple simple and compound	Multiple simple (and compound?)	Multiple simple?	Simple?	Multiple simple and compound	Multiple simple and compound
Determinate/ Indeterminate	Determinate	Determinate	Determinate	Determinate	Determinate	Determinate
Sterile relative size morphology	Smaller, leaflike immediate below male region to none	Smaller, leaflike to none	Smaller, bract	Smaller, leaflike	Smaller, leaflike to none	Smaller, bract
Phyllotaxy of ultimate reproductive axis on reproductive axis	Alternate	Alternate	Opposite, dichasial	Alternate, opposite	Alternate	Alternate (monochasial), opposite (dichasial, terminal oldest)
Arrangement of URA	Male proximal, female distal	Male proximal, female distal	Male proximal, female distal	—	Male proximal, female distal	Female
Number of URAs	Many	Many	Three	—	Many	Few
Ratio of URAs of similar size (age)	High	High	—	—	High	High
Characteristics of Ultimate Reproductive Axis						
Unit	Male, female, bisexual	Male, female	Male, female	—	Male, female, bisexual	Female
Sterile type	None	None	Tepal	Tepal	None	Tepal
Sterile number	NA	NA	4–5	2–4	NA	2
Sterile arrangement	NA	NA	Opposite	Opposite	NA	Opposite
Structure			Hypogynous	Hypogyous		Hypogynous
Characteristics of Reproductive Organs						
Stamen				—		—
Number	1–3	1–3	4–5		1–2	
Phyllotaxy	Single, cluster	Single, cluster	Opposite		Single, cluster	

Pluricarpellatia	Xingxueina	Koonwarra plant	Araripia	Endressinia	Caspiocarpus	Hedyosmum-like	Platanus-like
Aptian–Albian	Aptian	Aptian	Aptian–Albian	Aptian–Albian	Albian	Barremian–Aptian	Albian
						—	—
5 cm	4–5 cm	1 cm	3 cm	4 cm	3–4 cm dispersed		
Long	Long	Long	None	Short	Long		
Alternate	Alternate	Alternate	Opposite	Alternate	Opposite		
Apical, axillary	—	—	Apical, axillary	Axillary?	—		
Straight	—	Deflected	Straight	Deflected	Straight		
Axillary	Terminal?	Axillary	Axillary	Terminal	Terminal	—	—
None	None	None	Dichasial (opposite)?	Monochasial (alternate)?	Dichasial (opposite), terminal oldest	NA	Monochasial (alternate)
Simple	Multiple simple	Multiple simple	Multiple compound	Multiple compound	Multiple compound	Simple	Multiple compound
Determinate	Determinate	Determinate	Determinate?	Determinate	Determinate	Determinate	Determinate at least in part
NA	Smaller, bract	Smaller bract	None?	None?	Smaller, leaflike	None?	None?
NA	Alternate	Alternate	Opposite, dichasial, terminal oldest	Opposite, dichasial, terminal oldest	Alternate	Whorl	Alternate
Female	—	Female	—	Bisexual		Male	Bisexual
One	Many	Many	Three	Three	Many	Many	Many
High	High	High	High	High	High	—	—
	—						
Female	Male	Female	—	Bisexual?	Female	Male	Male, female
None		Bracteole	Tepal	Tepal	None	None	Tepal
NA		2	Many	Many	NA	NA	5 to many
NA		Opposite	Alternate	Alternate	NA	NA	Alternate to whorled
		Hypogynous	Hypogynous	Hypogynous			Hypogynous
—	—	—	—		—		
				Stamenoids			
				Many		Many	Many
				Alternate		Whorl	Alternate to whorled

Table 7.1 (continued).

Taxon	Archaefructus eoflora	Archaefructus liaoningensis	Spanomera	Bevhalstia	Archaefructus sinensis	Hyrcantha (Sinocarpus)
Size variation			No			
Connation	Free	Free	Free		Free	
Filament length	Short	Short	Short		Short	
Filament continuation	No joint	No joint	No joint		No joint	
Anther attachment	Basifixed	Basifixed	Dorsifixed		Basifixed	
Number of sporangia	Tetrasporangiate	Tetrasporangiate	Tetrasporangiate		Tetrasporangiate	
Number of theca	Dithecal	Dithecal	Dithecal		Dithecal	
Connective	Apical tip	Apical tip	Apical tip		Apical tip	
Pollen	—	—			—	
Carpel/fruit				—		
Number	1–2	1–2	2		1–2	2–4
Phyllotaxy	Single, cluster	Single, cluster	Opposite		Single, cluster	Opposite
Size variation	Yes	Yes	No		Yes	Yes
Connation	Free	Free	Free		Free	Basally syncarpous
Stigma	Yes, apiculate	Yes, apiculate	Long		Yes, apiculate	—
Style	No	No	No		No	No
Stigmatic tissue	Apex	Apex	Along closure		Apex	—
Closure	Upper third	—	Lateral		—	—
Type	Ascidiate	—	Ascidiate		—	—
Ovule				—		
Number	4–8	2–4	—		8–12	10–16
Attachment of chazala	Opposite micropyle	Opposite micropyle	—		Opposite micropyle	?
Angle	Orthotropous	Orthotropous	—		Orthotropous	Anatropous?
Number of integuments	—	—	—		—	2

Note. Most terms are in common usage; those that are not are defined in the Methodology section.

Pluricarpellatia	Xingxueina	Koonwarra plant	Araripia	Endressinia	Caspiocarpus	Hedyosmum-like	Platanus-like
						Variable	—
						Free	—
						Short	Long
						No joint	No joint
						Basifixed	Basifixed
						Tetrasporangiate	Tetrasporangiate
						Dithecal	Dithecal
						Simple tip	Apical tip
	Inaperturate, reticulate					Monosulcate, reticulate	Tricolpate, reticulate
	—					—	
6–many		Few		Many	Many		5
Alternate		Whorled?		Alternate	Alternate		?
No		No		Yes	—		No
Free		Free		Free	Free		Free
Yes, apiculate		Yes, small		Yes, apiculate	Yes		Yes, apiculate
No		No		No	No		No
—		Apex		Apex	Apex		Apical
—		—		Lateral	—		—
?Ascidiate		—		—	—		—
		—				—	
				—	Up to 3		1
Next to micropyle				—	—		Opposite micropyle
Anatropous				—	Orthotropous?		Orthotropous
—				—	2		—

Archaefructus eoflora (Fig. 7.1A) is one of the three species from China (Table 7.1) and is based on a unique specimen that has multiple RAs of variable age attached to an entire plant with roots and leaves (Ji et al. 2004). The basal portion is a swollen stem (rhizome) with attached adventitious roots. The RAs have a proximally naked peduncle, attached leaves with long petioles, and usually smaller leaves without petioles before the attachment of the alternately arranged reproductive organs. The male region has URAs composed of clusters of up to three stamens. The URAs of the distal female region have usually single but occasionally paired carpels. In one case, there is a bisexual cluster composed of an anther and two carpels between the two regions. The main axis is terminated by an RA, with progressively smaller and younger RAs toward the base in a monochasial pattern. The two nodes below those having RAs have lateral vegetative buds; the vegetative branching is thus sympodial.

The preservation of the entire plant in different stages of reproduction provides a unique opportunity to examine the development of the organs (Table 7.2). Four stages (in Ji et al. 2004) were chosen: bud (fertile axis 8), early maturation (receptive stigmas, fertile axis 4), late maturation (maturing carpels, fertile axis 2), and mature (mature stamens, fertile axis IV). Clearly, there are two different growth trajectories in the stamens and carpels. In the bud, the two organs are of similar size. From there on, the male fertile region increases steadily, tripling in size during the course of two stages until it reaches maturity, while the stamen length increases slowly until its mature stage, when it has doubled in size and the stamens have opened. In contrast, the female region increases initially sixfold, then slowly continues to increase until its mature stage, while the carpels initially double or triple in size during the intermediate stages and develop a stigma. Note the protogynous order of maturation of the organs. The sterile region greatly increases at the mature stage, while the modified leaf steadily increases to about double its size.

The first species of *Archaefructus* described from China is *A. liaoningensis* (Table 7.1; Fig. 7.1B). Although the species was initially described from one specimen at the fruiting stage (Sun et al. 1998), later publications have documented many additional specimens that show better details of the structure of the RAs, younger stages, and the attachment of leaves (Sun and Dilcher 1997; Sun et al. 2001, 2002). In addition, other interpretations have been made on the basis of examinations of the material (Friis et al. 2003; Ji et al. 2004).

In this species, the RAs may be terminal with an apparent vegetative axillary bud, or monochasial with an additional smaller or younger axis originating from the axil of the last vegetative leaf. The female organs are distal on the RAs, and the male proximal; occasionally there is a small, less dissected leaf toward the base of the axis. The URAs, composed of carpels or stamens, are spirally to suboppositely arranged, and the stamens can be single or in clusters of up to three.

Three stages (Table 7.3) can be identified from fossil material from the same locality (Sun et al. 1998, 2001). The first, early maturation stage

Floral Structure or Characteristic	Bud	Early Maturation	Late Maturation	Mature
Internodal distance between reproductive structures	Short	Short	Moderate male, short female	Elongate male, moderate female
Sterile region	1.0 cm	1.0 cm	1.3 cm	4–11 cm
Male fertile region	0.33 cm	0.90 cm	2.7 cm	4.0 cm
Female fertile region	0.12 cm	0.80 cm	1.7 cm	2.0 cm
Sterile organ	0.74 cm	1.0 cm	Incomplete	2.0 cm
Stamen length	0.14 cm	0.20 cm	0.35 cm	0.60 cm
Stamen status	Unopened	Unopened	Unopened?	Open, dehiscent
Carpel length	0.16 cm	0.33 cm	1.1 cm	1.4 cm
Stigma	Not developed	Yes	Yes	Yes

Table 7.2. Developmental Stages in *Archaefructus eoflora*

has RAs with short sterile sections, short nodes, apparently unopened stamens, and small carpels that are tightly packed and projecting upward at about 45 degrees (Sun et al. 2001; specimens PB18951, PB18952). The stigmas are well developed, suggesting a protogynous order in the maturation of the reproductive organs. The mature stage (Sun et al. 2001, 2002, specimen B2000) is shown by fossils that have elongate RAs, but still have clusters of attached anthers and enlarged carpels. The elongation of the axis includes both the sterile base and the portion with attached reproductive organs. The fruiting stage is represented by specimens that have long internodes, the bases of the clusters of stamens, and distal fruits with a thickened closure and irregularly sized seeds (Sun et al. 1998, 2001, specimen PB18938). The sterile portion of the two axes ranges from 1.3 to 2.3 cm, while the fertile portion ranges from 5.0 to 6.0 cm. The carpels remain about the same size.

The species of *Spanomera* (Table 7.1; Fig. 7.1C) from the United States also provide interesting data on flowers, inflorescences, and developmental stages (Drinnan et al. 1991; Friis et al. 2006). The fossil species are described from a large variety of remains that allow for a good reconstruction of the terminal inflorescence unit, and the male and female flowers at different stages. The determinate dichasium is composed of a terminal female flower and two lateral male flowers, subtended by lateral bracts. One specimen appears to show these units subtended by a bract on an axis with alternate phyllotaxy. Unlike typical cymes, the terminal flower seems to mature after the lateral male flowers, and the stamens dehisce before the female is fully developed. The male flowers have four or five oppositely and decussate bracts that are interpreted as tepals and subtend the four or five stamens. In some flowers, there appears to be a small and presumably aborted gynoecium; the tepals initially protect the stamens and have valvate dehiscence. The terminal flowers have

Table 7.3. Developmental Stages in *Archaefructus liaoningensis*

Floral Structure or Characteristic	Early Maturation	Mature	Fruiting
Internodal distance between reproductive structures	Short	Long	Long
Sterile region	0.5 cm	1.6 cm	1.3 to 2.3 cm
Fertile region	2.5 cm	At least 4.8 cm	At least 5.0 to 6.0 cm
Stamen length	0.4 cm	0.4 cm	Dehisced
Stamen status	Unopened	Opened	NA
Carpel length	0.3 cm	0.7 cm	0.7 cm
Stigma	Developed	Developed	Developed

four bracts in opposite and decussate pairs (the last two are interpreted as tepals), which subtend two carpels that may be connate at the base.

The fossil specimens (Drinnan et al. 1991) seem to show a developmental series (Table 7.4). Unfortunately, this series could include organs that are aborted, as opposed to being immature. Nevertheless, in the attached flowers, small anthers (0.3 mm) are found in which the pollen sacs are poorly developed and differentiated from the filament, and the tepals are small. Numerous dispersed, large (1.0 mm), dehisced anthers are found with apical extension of the connectives and longitudinal dehiscence. Dispersed male flowers without anthers have tepals twice the size and with a small ovary that is interpreted as a pistillode. Thus, there appears to be a developmental sequence from flowers with small immature stamens and small tepals, to large, dehisced stamens with open anthers, mature pollen, and large tepals. The carpels in the flowers attached to the inflorescence are small (0.3 mm long) and lack structure. Some of these flowers are attached to inflorescences bearing male flowers without stamens. Small dispersed gynoecia (0.7–1.0 mm long) have a long closure. Larger specimens (over 1.2 mm and wider) have an elongate stigma that surrounds the closure with a papillate stigmatic surface. Finally, the largest are fruit fragments over 2.2 mm long with thick walls and are open without seeds. Altogether, it appears that female flowers are delayed in respect to male flowers (so the inflorescence may be protandrous; but see reconstruction in Friis et al. 2006), and the carpel is about 1.2 mm long when sexually mature and increases substantially in size during the development of the fruit.

These species show that considerable developmental data are available from fossils and, more importantly, that it is possible to determine the details of growth dynamics. The dynamics of the RAs in *Archaefructus* are not typical of a receptacle but instead much more similar to an inflorescence, as has been previously suggested (Friis et al. 2003; Ji et al. 2004). In addition, these data aid in the understanding of whether the RAs were protogynous or protandrous. Clearly, the *Archaefructus* species were protogynous with the development of the stigmas before the anthers opened. Last, all show evidence of fruit development after fertilization,

Table 7.4. Developmental Stages in *Spanomera*

Floral Structure or Characteristic	Monoecious Inflorescence Early	Monoecious Inflorescence Late	Dispersed Stamen	Dispersed Carpel Small	Dispersed Carpel Large	Dispersed Fruit
Male flower						
Tepals	Small	Large				
Stamen	0.3 mm	Dehisced	1.0 mm			
Female flower						
Tepals	Small	Small				
Carpel	0.3 mm	0.3 mm		0.7–1.0 mm	1.2 mm	2.2 mm
Stigma	Simple	Simple		Simple	Papillate	NA

Fig. 7.2. Diagram of the RAs (inflorescences) and URAs (flowers) of *Bevhalstia* (A), *Archaefructus sinensis* (B), *Hyrcantha/Sinocarpus* (C), and *Pluricarpellatia* (D). The number of RAs and URAs and organs are reduced in number for clarity. See the key in Figure 7.1 for organs and oval patterns designating the additional URAs and RAs. The relative sizes of the ovals show relative maturity: the larger the size, the greater the maturity.

an angiosperm character that appears to make them more efficient than gymnosperms (Taylor and Hickey 1996).

Species Preserved as Whole Plants

It is evident that one of the previously described fossil species, *Archaefructus eoflora*, is preserved as a whole plant. Yet there are other whole-plant species as well. It is notable that many these do not have reproductive organs at very different ages, suggesting the flowering was synchronous.

Bevhalstia pebja (Table 7.1; Fig. 7.2A) is an enigmatic plant from England that represents a probable aquatic angiosperm (Hill 1996). It has the type of variable morphology that is typically found in aquatics that are partly submerged and partly emergent. The stems have either alternately or oppositely arranged leaves or bracts. The leaf organs can be palmate, lobed, and petiolate to simple or fimbrate, and sessile or clasping, but always with simple, dichotomous venation. Some leaves have a cuticle and either hair bases or oil cells. The branching appears sympodial and is similar to the growth pattern observed in water lily axes from the Albian (Taylor et al. 2008). In some plants, up to three branches or short branches can arise at the nodes that bear opposite leaves (one of which is reduced), originating from the two axillary buds and the main apical bud. Vegetative buds or an RA can occur terminally.

Archaefructus sinensis (Table 7.1; Fig. 7.2B) from China is preserved in its entirety, but it shows little size variation in its RAs (Sun et al. 2002). This species has been described on the basis of multiple specimens (Sun et al. 2002), as well as redescribed on the basis of other specimens (Friis et al. 2003) and reexamination of the type specimen (Ji et al. 2004). Some specimens show that the distinctive reproductive organs are attached to a slightly swollen axis with lateral roots. The type specimen has several overlapping axes that have been considered fertile. Yet careful examination shows that the axis to the left originates from the axis of a leaf (sympodial branching) and continues without any branches or reproductive structures. In addition, although most of the RAs originate from a single axis, at least one has an axillary RA as well. Like all species of *Archaefructus*, the RAs in *A. sinensis* have a sterile peduncle to which smaller and simpler leaves are usually attached, the URAs in the male region have stamens either singly or in clusters of two, and the female region has carpels either singly or in clusters of two (Friis et al. 2003; Ji et al. 2004). As in *A. eoflora*, one bisexual URA cluster is found between the two regions (Ji et al. 2004).

Two related plants (Table 7.1, Fig. 7.2C) that have been described multiple times are *Hyrcantha karatscheensis* from the Albian of Kazakhstan (Krassilov et al. 1983; Krassilov 1997; Dilcher et al. 2007) and *Hyrcantha (Sinocarpus) decussata* from the Barremian to Aptian of China (Leng and Friis 2003, 2006; Dilcher et al. 2007). These fossil species appear to be congeneric (Dilcher et al. 2007), but the differences in the material and interpretations make understanding the plant somewhat difficult. Some specimens appear to have simple roots from which arise from a single (or double) axis to which alternately arranged bracts (or leaf bases?) are attached (Leng and Friis 2006; Dilcher et al. 2007). From the axils of the bracts, RAs arise that may bear a single URA (Krassilov 1997; Leng and Friis 2003, 2006; Dilcher et al. 2007), two URAs (Krassilov 1997; Leng and Friis 2003, 2006; Dilcher et al. 2007), or three URAs (Leng and Friis 2003, 2006) even on the same plant. In the three-URA specimens, the two lateral, oppositely arranged URAs may be smaller and at an earlier developmental stage (Leng and Friis 2003, 2006). The suggestion that there might be stamens (Krassilov 1997) does not seem to be supported (Dilcher et al. 2007).

A well-preserved nymphaeaceous aquatic plant, *Pluricarpellatia peltata* (Table 7.1; Fig. 7.2D), has been described from the Aptian–Albian of Brazil (Mohr and Friis 2000; Mohr et al. 2008). The plant includes a thin rhizome, adventitious roots (mostly at the nodes), variable eccentric–cordate to centrally peltate leaves, and RAs originating from the axil of the leaves. Vegetative growth continues at the apex and with axillary branches. The RAs have a long, naked axis that is terminated with a single cluster of reproductive organs. The URAs do not appear to have perianth or stamens, and the carpels are six to many. The open and closed carpels are about the same size.

Species with All Organs Except Roots

In one of the species described above, *Archaefructus liaoningensis*, all of the organs are preserved except the roots. The rest of the following fossils have at least one axis with a large vegetative leaf and reproductive structures attached.

The single fossil of *Xingxueina heilongjiangensis* (Table 7.1; Fig. 7.3A) from China has a short axis with what appears to be a single foliage

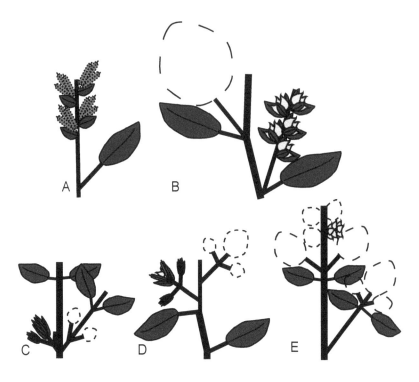

Fig. 7.3. Diagram of the RAs (inflorescences) and URAs (flowers) of *Xingxuenia* (A), Koonwarra plant (B), *Araripia* (C), *Endressinia* (D), and *Caspiocarpus* (E). The number of RAs and URAs and organs are reduced in number for clarity. See the key in Figure 7.1 for organs and oval patterns designating the additional URAs and RAs. The relative sizes of the ovals show relative maturity: the larger the size, the greater the maturity.

leaf and an RA (Sun and Dilcher 2002). The RA has poorly preserved URA "florets" subtended by spirally arranged bracts along the distal half of the axis. Pollen has been described from these florets. No other details are described or observable.

The Koonwarra plant (Table 7.1; Fig. 7.3B) comes from Australia and has axillary RAs (Taylor and Hickey 1990). The RAs appear to be composed of spirally arranged bracts with opposite bracteoles that subtend groups of free carpels (URAs).

The fossil *Araripia florifera* (Table 7.1; Fig. 7.3C) comes from Brazil and has both vegetative axes and RAs (Mohr and Eklund 2003). There appears to be opposite pairs of axes at the base of the plant. The authors suggest that there is a fertile branch with basal flowers (URAs) and distal leaves, as well as a vegetative branch. My interpretation is that there are two oppositely arranged fertile branches, and the distal end of the main axis is vegetative. At least one of the fertile RAs has two URAs of different sizes and possibly the peduncle of a third. The opposite axis may be bearing a pair (or more) of URAs, but they are poorly preserved. In either interpretation, at least some of the stems with lateral fertile axes continue vegetatively with opposite leaves.

The Brazilian fossil *Endressinia* (Table 7.1; Fig. 7.3D) has an axis with leaves that terminates in an RA branching system (Mohr and Bernardes-de-Oliveira 2004). The RAs have a larger terminal URA and two smaller and developmentally younger lateral URAs that are opposite. These units may be part of a larger branching system but the preservation is not sufficient to tell. The larger URA is not open and has tepals followed by staminodes and finally carpels with the distal organs developmentally younger.

A last fossil with leaves and reproductive organs is *Caspiocarpus* (Table 7.1; Fig. 7.3E) from Kazakstan, which has pairs of small, petiolate, palmately lobed, toothed leaves that subtend branched reproductive organs (Krassilov 1984, 1997). At the terminus of the main branch or lateral branches, there appears to be a node with a pair of small leaves below an apex that develops into a branched RA with a large central RA and the two lateral buds that develop into two smaller and more compact RAs. On these are spirally arranged URA "racemes" of variable length with attached fruits that lack any perianth or stamens. Within the fruits are seeds, but the attachment site of the funiculus is unclear. The interpretation is that the subtending leaves are actually modified bracts because they are smaller than typical *Cissites* leaves. Note that on the main axes, the terminal cymose RA is similar in size and stages of development to those of the lower lateral cymose structure. Whether the reproductive structure is an inflorescence or a flower cannot be determined, which again shows the difficulty in distinguishing between the two.

Species with Flowers and Inflorescences

Reproductive organs similar to living *Hedysomum* (Table 7.1) have been reported from Portugal (Friis et al. 2006). The probably determinate male RA has a long, naked axis and whorls of stamens along the end of the axis, with the distal stamens appearing smaller. Other fossils similar to living species are the platanoid fossils (Table 7.1) from the Albian of the United States (Crane et al. 1993; Friis et al. 2006). The compound RAs have widely spaced, alternately attached, determinate, sessile axes that are nearly spherical heads. Attached to these are sessile flowers with carpels subtended directly by five long sepals and up to five additional shorter tepals. The male flowers, which are not well preserved, have tepals and stamens with pollen grains similar to those found on the stigmas.

Selected Dispersed Flowers Species that Show Timing of Floral Characters

These fossil URAs can be considered flowers because they have short internodes and typical floral structures; all occur during the Early Cretaceous. They can be assigned to three major groups of angiosperms: the basal dicot grade (not a monophyletic group), the monocot clade, and the eudicot clade.

Dispersed flowers of basal dicots (Table 7.5), as well as those attached to RAs (*Spanomera*, *Pluricarpellatia*, *Hedyosmum* type; Table 7.1), have been described. The oldest is an *Amborella*-like flower from Portugal that is staminate and bears alternate tepals and many stamens (Friis et al. 2000b). Another, geologically younger, flower from Portugal is similar to several members of the ANITA grade (Friis et al. 2001), the basal living taxa that include the Amborellaceae, Nymphaeales, Illiciales,

Table 7.5. Summary of the Structure of Selected Dispersed Flowers of Early Cretaceous Age

Taxon	Amborella-like	ANITA-like	Potomacanthus	Virginianthus	Mayoa	Monocot Unnamed A	Monocot Unnamed B	Pennistemon/ Pennicarpus/ Pennipollis	Lusistemon– Lusicarpus	Teixeiraea
Geological age	Barremian– Aptian	Aptian–Albian	Albian	Albian	Barremian– Aptian	Barremian– Aptian	Barremian– Aptian	Aptian–Albian	Aptian–Albian	Aptian–Albian
Characteristics of Ultimate Reproductive Axis										
Unit	Male	Bisexual?	Bisexual	Bisexual	Male?	Bisexual	Bisexual	Male, female	Male, female	Male
Sterile type	Bracts and tepals	Tepals?	Tepals	Bracts, tepals		Tepals	Tepals	None?	Bractlike	Bracts to tepals
Sterile number	Few	Many	3 + 3	Many		3	6	—	1 to 6?	Many
Sterile arrangement	Alternate	Alternate?	Whorled	Alternate to irregular		Whorled	Whorled	—	Alternate	Alternate
Structure			Hypogynous	Hypanthium		Hypogynous	Epigynous			
Characteristics of Reproductive Organs										
Stamen										
Number	Many	Many	3 + 3	Many		9	6	Few	6	Many
Phyllotaxy	Alternate?	Whorled	Whorled	Alternate		Whorled	Whorled?	Alternate	Alternate	Alternate
Size variation	Yes, staminoids?	—	No	Yes, staminoids		No	No?	No	Yes	Yes
Connation	No	—	No	No		No	—	No	No	No
Filament length	Long	—	Long	Sessile		Short	—	Short	Short	Short
Filament continuation	No joint?	—	No joint	No joint		No joint?	—	No joint	No joint	No joint
Anther attachment	Basal	—	Basal	Basal		Basal	—	Basal	Basal	Basal
Number of sporangia	Tetrasporangiate	—	Bisporangiate	Tetrasporangiate		?Tetrasporangiate	—	Tetrasporangiate	Tetrasporangiate	Tetrasporangiate
Number of theca	Dithecal	—	Dithecal	Dithecal		Dithecal?	—	Dithecal	Dithecal	Dithecal
Connective	Apical tip	—	Apical tip	Apical tip			—	Apical tip	Apical tip	Apical tip
Pollen	Trichotomono- colpate, verrucate- rugulate	Monosulcate, reticulate	Inaperturate, microreticulate	Monosulcate, reticulate	Inaperturate, striate	Monosulcate, microreticulate		Monosulcate, reticulate	Tricolpate, striate	Tricolpate, perforate

Table 7.5 (continued).

Taxon	Amborella-like	ANITA-like	Potomacanthus	Virginianthus	Mayoa	Monocot Unnamed A	Monocot Unnamed B	Pennistemon/ Pennicarpus/ Pennipollis	Lusistemon– Lusicarpus	Teixeiraea
Carpel/fruit	—						—			—
Number		Many	One	Many		Three		—	Two	
Phyllotaxy		Whorled	NA	Alternate		Whorled		—	Opposite	
Size variation		No	NA	No		No		—	No	
Connation		Yes, lateralmost	NA	Free		—		No	Yes, four-fifths	
Stigma		Yes	Yes	Yes		Yes		Yes	Yes	
Style		No	Yes	No		No		No	No	
Stigmatic tissue		Along closure	Along closure?	Terminal?		Apical?		Apex	Terminal portion of closure	
Closure		Long from stigma to base	Style plicate	NA		—		—	Style plicate	
Type		—	Ascidiate	—		—		—	-	
Ovule						—		—		
Number		Several	One	Two				Single		
Attachment of chalaza		—	Next to micropyle	—				Opposite		
Angle		—	Anatropous	—				Orthotropous?		
Number of integuments		—	—	—						

Note. Most terms are in common usage; those that are not are defined in the Methodology section.

Trimeniaceae, and Austrobaileyaceae. It was probably bisexual with many alternate tepals, many whorled stamens, and laterally connate carpels. The lauraceous *Potomacanthus* comes from the United States and has two whorls of three tepals and two whorls of stamens (von Balthazar et al. 2007). Another flower from the United States is *Virginianthus*. It has many alternately to irregularly arranged tepals, staminoids, and stamens, and alternately arranged fruits (Friis et al. 1994).

A number of fossil flowers are thought to be members of the monocots (Table 7.5) on the basis of floral structure or pollen (Friis et al. 2006). *Mayoa* comes from the Barremian–Aptian of Portugal and is preserved with little structure other than inaperturate, striate pollen (Friis et al. 2004). Two other unnamed flowers (Friis et al. 2006) show some of the variation in fossil monocots. One has three whorled sepals, nine whorled stamens, and three carpels. The other has six whorled epigynous tepals, six whorled stamens, and an inferior ovary. The last putative monocot species, *Pennistemon/Pennicarpus/Pennipollis* (Friis et al. 2000a), has unisexual flowers without a perianth. The male flowers has few stamen, while the female flowers have free carpels and an ovule that probably has a chalaza opposite the micropyle, and an orthotropous angle. Similar pollen grains occur in the anthers and on the stigmas; they are also found dispersed.

Several fossil flowers have tricolpate type pollen and thus are members of the eudicots (Table 7.5). A *Platanus*-like flower has been described previously in this chapter. *Lusistemon striatus*–*Lusicarpus planatus* (Pedersen et al. 2007) have the same pollen in the anther and on the stigma, respectively. The male flowers have six tepals and six, probably spirally attached, stamens. The female flower is bicarpellate, syncarpous, and hypogynous, with a smooth, elongate stigmatic region for each carpel and short style. *Teixeiraea lusitanica* (von Balthazar et al. 2005) is based on a male flower with free bracts, sepallike tepals, and petallike tepals of increasing size arranged spirally on a slightly elongate receptacle. The stamens are spirally arranged, but the innermost are shorter. Other species show variations, such as tricarpellate, syncarpous ovaries and differences in the size of the stigmatic regions (Pedersen et al. 2007)

Summary

The fossils described so far (Tables 7.1, 7.5) show a number of trends in the morphology of early angiosperms. First, there are some characteristics that are quite variable (e.g., organ number). Other characters are uniform across taxa and appear to be canalized in their expression (e.g., stamen morphology). Finally, a few of the characters appear canalized early, but show later elaborations to form new states (e.g., filament length).

Several vegetative characters are found in the majority of taxa. These are a uniformly small leaf size (1 to 4 cm) and a long petiole size (short petioles appear by the Aptian–Albian). The RAs or inflorescences also appear to be uniformly determinate, as was previously suggested as ancestral by Taylor and Hickey (1996), and if the URAs or flowers are unisexual, the female are distal and the male proximal. Within the URAs, the female organs are distal and male organs are proximal, and they are hypogynous (epigyny appears by the Barremian–Aptian). The sterile organs are tepals with differentiation into multiple whorls by the Albian and sepals and petals by the Cenomanian (Basinger and Dilcher 1984). Beyond these characters, overall, there is not much similarity in the structures of the RAs or URAs.

Specific organs show great similarity. Specifically, all the perianth organs are simple tepals. The stamens have a broad connective that ends in an elongate apical tip. These should not be considered laminar because they are not embedded in a leaflike tissue. Endress (1986; Endress and Hufford 1989) has successfully debunked the idea that living basal dicots have a similar type of laminar stamens in which the sporangia are embedded in a leaflike organ. In addition, the stamens are tetrasporangiate and dithecal with a short filament (bisporangiate appears later). The pollen is monosulcate (or a related morphology), with tricolpate pollen appearing during the Aptian–Albian. Carpels are uniformly free (syncarpy shows up later), have no style, and have a short apical stigma (styles appear by the Barremian–Aptian and elongate stigmas by the Albian). Finally, ovules appear to be bitegmic.

In contrast, several groups of characters have variable states. These include vegetative and reproductive phyllotaxy and branching, URA (or flower) structure and sexuality, and organ size and number. Clearly, these characteristics related to floral and inflorescence structures are in evolutionary flux during the Early Cretaceous and became more canalized later in different lineages. In several species, such as *Archaefructus* and *Caspiocarpus*, it is difficult to clearly distinguish between RAs and flowers. The RAs shows the typical structure of a flower with distal female organs and proximal male organs, but the organs can occur in clusters that may even be bisexual. In any case, these species do not show full canalization of the typical monocot or eudicot flower. Other examples of this wide variation are the many combinations of the phyllotaxy of the vegetative axes, with that of the branches on the RAs, the flowers on the RAs, and the floral parts on the same plant. In some, like *Araripia*, the phyllotaxy of all units are opposite, while in others, such as *Archaefructus*, they are all alternate. Yet there are other combinations. For example, *Endressinia* is alternate in all units except the arrangement of the URAs on the RA, which is opposite, while *Caspiocarpus* is opposite in all units except the URAs on the RA, which are alternate.

This variation of canalized and variable characters illustrates the problem of determining what the earliest flower might have looked like, a problem that has been pointed out by others (e.g., Taylor and Hickey 1996; Soltis et al. 2005a; Endress 2006). On the basis of these data, we can be fairly certain about the morphology of the basic organs (similar-looking tepals, stamens, carpels, and maybe ovules), but much less about the arrangement, number, and structure of the flowers and inflorescences. This obviously has implications for development and is discussed in greater detail below.

Review of Development Controls and Discussion of their Evolution

The transition to flowering—the production of determinate floral meristems—is due to the interactions of several groups of genes (Fig. 7.4). First are the meristem identity genes (MIGs), which include controls for producing vegetative apices, inflorescence apices, and floral apices. The vegetative MIGs maintain the shoot apical meristem (SAM), which permits production of leaves and axillary buds, and the root apical meristem. Inflorescences can be considered transitional forms, essentially indeterminate floral meristems (FM), which may stay indeterminate or may become determinate by converting to FMs. Floral apices are determinate FMs. The interactions between the MIGs are influenced indirectly by floral timing genes (FTGs) through integrator genes (IGs), which directly influence the MIGs. The interactions between the MIGs, especially in the way inflorescences are produced and maintained, are beginning to be understood in eudicots (Benlloch et al. 2007). The MIGs also directly or indirectly control the floral organ identity genes (FOIGs) that define the major whorls of the flower. In addition to these genes are a variety of separate genes that affect boundary expression of the FOIGs called the cadastral genes, while other genes affect floral symmetry, size, organ number, and organ polarity. Finally, the FOIGs directly control the organ-building genes (OBGs) that result in the formation of floral organs such as sepals, petals, stamens, carpels, and ovules.

Details of these controls can be observed in the transition to flowering in *Arabidopsis*. The following summary is possible owing to many excellent reviews (e.g., Battey and Tooke 2002; Theissen et al. 2002; Zik and Irish 2003; Buzgo et al. 2005; Krizek and Fletcher 2005; Soltis et al. 2005a; Floyd and Bowman 2007; Theissen and Melzer 2007). It is clear that the discovery of new genes and the better understanding of the expression and distribution of these new genes is occurring every day. As a result, this review will concentrate on the more extensively studied genes, and this basic information will undoubtedly be supplemented by additional knowledge in the near future.

Meristem Identity Genes (MIGs)

FUNCTION OF VEGETATIVE GENES

The SAM is maintained by a group of genes that define the extent and number of cells. Genes such as CLAVATA (CLV) and AGAMOUS (AG), which control stem cell number and development of carpels, respectively, interact with WUSCHEL (WUS), which controls apical meristem structure (Fletcher 2002). WUS directly regulates cytokinin regulators (Leibfried et al. 2005) and is also necessary for FMs; it is only completely

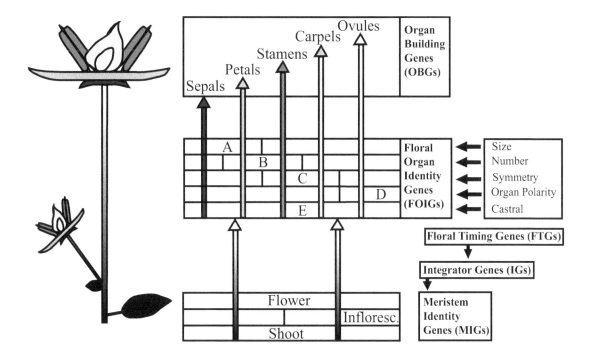

Fig. 7.4. An overall model showing the major groups of developmental controls and the basic organs they control.

repressed at the final stages of floral development with the production of the carpels (Krizek and Fletcher 2005). CURLY LEAF (CL) is necessary for normal leaves and may repress AG (Schubert et al. 2005). It may also play a role with EMBRYONICFLOWER2 (EMF2) in the repression of flowering (Chanvivattana et al. 2004). Also necessary are KNOTTED1-like genes (Floyd and Bowman 2007) such as SHOOTMERISTEM-LESS (STM), which promote the establishment of the SAM (Lenhard et al. 2002; Guillet-Claude et al. 2004; Hake et al. 2004).

EVOLUTION OF VEGETATIVE GENES

The evolution of vegetative MIGs is not well understood (Friedman et al. 2004). A protein BLAST search of CLV shows a similar gene in conifers, mosses, and liverworts, indicating a preangiosperm distribution. Genes similar to WUS are found in monocots and eudicots, suggesting a possible all-angiosperm distribution, but it originates from the large WOX-like gene family found in the conifers. The structural function of WUS is better understood with evidence of dimer formation (Nagasaki et al. 2005). These authors suggest this system may have evolved from a homodimer to a heterodimer condition through a duplication event. CL is not only reported in monocot and eudicot angiosperms, but a similar expressed sequence tag is reported in *Ginkgo* (Brenner et al. 2005), and a protein BLAST search shows similarities to a moss sequence indicating a preangiosperm distribution. KNOT-like genes are better studied. Similar genes are widespread throughout monocots, basal eudicots, and eudicots, but they are also found in gymnosperms and others (Guillet-Claude et al. 2004; Hake et al. 2004), and AG is also widely distributed and is found in nonangiosperms. This suggests that both had preangiosperm distributions. Fossil plants show the existence of vegetative meristems that may have been potentially large in the rhizomes and smaller in axillary branches. This suggests a role for stem cell number that affects size and the fundamental structure of the apical meristem. In addition, the occurrence of fully determinate RAs in fossil plants indicate the existence of repression of WUS by probably an AG type gene as is found in living plants (Lohmann et al. 2001) because it is related to carpel formation. Thus, it is of little surprise that there is fossil evidence for WUS, CL, CLV, AG, and KNOT-like genes.

FUNCTION OF REPRODUCTIVE GENES

The SAMs are maintained and prevented from becoming fertile by floral repressors. Some of the repressors, such as TERMINALFLOWER1 (TFL1), act directly on floral MIGs (Ratcliffe et al. 1998; Amaya et al. 1999; Ahn et al. 2006; Benlloch et al. 2007; Prusinkiewicz et al. 2007). Yet others, such as FLOWERING LOCUS C (FLC), act on IGs while EMBRYONICFLOWER1 (EMF1) acts on FOIGs. It has been suggested that flowering may be the default and the vegetative stage as a result of that repression (Boss et al. 2004). This idea would fit in with the longer-held view that the evolution of land plants occurred through the elaboration of the sporophyte (predominate vegetative form) through the delay of meiosis (Niklas 1997).

The transition to flowering is complex, especially the production of inflorescences (Singer 2006; Benlloch et al. 2007). There appear to be redundancies that both permit reproduction even after the loss of certain genes and allow the variety of inflorescence that distinguish angiosperms from other seed plants. Most agree that it is an antagonistic interaction among SAM genes such as WUS, floral MIG repressors (e.g., TFL1), floral MIG promoters such as LEAFY (LFY), and FOIGs such as AG. There are a number of repressors; for example, TFL1 directly interacts with LFY and FRUITFULL (FUL; =AP1) to prevent the establishment of the FMs (Zahn et al. 2006a). Clearly, flowering is under negative regulation by EMF1 and EMF2 (Moon et al. 2003). TFL1 may also be under the control of EMF1 (Espinosa-Soto et al. 2004; Gambin et al. 2006). Other genes are promoters, e.g., LFY, which is central for the formation of flowers from inflorescence meristems, and there are many FOIG targets (Maizel et al. 2005; William et al. 2004). FUL and AP1 are two closely related genes; one is an A-class gene (AP1), while the other, FUL and its orthologs, helps regulate flowering (Litt and Irish 2003).

Depending on the level of expression, the apex can be vegetative, reproductive, and indeterminate (a growing inflorescence axis), or floral and determinate (Ratcliffe et al. 1998; Amaya et al. 1999). The floral MIGs, along with AGAMOUS-LIKE 24 (AGL24), establish the inflorescence meristem. It also appears that AGL24 promotes and maintains inflorescence identity and is also negatively controlled by LFY and FUL (=AP1; Yu et al. 2004). The elaboration of inflorescence branching, the length of time before a lateral meristem becomes floral, and whether an inflorescence is determinate (ending in a flower) is likely contingent of the degree of dominance of the repressor or promoter (Ratcliffe et al. 1999; Amaya et al. 1999; Battey and Tooke 2002; Benlloch et al. 2007). WUS, LFY, and AG also interact to make the transition (Zik and Irish 2003). In this role, AG is important in making determinate axes, as is ULTRAPETALA1 (ULT1), which represses WUS and STM (Carles et al. 2004). APETALA2 (AP2) interacts with AP1, LFY, and possibly TFL1, and has negative reciprocal interactions with AG (Kim et al. 2006). Another gene with multiple interactions and effects is UNUSUAL FLOWER ORGANS (UFO), which affects the regulation of floral identity and floral organ development (Ni et al. 2004), especially B-class genes. UFO, together with LFY, appears to promote FMs and negatively regulate bract growth (Hepworth et al. 2005).

The importance of regulators is shown by genomewide expression studies in *Arabidopsis* (Zhang et al. 2004). In young inflorescences and young flowers, it is conservatively estimated that 28% of the genes that are expressed and unique to these tissues are involved with transcription regulation. This compares to 6% in the entire genome. Cluster analysis suggests these genes are most similar to those expressed in the shoot apex and, secondarily, are more distantly related to floral genes.

Other floral MIGs play a role in the transition and maintenance of the FM, but not in inflorescence formation (Parcy et al. 2002). FUL and CAULIFLOWER (CAL, unique to the Brassicaceae; Litt and Irish 2003) act together to promote LFY, while AP2 and LFY also interact positively. It is evident that LFY, which is fundamental to the flowering process, is at the core of this. LFY plays a direct role in the activation of MIGs such as FUL (=AP1) and CAL (William et al. 2004). It also activates FOIGs, including A-, B-, C-, and E-class genes (Maizel and Weigel 2004). Yet other repressors also directly regulate FOIGs, such as EMF1 and FER-TILIZATION INDEPENDENT ENDOSPERM (FIE; Katz et al. 2004). It appears that 60% of the genes in *Arabidopsis* activated by LFY are transcription factors (William et al. 2004), which again shows the degree that regulation plays in this transition.

EVOLUTION OF REPRODUCTIVE GENES

Phylogenetically, TFL1 is closely related to FLOWERING LOCUS T (FT) and MOTHER OF FT AND TFL1 (MFT), which are widespread in monocots and eudicots (Ahn et al. 2006), and genes and expressed sequence tags similar to FT are also found in gymnosperms (Brenner et al. 2005), suggesting a preangiospermous distribution, although expression has not been shown in the RAs. Most recently, this class of genes has been found in bryophytes and lycopods; their phylogenetic data suggest FT and TFL1 may be related to the evolution of seed plants and angiosperms

(Hedman et al. 2009). A protein BLAST search has found eudicot homologs for EMF1. On the other hand, a BLAST search for EMF2 shows polycomb-group proteins of high similarity in monocots and eudicots, suggesting a possible all-angiosperm distribution. Orthologs of LFY have been found widespread in plants, including ferns. Yet gymnosperms appear to have had two copies, and their function appears less specialized (Maizel et al. 2005; Frohlich 2006a). FUL (=AP1) is found in magnoliids, eudicots, and monocots, indicating an all-angiosperm distribution (Litt and Irish 2003). A protein BLAST search of AGL24 shows that it appears to be a unique subgroup within the SHORT VEGETATIVE PHASE (SVP) family, which includes monocots, eudicots, and other seed plants. Because it is a subgroup, it should be considered to have a eudicot distribution until there is additional research. AP2 has an all-angiosperm distribution because it is found in the basal angiosperm *Nuphar*, as well as in monocots and eudicots. The gene and orthologs (called euAP2) are part of a group that include AINTEGUMENTA (ANT; also found in other seed plants) and ETHYLENE RESPONSE FACTOR–like genes (ERF, also found in mosses; Kim et al. 2004), and could be another example of a duplication related to flowering. UFO and ULT1 are found in monocots and eudicots, suggesting a possible all-angiosperm distribution. FIE-like sequences are found in monocots and eudicots, and similar sequences occur in conifers and mosses, suggesting a preangiosperm distribution.

Early Cretaceous fossils show that the RAs are determinate. First, the URAs (flowers) are determinate with the production of carpels, or in unisexual flowers, the production of stamens. Second, fossils with inflorescences are also determinate. Most of these are either monochasial or dichasial, with inflorescence axes ending with a terminal flower that matures before the more proximal flowers. This is consistent with hypotheses that cymose inflorescences were ancestral (Taylor and Hickey 1996). Thus, the control genes of TFL1, LFY, and FUL (=AP1) appear to be actively controlling the determinate nature of the apex. Yet the fact that the stamens in the RA in *Archaefructus* can be replaced by clusters of stamens or by even bisexual clusters, and that the carpels may occur in clusters or as singles, raises questions about whether the control observed in extant plants was as fully canalized then. Although this plant may be specialized for an aquatic habit, these intermediates are similar to the evolution of the flower proposed by Hickey and Taylor (1996), in which the carpels and anthers are reduced axes, and the determinate nature of the flower is due to the evolution of the carpel. Interestingly, a number of fossils have bracts or transitions from leaves to bracts, suggesting control by genes like UFO. Clearly, indeterminate inflorescences are not found until the Campanian (Friis et al. 2004).

Floral Timing Genes and Integrator Genes

The conditions under which flowering is allowed are controlled by FTGs. There are many genes that control this, and the pathways can be categorized as gibberellic acid, autonomous, vernalization, and light dependent (Boss et al. 2004; Komeda 2004; Tremblay and Colasanti 2006; Devlin 2007). The signals from the autonomous pathway are integrated by the repressor FLC, whose effects can be inactivated by members of the vernalization pathway. The light-dependent pathway signals are integrated by the gene CO, while the gibberellic acid pathway is independent and acts directly on LFY. The signals from most of these genes are then integrated by promoters such as SUPPRESSOR OF OVEREXPRESSION OF CO1 (SOC1), FT, and F-class genes (AGL20, TM3; Nam et al. 2003). FLC has a direct role in repressing both SOC1 and FT. These integrator genes (Komeda 2004) then act on MIGs such as LFY, AP1, FUL, and CAL (Theissen et al. 2002; Soltis et al. 2005a).

The phylogenetic relationships of many of these genes have not been examined. SHORT VEGETATIVE PHASE (SVP), a T-class gene (Nam et al. 2003), is thought to be a floral repressor (Gregis et al. 2006) and might form part of the autonomous pathway (Theissen et al. 2002) or light-dependent pathway (Boss et al. 2004). The orthologous genes of SVP and CO are reported from gymnosperms (Nam et al. 2003) and thus have a preangiosperm distribution. The same holds for FT and F-class genes (AGL20/TM3; Nam et al. 2003), so they are preangiospermous too (Brenner et al. 2005). PHYTOCHROME A (PHYA) is part of the light-dependent pathway and appears to be only found in angiosperms (Mathews et al. 2003). Data exist for several integrators: FT, SOC1, and F-class genes have preangiospermous orthologs, while FLC is found in monocots and dicots, suggesting an all-angiosperm distribution.

There is fossil evidence for these genes. Many of the Early Cretaceous fossils show synchronous flowering, and some exhibit evidence of alternation between vegetative and floral stages. This could be due to either day length (light-dependent pathway), as they are found in localities with paleolatitudes above or below $25°$ north or south, or due to the autonomous pathway. The angiosperm fossils from Brazil also appear to synchronously flower, although in these cases, they would have been at paleoequatorial latitudes. Thus, on the basis of the fossil evidence, both FTGs (e.g., CO) and IGs (e.g., FLC) were likely to have existed in the Early Cretaceous.

Floral Organ Identity Genes

FOIGs have been categorized by the ABC model and later modifications into five groups of genes (Coen and Meyerowitz 1991; Soltis et al. 2006; Irish 2006). They regulate the formations of the four whorls—sepals, petals, stamens, and carpels—as well as with ovules within the carpels. The fifth class (E) is expressed in all whorls. Most of these are MADSII (MIKC type; from N- to C-terminus, a MADS [M-], intervening [I-], keratin-like [K-], and C-terminal [C-]) class domain proteins, and strong evidence suggests they form quartets (Egea-Cortines et al. 1999; Theissen et al. 2002; Krizek and Fletcher 2005; Melzer et al. 2006; Theissen and Melzer 2007). For example, in *Arabidopsis*, sepal formation is regulated by a quartet formed of two A-class and two E-class components; petals by one A-class, one E-class, and two (different) B-class components; stamens by one E-class, two (different) B-class, and one C-class component; and carpels by two E-class and two C-class components. Phylogenetic analysis clearly shows separate clades composed of E-, A-, C-, and D-class components, and a two-part clade for the B class (Kim et al. 2005). Several duplications are important for the functioning of the genes (Kramer et al. 2004; Kramer and Zimmer 2006). In addition, it is clear that the ABC model is not sufficient to explain the flowers of noneudicots (Kim et al. 2005; Scutt et al. 2006; Soltis et al. 2006). Modifications show that the functions of these classes of genes were less canalized and more flexible in their expressions than in eudicots (Kim et al. 2005; Scutt et al. 2006; Soltis et al. 2006; Zahn et al. 2006a), forming a sliding or fading expression (Soltis et al. 2006). The regulation of these FOIGs is also complicated and is still under study (e.g., Kramer 2006; Zahn et al. 2006a).

FUNCTION AND EVOLUTION OF A CLASS

The ABC model suggests that the A-class genes were necessary for the outer sepal whorl in flowers. A-class genes such as AP1 definitely exist for *Arabiodopsis*, but related genes (Parcy 2002) such as FUL and CAL are important for initiating floral development. Yet analysis of these genes (Litt and Irish 2003) shows that although FUL is widespread in all angiosperms, AP1 is restricted to sepals and petals in eudicots, which Litt and Irish have named euAP1. A protein BLAST search shows that possible homologs are found in basal eudicots. This indicates that perianth is not homologous in all angiosperms. Thus, the molecular evidence does support evidence of multiple origins for the perianth (as reviewed in Soltis et al. 2005b; Ronse de Craene 2008). AP2 is another A-class gene, yet it too plays a role in flower initiation and gene regulation (Kim et al. 2005). It appears that sepals and petals are not homologous between eudicots and monocots (Soltis et al. 2005b; Endress 2006; Ronse de Craene 2008). The fossil record supports this, as all the Early Cretaceous flowers with tepals in two whorls do not appear until the Albian. Truly differentiated sepals and petals are not found in fossils from the Early Cretaceous;

clearly larger and differentiated petals first occur during the Cenomanian (Basinger and Dilcher 1984). The first occurrence of the eudicots, as evidenced by the development of tricolpate pollen—a synapomorphy for the eudicots—appears during the Aptian to Albian (Brenner 1996; Table 7.5), suggesting that true petals evolved within eudicots (Litt and Irish 2003; Rasmussen et al. 2009).

FUNCTION AND EVOLUTION OF B CLASS

There has been considerable research on B-class genes (Zahn et al. 2005b; Hernández-Hernández et al. 2007; Hileman and Irish 2009). These play a fundamental role in petal and stamen identity. The best known are PIS-TILLATA (PI) and AP3 in *Arabidopsis*. All angiosperms studied thus far, including basal angiosperms, monocots, and eudicots, have two copies that appear to be due to a duplication event. In contrast, gymnosperms have a single copy that has a similar function in the identity of male reproductive structures. Thus, the gene type is preangiospermous, but the duplication appears in all living angiosperms. In addition, other duplications have been found, including one that seems to be a synapomorphy for eudicots and their petals (Hernández-Hernández et al. 2007). All the fossils have well-developed tetrasporangiate and dithecal stamens with short filaments, but a wide range of floral morphologies. In all cases, the bisexual flowers have the stamens arranged between the perianth and carpels. This is true in the unusual RAs in the genus *Archaefructus*, where the stamens and clusters of stamens are proximal on the RA and in the bisexual URA. Thus, the fundamental order is retained whether they are "flowers" or clusters of "flowers." The fossils have well-developed stamens, but their placement implies that fossils are older than the duplication of B-class genes, that the arrangement was not fully canalized, or that it represents an unusual, specialized situation.

FUNCTION AND EVOLUTION OF C CLASS

The C-class genes are well studied and play a role in the formation of stamen, carpels, and ovules (Kramer et al. 2004; Zahn et al. 2006b; Scutt et al. 2006). A-class genes mutually inhibit C-class genes, so they are never coexpressed. The best known is AG and, along with SHATTERPOD (SHP1, SHP2) in *Arabidopsis*, also plays a role in flowering (Kramer et al. 2004). AG not only specifies stamens (in combination with B-class genes), carpels, and ovules (in combination with D-class genes), but also makes the floral axis determinate (Zik and Irish 2003; Kramer et al. 2004; Zahn et al. 2006b). Many orthologs that have been sequenced, so it is possible to examine the phylogenetic relationships of the sequences (Kramer et al. 2004; Kim et al. 2005; Zahn et al. 2006b). These genes only appear to be found in seed plants (Kramer et al. 2004) and clearly form three major clades: gymnosperm orthologs, D-class orthologs, and C-class orthologs. In conifers, female reproductive organs express B- and C-class genes, while male organs express C-class genes. Angiosperms show a major duplication with the formation of ovule D-class genes and carpel/ovule/floral apex C-class genes (Kramer et al. 2004; Scutt et al. 2006), although there is some doubt about the monophyly of the D-class clade (Zahn et al. 2006b). Beyond the D-class clade, there appears to be a clade composed of basal dicots including ANITA, magnoliids, and monocots, and a clade of "basal tricolpates," which is followed by a second duplication within the eudicots (excluding the basal tricolpates), forming the AG and SHP lineages (Kramer et al. 2004; Zahn et al. 2006b). It has been suggested that this resulted in true floral determinacy (Zahn et al. 2006b). Thus, the gene class is clearly preangiospermous. Floral determinacy is clear in the fossils, but most of them lack a syncarpous ovary, and the last carpel is not actually terminal. Perhaps the duplication is related to this morphology. The fossils clearly show variability in the number, size, and arrangement of the carpels.

FUNCTION AND EVOLUTION OF D CLASS

The D-class genes are less well understood than the previous three classes (Kramer et al. 2004). The class was originally proposed by Colombo et al. (1995), and evidence for the class continues to grow. In *Arabidopis*, SEEDSTICK (STK) is expressed in the ovule, albeit in combination with C-class genes. A BLAST search shows that similar genes are found in monocots, suggesting a possible all-angiosperm distribution. The most significant evidence has been the findings that some MADSII class genes with expression in ovules are found in a class separate from the C-class genes, some of which are also expressed in the ovule (Kramer et al. 2004), although this is disputed by Zahn et al. (2006b). This clade of genes has clear structural differences and has been interpreted as subfunctionalization for ovules (Kramer et al. 2004). The fact is that angiosperm ovules are different from gymnosperm ovules in their shape and number of integuments (Taylor 1991); D class could have to do with this evolutionary change. The fossil record shows evidence of both orthotropous and anatropous ovules. In addition, some appear to be bitegmic, but orthotropous, unitegmic ovules are known from living and fossil seed plants. Certainly the presence of ovules is preangiospermous, but the presence of bitegmic ovules is all-angiospermous. Considerably more work is needed on the development of these organs. Because the homologies between ovules are clearer and there is a good fossil record, the evolution and development of these plant parts are probably likely to be better understood than other organs that do not have clear morphological homologs, namely stamens and carpels.

FUNCTION AND EVOLUTION OF E CLASS

E-class genes are required for the identity of all whorls and for meristem determinacy (Zahn et al. 2005a). In *Arabidopsis*, SEPALLATA1, 2, and 3 (SEP1–3, formerly AGAMOUS-LIKE 2, 4, and 9, respectively) are necessary for the B, C, and D whorls, while SEP4 (formerly AGL3) is apparently necessary for all whorls. SEP3 is in a clade separate from the remaining SEP genes and is probably due to duplication before the evolution of the living crown group of angiosperms. There are additional duplications in both the monocot and eudicot clades. Although SEP genes are phylogenetically related to the AGL_6 gene with both angiosperm and gymnosperm homologs, there are no identified preangiosperm SEP homologs. In bisexual fossil reproductive organs, there is evidence of three of the whorls in an orderly progression, suggesting that all fossil plants have these genes.

FUNCTION OF OTHER FOIG GENES

In addition to these genes, there are other types of genes that influence the expression of the flower structure. These genes influence floral size, number of organs, floral symmetry, organ polarity, and location of whorls (castral genes) in the flowers (Zik and Irish 2003; Weiss et al. 2005; Fulton et al. 2006). Many of these characteristics are influenced by FOIGs and MIGs, while others are specific. Many genes, such as CLV, WUS, and ULT1, influence overall size (Krizek and Fletcher 2005; Weiss et al. 2005; Fulton et al. 2006). This is also true of organ number, in that some genes increase the number of organs, some decrease them, while others change the relative organ number (Weiss et al. 2005). One of these is an auxin response factor ETTIN, which affects the latter. A much smaller number of genes help control floral symmetry, including CYCLOIDEA (CYC), DICHOTOMA (DICH), and RADIALIS (RAD) in *Antirrhinum* (Cubas 2004; Costa et al. 2005). Organ polarity has also been investigated (Zik and Irish 2003), and the important genes include the more general YABBY genes (Siegfried et al. 1999), as well as more specific genes KANADI (KAN), REVOLUTA (REV), and PHANTASTICA (PHAN) in *Antirrhinum* (Zik and Irish 2003; Weiss et al. 2005). PHAN may interact with BLADE-ON-PETIOLE1 (BOP1) to regulate symmetry in both flowers and leaves (Hepworth et al. 2005). Finally, castral genes include SEUSS (SEU), LEUNIG (LUG), and SUPERMAN (SUP).

EVOLUTION OF OTHER FOIG GENES

Although no specific phylogenetic analyses have been published, homologous genes have been found in other groups. ETTIN changes the relative organ numbers and is also similar to genes found in monocots, suggesting a possible all-angiosperm distribution. Symmetry genes CYC and DICH are closely related to DICH-like genes, forming a clade within the CYC family; this is based on a protein BLAST search of all monocot and eudicot representatives and also indicates a possible all-angiosperm distribution. The same is found for RAD. The organ polarity gene REV is a plant gene family with several duplications that are found in nonangiosperms, as well as several apparent duplications in angiosperms. This indicates a preangiosperm distribution. PHAN is also found in eudicots and monocots, indicating a possible all-angiosperm distribution. The

SUP and SEU genes may have a possible all-angiosperm distribution, but the sampling is small. LUG is similar, but there is a moss gene that is also similar that could suggest a possible preangiosperm distribution. The fossils show variation in size and number of organs, although the symmetry and organ polarity seem established. The fact that the number of organs in the fossil species seem to be variable suggests that these genes may not be fully canalized. In some fossil species, the demarcation between the whorls is not well established with transitional organs, suggesting the castral genes may not be fully canalized. Bilateral symmetry does not appear in fossils until the Turonian (Crepet 2008).

Organ-Building Genes

Considering the importance of the uniqueness of the floral organs in angiosperms, it is surprising how little we understand about their development. Perhaps this has to do with the potentially large number of genes that may be involved (Krizek and Fletcher 2005), which is reflected in microarray studies that show that in *Arabidopsis*, the number of cDNA and oligonucleotides unique to sepals and petals are about a dozen, to stamens about 1,000 (with approximately a third related to pollen development), and to carpels 200 (Wellmer et al. 2004). The presence of large numbers of genes is also shown by down stream activation by AG (Gómez-Mena et al. 2005). Yet only a couple of dozen genes have been described for organ morphogenesis.

FUNCTION AND EVOLUTION OF THE PERIANTH

A number of genes appear to control the specifics of prophyll and perianth development. These include INCOMPOSITA (INCO), BIGPETAL (BPE), and JAGGED (JAG), with the last one controlling the general shape (Dinneny et al. 2004; Masiero et al. 2004; Szécsi et al. 2006). Another is RABBIT EARS (RBE), which is required for the proper formation of the petal primordia (Takeda et al. 2003). Protein BLAST searches indicate that most of these genes do not have obvious homologs, but BPE appears to be found in both eudicots and monocots, suggesting a possible all-angiosperm distribution. On the basis of fossil evidence, modified leaves in the form of bracts and tepals are well established. These are fundamentally different than those in other seed plants, so additional study in identifying the developmental genes unique to angiosperms would be valuable. True sepals and petals do not show up until later, when two whorls appear in the Albian and true eudicot petals in the Cenomanian.

FUNCTION AND EVOLUTION OF STAMENS AND POLLEN

Several genes have been identified that affect stamen, anther, and pollen development (Scott et al. 2004). SPOROCYTELESS (=NOZZLE [NZZ]) is regulated by AG but can cause microsporogenesis on its own

(Ito et al. 2004, 2007). Morphological development in stamen, including filament formation, seems to be influenced by AG in conjunction with the phytohormone jasmonic acid and DEFECTIVE-IN-ANTHER-DEHISCENCES (DAD1). Although both B-class and C-class genes are needed to initiate the formation of stamens, C-class genes appear to be needed longer. JAG, along with related NUBBIN (NUB), also influences shape in stamens (Dinneny et al. 2006). Protein BLAST searches have provided few data on homologous genes. SPL is obviously important, but only scattered homologies are found that include mosses. DAD1 clearly has a monocot and eudicot distribution, suggesting a possible all-angiosperm pattern. The lack of data is unfortunate because the fossils clearly show that basifixed, tetrasporangiate, dithecal stamens with a short filament, and an apical connective extension are found during the Early Cretaceous. Clearly, the genes that control for this stamen type would occur in all these angiosperms and were well established early on in angiosperm evolution.

Pollen also has a number of genes regulating its development, such as those related to physiology, transduction, and structure (Twell 2002). Some of these are under the control of the sporophyte and regulate exine development, such as MALE STERILE (MS) and POLLEN WALL (POW), while others are controlled by the gametophyte, including the intine and microgametophyte itself, such as SIDECAR POLLEN (SCP), GEMINI POLLEN (GEM), and TWO IN ONE POLLEN (TIO).

FUNCTION AND OF EVOLUTION OF CARPELS

Only a few genes have been identified that play a role in carpel morphogenesis (Gómez-Mena et al. 2005; Fourquin et al. 2005; Scutt et al. 2006). The most important gene is the YABBY transcription factor CRABS CLAW (CRC; Alvarez and Smyth 1999; Bowman and Smyth 1999; Fourquin et al. 2005, 2007) that appears to control carpel fusion and perhaps carpel closure. Another gene with similar effects is TOUSLED (TSL). Both JAG and NUB also play a role in carpel morphology (Dinneny et al. 2006). SPATULA (SPT) appears to control style, stigma, and septum development (Alvarez and Smyth 1999), as do LUG and ANT (Liu et al. 2000). SEU, ANT, and LUG probably play a role in polarity (Azhakanandam et al. 2008). Other genes of importance are REV, CUP-SHAPED COTYLEDON (CUC), and PHAVOLUTA (PHB), which are expressed in the placenta (Chevalier et al. 2002; Skinner et al. 2004; Azhakanandam et al. 2008). Scutt et al. (2006) has reviewed the evolution and developmental controls of carpels, suggesting both CRC and TSL have an all-angiosperm distribution because orthologs have been reported from ANITA-grade species as well as monocots (Fourquin et al. 2005, 2007).

Many of the genes that influence carpel development are also active in ovule formation (Chevalier et al. 2002; Skinner et al. 2004; Azhakanandam et al. 2008). In particular, there are genes that help form the placental areas in the carpel, which are required for ovule initiation. These include ANT, LUG, CRC, TSL, SEU, and REV, with the addition of CUC and similar HUELLENLOS PARALOG (HLP; Skinner et al. 2004; Azhakanandam et al. 2008). PRETTY FEW SEEDS (PFS) also regulates ovule patterning or differentiation (Park et al. 2005). Expansion of the placenta appears to be controlled by TSO, DICER-LIKE (DCL), and SHORT INTEGUMENTS (SIN), although the final stages are regulated by LUG, novel nuclear protein TSO, and SEU. Several stages and interactions occur after primordia initiation (Sieber et al. 2004; Skinner et al. 2004). First, PHB exists in the primordia and then, with WUS and NZZ/SPL, interacts with ANT in the late primordia to begin embryo sac and integument formation. Integument formation continues with the expression of homeodomain protein BELL (BEL) and regulation by YABBY gene INNER NO OUTER (INO initiates the outer integument) and SUP. Finally, sporogenesis is controlled by NZZ/SPL, as in the anther. BEL-like genes have been identified in other seed plants (Becker et al. 2002), and a BLAST search shows similar distribution for ANT, suggesting a preangiosperm distribution. This is supported by the presence of ovules in all seed plants and the fossil record of angiosperms. Whether anatropous or orthotropous ovules are ancestral (Taylor 1991; Soltis et al. 2005b; Endress 2006) is still under debate, and there are no developmental data on this question.

Discussion of Early Reproductive Development

The data in Figure 7.5 provide evidence that all the groups of floral developmental genes were functional in Early Cretaceous fossils. The fossils are in broad agreement with the data on distribution in living plants. The genes that are found in angiosperm sister groups with a preangiospermous distribution are also found in early fossil angiosperms or gymnosperms. The genes that are uniquely found in angiosperms with all-angiosperm or possibly all-angiosperm distributions are found in a range of older and younger fossils. As expected, the vegetative MIGs are preangiospermous, as are most IGs and FTGs. The major exception is the PHYA gene, which has an all-angiosperm distribution. In contrast, the fertile MIGs include genes with preangiospermous distributions such as LFY, FIE, AG, WUS, and CLV, but also appear to include all-angiosperm genes AP2, FUL=AP1, TFL1, UFO, ULT1, and the possible all-angiosperm gene EMF2. EMF1 and AGL24 may be restricted to eudicots. The all-angiosperm distribution of many of these genes is evidence that inflorescences are also a unique characteristic for angiosperms.

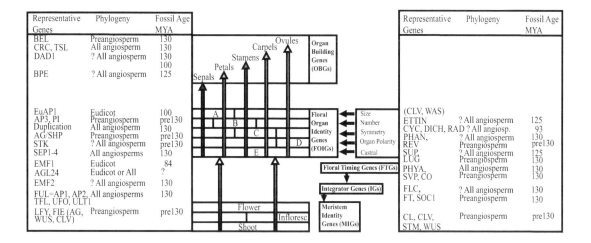

Representative Genes	Phylogeny	Fossil Age MYA
BEL	Preangiosperm	130
CRC, TSL	All angiosperm	130
DAD1	? All angiosperm	130
		100
BPE	? All angiosperm	125
EuAP1	Eudicot	100
AP3, PI	Preangiosperm	pre130
Duplication	All angiosperm	130
AG/SHP	Preangiosperm	pre130
STK	? All angiosperm	pre130
SEP1-4	All angiosperms	130
EMF1	Eudicot	84
AGL24	Eudicot or All	?
EMF2	? All angiosperm	130
FUL=AP1, AP2, TFL, UFO, ULT1	All angiosperms	130
LFY, FIE (AG, WUS, CLV)	Preangiosperm	pre130

Representative Genes	Phylogeny	Fossil Age MYA
(CLV, WAS) ETTIN	? All angiosperm	125
CYC, DICH, RAD	? All angiosp.	93
PHAN,	? All angiosperm	130
REV	Preangiosperm	pre130
SUP,	? All angiosperm	125
LUG	Preangiosperm	130
PHYA,	All angiosperm	130
SVP, CO	Preangiosperm	130
FLC,	? All angiosperm	130
FT, SOC1	Preangiosperm	130
CL, CLV, STM, WUS	Preangiosperm	pre130

Fig. 7.5. The major groups of developmental controls (with representative genes) integrated with the phylogenetic evidence of their origin and age of the fossils showing evidence of their presence. Preangiospermous fossil evidence is based on the general literature of fossil and living gymnosperms; see Tables 7.1 and 7.2 for fossil angiosperm evidence.

The strictly floral genes have a more complex distribution (Fig. 7.5). The FOIGs show a variety of patterns. Those related to reproduction, such as classes B, C, D, organ polarity, and castral, all have members with preangiospermous distributions (AP3, PI, AG, SHP, STK, REV, and LUG). Yet these patterns are complicated by duplications that appear to have all-angiosperm patterns (Kramer et al. 2004). Thus, existing genes have been modified and elaborated to make angiosperm structures and the diversity of structures in living angiosperms. Importantly, others appear to be only angiospermous. A class (euAP1) is restricted to eudicots, while genes for E class (SEP1–4), number, and symmetry have all-angiosperm distributions. Thus, it appears that new genes evolved that are unique to the angiosperm reproduction.

If we look at genes for specific organs, these data are less clear as a result of the lack of research. For the perianth, a possible all-angiosperm distribution is found in BPE, as would be expected with the presence of tepals in most of the fossils. True sepals and petals in eudicots have narrower distributions and are not found in the fossil record until the Cenomanian, and there is considerable evidence of multiple origins (see Endress 2006; Ronse de Craene 2008). A couple of stamen (DAD1) and carpel (CRC, TSL) genes have possible all-angiosperm and all-angiosperm distributions, respectively, as would be expected from the fossil record. Finally, BEL is an ovule gene with a preangiosperm distribution.

The presence of all the major groups of developmental genes is supported by the fossil record; what is surprising is evidence that some morphological characteristics are uniform while others are variable (Table 7.6). This might suggest that the genes controlling these structures are already canalized in some cases, although in others, organs were still in the process of evolving. Of course, variation in characteristics could be due to the elaboration of a previously canalized gene that would be expressed by multiple character states, or that it was possible to vary multiple morphological states with a simple molecular switch. Yet the variation in the fossils is frequently found within a single plant, or in multiple combinations not frequently found in living angiosperms.

What is immediately obvious is that certain gene groups show more variability than others (Table 7.6). The RA or inflorescence characteristics are quite variable in the fossils, suggesting the fertile MIGs are not fully canalized. Equally uncanalized are the structures of URAs or flowers. They show much variation, presumably as a result of the lack of canalization of the development from FOIGs. Other FOIG genes with less canalization control number, size, and arrangement (phyllotaxy). This is consistent with molecular developmental data that suggest sliding boundaries or fading borders (summarized in Soltis et al. 2006; Theissen and Melzer 2007). Still, the evolution of some of these gene classes has occurred. The Early Cretaceous fossils show strong evidence of synchronous flowering, as would be expected if FTGs and IGs were well developed and canalized. In addition, all the reproductive structures, whether equivalent to inflorescences or flowers, are clearly determinate and always have the male proximal and the female distal, both presumably controlled by MIG and FOIG genes. Finally, the OBGs for specific organs, especially tepals, stamen, carpels, and bitegmic ovules, appear to show little variation. Apparently selection has canalized these developmental genes to form morphologically similar organs. Interestingly, these organs are later elaborated into other states.

Evo-Devo Models of Flowers

In recent years, there have been a number of implicit and explicit evo-devo models developed for the evolution of the angiosperm flowers (Frohlich 2006a; Frohlich and Chase 2007). Although morphologically implicit models, such as my own (Taylor and Hickey 1990, 1992, 1996; Hickey and Taylor 1996; Ji et al. 2004), are useful in advancing ideas, they are difficult to test without new and extensive fossil data, and many were based on an earlier understanding of angiosperm sister-group relationships. Parsimony models based on current understanding of relationships based on molecular characters (Buzgo et al. 2004; Soltis et al. 2005b; Endress 2006; Endress and Doyle 2009; Soltis et al. 2009) also have problems owing to the morphological variation in the likely group of ancestral angiosperms (ANITA grade; Rudall et al. 2009). Recently, a series of related models has been developed that makes explicit hypotheses that can be tested by developmental or molecular phylogenetic data (reviewed in Frohlich 2006a; Frohlich and Chase 2007). The benefit of these models is that they can be tested with both living and fossil data (Frohlich 2006b). The most important of these are the Mostly Male, Out of Male/Out of Female, and Enhanced Out of Male theories, as well as the narrower Sliding Boundaries or Fading Borders models.

The Mostly Male model concentrates on the development of the bisexual flower (Frohlich and Parker 2000; Frohlich 2001, 2002, 2003,

Characteristic	Character Status	State	Gene Class
Reproductive Axis			
Placement	Variable	Axillary, terminal	MIG
Branching arrangement	Variable	None, monochasial, dichasial	MIG
Morphology	Variable	Single, multiple simple, multiple compound	MIG
Determinate/ indeterminate	Uniform	Determinate	FOIG class E
Sterile relative size morphology	Uniform	Smaller to none	FOIG size
Phyllotaxy of URA on reproductive axis	Variable	Alternate, opposite	MIG
Arrangement of URA	Uniform	Male proximal, female distal	MIG, FOIG
Number of URA	Variable	Variable	FOIG number
Ratio of URA of similar age	Uniform	Many flowers of similar age	IG, FTG
Ultimate Reproductive Axis			
Unit	Variable	Male, female, bisexual	FOIG
Sterile type	Uniform	Tepals	FOIG class A, B
Sterile number	Variable	Variable	FOIG number
Sterile arrangement	Variable,	Alternate, opposite, whorled	FOIG
Structure	Uniform	Hypogynous	OBG
Reproductive Organs			
Stamen			
Number	Variable	Variable	FOIG number
Phyllotaxy	Variable	Alternate, opposite, clustered	MIG
Size variation	Variable	Variable	FOIG size
Connation	Uniform	Free	OBG
Filament length	Uniform	Short	OBG
Filament continuation	Uniform	No joint	OBG
Anther attachment	Uniform	Dorsifixed	OBG
Number of sporangia	Uniform	Tetrasporangiate	OBG
Number of theca	Uniform	Dithecal	OBG
Connective	Uniform	Broad, apical tip	OBG
Pollen	Uniform	Monosulcate	OBG
Carpel/fruit			
Number	Variable	Variable	FOIG number
Phyllotaxy	Variable	Alternate, opposite, clustered	MIG
Size variation	Variable	Variable	FOIG size
Connation	Uniform	Free	OBG
Stigma	Uniform	Short, apiculate	OBG
Style	Uniform	No	OBG
Stigmatic tissue	Uniform	Apex	OBG
Closure	Variable	Variable	OBG

Table 7.6. Summary of Fossil Data Showing Whether a Characteristic Is Variable, as Well as the Likely Major Group of Developmental Controls that Would Have Control the Character's Development

Note. MIG = meristem identity gene; FOIG = floral organ identity gene; IG = identity gene; OBG = organ-building gene.

Table 7.6 (continued).

Characteristic	Character Status	State	Gene Class
Type	Variable	Ascidiate, uncertain	OBG
Ovule			
Number	Variable	Variable	FOIG number
Attachment of chalaza	Variable	Opposite or next to micropyle	OBG
Angle	Variable	Orthotropous, anatropous	OBG
Number of integuments	Uniform	2	OBG

2006a; Frohlich and Chase 2007). It suggests that the loss in angiosperms of the NEEDLY copy of LFY—both found in gymnosperms—led to a fundamentally male axis with ectopic expression of carpels/ovules on the axis. Frohlich (2006a; Frohlich and Chase 2007) has most recently reviewed the evidence for and against the model. I would add that from my personal observations of mutant bisexual cones in pine (ovulate distal, microsporangiate proximal), because the male axes are adapted to dehisce after shedding pollen, the ovules have no time to develop before the axis falls from the tree. Thus, any model that uses the male axis as fundamental must add a stage of evolution in which the male axes are retained on the plant. In addition, Frohlich has discussed the evolution of carpels, but it is unclear how the timing of carpel characteristics tie with the timing of the evolution of the flower. It is interesting that LFY expression is graded (Albert et al. 2002), and these authors have developed differential expression of LFY model for the origin of the bisexual flower.

The Out of Male/Out of Female model also examines the origin of the bisexual flower (Theissen et al. 2002; Theissen and Becker 2004; Theissen and Melzer 2007). They suggest a homoeotic transformation of the distal male organs in a male gymnosperm strobilus into female organs (or vice versa). This would occur because of changes in FOIG B-class genes. They consider this bisexualization as the fundamental trait of angiosperms and insect pollination as the selective force. This has been enhanced by Baum and Hileman (2006), who suggest increasing levels of LFY affects the combination of MADS quartet of genes found in the ABC developmental model, resulting in the male proximal and the female distal. Again, the first step (ignoring carpel and stamen origin) is bisexualization as a result of selection by pollinators, followed by determinacy and then sepal–petal elaboration. Animal pollination is frequently considered to be correlated with bisexual organs (e.g., Theissen et al. 2002; Theissen and Becker 2004; Baum and Hileman 2006; Theissen and Melzer 2007), yet there are plants with unisexual flowers, such as the Chloranthaceae, that are insect-pollinated. Selection for insect pollination could have occurred at this level, with rewards consisting of pollen and stigmatic drops (Frohlich 2006a). Most recently, this model has been enhanced (Theissen and Melzer 2007) to integrate data with the fading and sliding boundaries. Intermediates might be expected to be found between the male portion and the female portion. If this model is applied to inflorescences, one could regard the inflorescences of *Archaefructus* as showing this. Low LFY, mostly BBCE quartets, makes males (stamens and clusters of stamen); moderate LFY and a mix of BBCE and CCCE make transition clusters that are bisexual; and high LFY and CCCE produce females (carpels and clusters of carpels).

A fundamental assumption of these models is that the bisexual organ called a flower is what fundamentally characterizes angiosperms and angiosperm success, rather than the structures that truly make angiosperms unique: the stamen and carpel. Most of these models do not discuss whether the bisexual organ evolved first or whether separate reproductive organs evolved first. This is partly because one of the potential fossil angiosperm sister groups is the bennettitaleans, which can be bisexual and bear flowerlike structures, suggesting that flowers are first. Yet there is good reason why angiosperms should be called angiosperms rather than flowering plants. Although there are other seed plant species with bisexual clusters of organs, none of them have carpels. Although this initially appears to be a semantic argument, a look at the results above strongly suggests a different model, one in which reproductive organs fully evolved and were developmentally canalized early, and the evolution of flowers was not only later but went through a period where there was little distinction between flowers and inflorescences. One could argue that this developmental lack of canalization allowed for the large diversity of flowers found today. Equally important is that the models do not address the inflorescence, another unique angiosperm characteristic. The fossils (like the molecular developmental data) also suggest that the structure of these flowers did not have the archetypical four whorls. Thus, a distinction between the presence of genetic controls and the canalization of those controls needs to be made.

The following are some of the problems that need to be addressed by molecular developmental data:

- Which came first, flowers or sexual floral organs?
- What causes determinacy in angiosperm flowers and inflorescences?
- How do controls in determinate and indeterminate inflorescences compare?
- What controls male proximal and female distal in flowers?
- What type or types of inflorescence is ancestral, and how does that relate to flowers?
- What are the developmental switches between basic morphological traits like alternate and opposite phyllotaxy, and anatropous and orthotropous angle of the ovules?
- What controls produce the carpel wall and the placenta?

Organs First, Flowers Later Model

The Organs First, Flowers Later model of the origin of angiosperm reproductive organs is outlined in Figure 7.6. It begins with the evolution of four structures—the tepal, stamen, carpel, and bitegmic ovule—on an RA under the control of FOIG B- and C-class genes (similar to gymnosperms). Like Stuessy (2004), I believe that flowers evolved after the evolution of the organs, although these organs could have been acquired sequentially. Thus, the first gene class to be angiospermous would be OBGs. There is simply insufficient developmental information to understand the evolution of these organs from a molecular view at this time (e.g., Scutt et al. 2006), although proposals on their evolution have been made (e.g., Taylor 1991; Taylor and Kirchner 1996; Taylor and Hickey 1996; Endress 2006).

Second, and important, was the evolution of determinacy (Fig. 7.6). I consider determinacy to be the using up of the apical meristem stem cells so that additional vegetative growth is not possible. Determinant RAs or inflorescences have apices terminated by flowers. Determinate URAs or flowers would have the FM used up by the carpels, or occasionally the stamens in unisexual male flowers. Evidence in fossils would be the presence of carpels in URAs or terminal URAs in RAs. In the fossils, we clearly have determinacy in all RAs and URAs, unlike gymnosperms. In living gymnosperms, the cones and other RAs are generally indeterminate with potentially indeterminate megasporangiate or microsporangiate structures. Although most agree that angiosperm flowers were determinate, this suggests that the inflorescences were initially determinate as well. Only later, after the canalization of the development of determinate URAs (flowers), is there the appearance of indeterminate inflorescences like racemes, although even these can show developmental evidence of their determinate origin (Sokoloff et al. 2006). The recent review by Benlloch et al. (2007) shows that LFY, FUL (=AP1), and TFL1 are primary for inflorescence control in both determinate and indeterminate eudicots. The primary difference is that in determinate inflorescences, TFL1 is restricted to branching and disappears from the inflorescence meristem to allow the formation of terminal flowers, while LFY and FUL (=AP1) expression is expanded into these areas. On this basis, candidates for the determinate inflorescence would be MIGs such as LFY, FUL(=AP1), UFO, ULT1, and related FOIGs such as AG. It is possible that genes like EMF1, EMF2, and TFL1 are related to the evolution of indeterminacy in *Arabidopsis*. Floral determinacy is likely to have been controlled by MIGs such as CAL and FUL (=AP1), and FOIGs like AG.

What is difficult is to know whether determinancy evolved contemporaneously with organ evolution or bisexualization, or whether it evolved as an intermediate stage. If it evolved along with the organs, it could be directly related to the enclosing of the ovules in the gynoecial appendage (carpel wall; Taylor 1991). This is directly related to the control

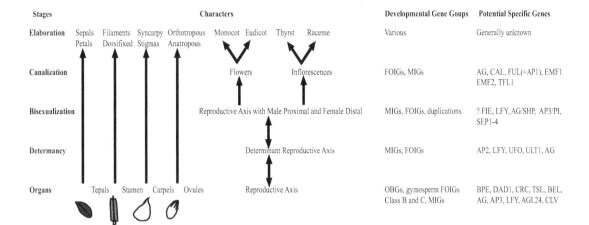

Stages	Characters							Developmental Gene Goups	Potential Specific Genes
Elaboration	Sepals Petals	Filaments Dorsifixed	Syncarpy Stigmas	Orthotropous Anatropous	Monocot Eudicot	Thyrst Raceme		Various	Generally unknown
Canalization					Flowers	Inflorescences		FOIGs, MIGs	AG, CAL, FUL(=AP1), EMF1 EMF2, TFL1
Bisexualization					Reproductive Axis with Male Proximal and Female Distal			MIGs, FOIGs, duplications	? FIE, LFY, AG/SHP, AP3/PI, SEP1-4
Determancy					Determinant Reproductive Axis			MIGs, FOIGs	AP2, LFY, UFO, ULT1, AG
Organs		Tepals	Stamen	Carpels Ovules	Reproductive Axis			OBGs, gymosperm FOIGs Class B and C, MIGs	BPE, DAD1, CRC, TSL, BEL, AG, AP3, LFY, AGL24, CLV

Fig. 7.6. Outline of the Organ First, Flowers Later model of angiosperm reproductive evolution. Included are the likely transitions in characters, developmental gene groups, and potential specific genes.

by AG, which has multiple roles as a MIG and a FOIG. Thus, determinacy could be directly related to the evolution of carpels. Alternatively, determinacy could be related to the evolution of bisexuality in which female systems are distal and male systems are proximal. Yet many early angiosperms are unisexual, and determinacy exists in both male and female flowers. Last, it could be independent and the result of potentially important MIGs such as UFO and ULT1. It is interesting that LFY expression is graded (Albert et al. 2002); this could result in variation in RAs and URAs simply by degree of expression (Baum and Hileman 2006; Theissen and Melzer 2007).

The third step would be the evolution of bisexuality (Fig. 7.6). The other models summarized above discuss the various possibilities. I suggest that although bisexuality is usually associated with flowers, the range of fossil morphologies indicates that it would be better to consider it as bisexualization of RAs and URAs. Although MIGs might play a role (especially AG and LFY), FOIGs, especially those of B, C, and E class, are important. Also important is the role of duplication (Kramer et al. 2004; Kramer and Zimmer 2006) of B- and C-class genes, as well as the evolution of E-class genes. The controls of bisexualization are unclear, but they are possibly are due to MIGs like LFY and FOIGs like AG/SHP, AP3/PI, and SEP1–4.

The next stage is canalization (Fig. 7.6). This is where selection optimized developmental pathways. This resulted in separate trajectories for flowers and inflorescences. Early Cretaceous fossils clearly show a lack of canalization in all flowers (including phyllotaxy and organ number) and inflorescences. The molecular data, especially the sliding boundaries or fading borders models, also suggest this (reviewed in Soltis et al. 2006). Research has suggested that in living basal dicots, the boundaries between whorls are not clear, and neither are their numbers. Essentially, FOIGs are not yet canalized. This suggests that several basic plans were canalized later and were responsible for the initial floral diversity found in basal dicots, the first monocots, and the basal eudicots.

Finally, after canalization, the basic plan became elaborated (Fig. 7.6). In regard to the organs, this would include the evolution of adnation and connation and the formation of discrete sepals and petals multiple times (Ronse de Craene 2008; but see Rasmussen et al. 2009; Warner et al. 2009). In addition, this would include the evolution of floral types within the basic plans described above. The large number of gene duplications, especially in MADS genes, would have allowed for the development of the vast array of morphologies found in living angiosperms. Variation in the ABC model is nicely summarized in Theissen and Melzer (2007) and shows the end points possible with expression of these genes. This would allow for the evolution of a wide range of determinate and indeterminate inflorescences, as well as elaboration of different floral types (Benlloch et al. 2007). The discrete nature of the FOIGs in whorl placement and floral part number is canalized in a number of directions that have allowed flowers to be good characters for taxonomy.

This models results in a number of explicit developmental expectations:

· OBGs will exist for each of the four organs, will have an all-angiosperm distribution, and will be independent of FOIGs for expression (not initiation).
· Relative clock comparisons will show these OBGs, such as carpel CRC, originated before all-angiosperm FOIG group genes.
· Control of determinacy for flowers and inflorescences will be unique to angiosperms and was fully developed early.
· If determinacy is fundamental to angiosperm flowers and inflorescences, genes or functions unique to angiosperms should affect both flowers and inflorescences.
· Indeterminate inflorescence axes will be shown to be elaborations of preexisting determinate types.
· In "flowerless" angiosperms (e.g., Chloranthaceae), OBGs will still be expressed, even if the full complement of FOIGs and other floral genes are not.
· Fully formed sexual organs are possible in unisexual flowers and should show that the full complement of FOIGs and other floral genes controlling bisexuality are not necessary.
· The fossil record should show the presence of dispersed tepals, stamen, and carpels (with ovules) that will not be associated with flowers, but rather with RAs.

To test this model and its implications, more research directed toward function and expression is necessary. We need much more work on OBGs than we currently have. We also need much more awareness that inflorescences and floral structure may be related. The model above does not exclude previously suggested models, but we do need to concentrate less on flowers and more on organs and inflorescences in living plants. Further comparisons between controls of determinate and indeterminate inflorescences are needed. In fossils, we need to concentrate less on finding flowers and more on finding earlier angiosperm organs that might not occur in typical flowers, as well as gymnosperms with possible relationships to angiosperms, such as gigantopterids (Taylor et al. 2006).

Conclusions

The Early Cretaceous was clearly a time of angiosperm diversification, but it was also a time that has recorded a diversity of angiosperm morphologies not found today. Although caution is necessary when considering the ancestral morphology of angiosperms, this review shows that some morphological states are uniform in the fossils, while others are variable. The organs of angiosperms are clearly similar in the fossils, while the number, size, and arrangement in flowers and inflorescences are quite variable. These fossil data and molecular developmental data suggest that all groups of developmental controls existed by the Early Cretaceous, but that some have a preangiosperm distribution and others have an all-angiosperm distribution. Yet these controls do not seem as canalized as they are today in living plants.

A number of explicit models of the evolution of floral development have been created. Many of these have value, but I do not believe that they are either broad enough or cover all areas that need to be covered. The new model, Organs First, Flowers Later, attempts to integrate these data and to propose a different course of evolution and development. The key differences are that the tepals, stamens, carpels, and bitegmic ovules are proposed to have evolved first, followed by determinacy, and then bisexualization of the flower and inflorescence. From that stage onward, there was selection to canalize the development of flowers and inflorescences observed today in living angiosperms, followed by elaboration to produce the diversity exhibited by flowering plants in the present day. This explicit model will, I hope, be tested in the future and provide new avenues of research.

Acknowledgments

It is a great pleasure to contribute to this volume in honor of Ted Delevoryas. My Ph.D. advisor was Bill Crepet, who was one of Ted's Ph.D. students, so I am a second-generation descendant of Ted's. Still, my pleasure comes from my interactions with Ted. I met him for the first time in the field in Texas looking for Tertiary fossil flowers. After that, Ted and I would speak at every meeting. To me, Ted exemplified friendliness; he always expressed concern for colleagues. He was interested in hearing how you were and how the research was going. But behind that friendliness was a sharp mind that was able to provide insight if asked.

This work is based on the huge and growing body of literature. I thank all my colleagues for their valuable work, and I apologize to my colleagues in the floral development field whose works were not cited because of the constraints of space. I also thank the editor, Carole Gee, for ongoing encouragement (and editing) on this project, and Mike Frohlich for interesting discussions on evolution and development. I thank Randy Hunt for evolutionary perspectives on the model. Last, I thank my research assistant, Erin McCorkle, for help compiling the references and Beth Boesche-Taylor and Shusheng Hu for help cleaning up the ideas and the writing.

References Cited

Ahn, J. H., D. Miller, V. J. Winter, M. J. Banfield, J. H. Lee, S. Y. Yoo, S. R. Henz, R. L. Brady, and D. Weigel. 2006. A divergent external loop confers antagonistic activity on floral regulators FT and TFL1. *The EMBO Journal* 25: 605–614.

Albert, V. A., D. G. Oppenheimer, and C. Lindqvist. 2002. Pleiotropy, redundancy and the evolution of flowers. *Trends in Plant Science* 7: 297–301.

Alvarez, J., and D. R. Smyth. 1999. CRABS CLAW and SPATULA, two *Arabidopsis* genes that control carpel development in parallel with AGAMOUS. *Development* 126: 2377–2386.

Amaya, I., O. J. Ratcliffe, and D. J. Bradley. 1999. Expression of CENTRORADIALIS (CEN) and CEN-like genes in tobacco reveals a conserved mechanism controlling phase change in diverse species. *Plant Cell* 11: 1405–1418.

Azhakanandam, S., S. Nole-Wilson, F. Bao, and R. G. Franks. 2008. SEUSS and AINTEGUMENTA mediate patterning and ovule initiation during gynoecium medial domain development. *Plant Physiology* 146: 1165–1181.

Basinger, J. F., and D. L. Dilcher. 1984. Ancient bisexual flowers. *Science* 224 (4648): 511–513.

Battey, N. H., and F. Tooke. 2002. Molecular control and variation in the floral transition. *Current Opinion in Plant Biology* 5: 62–68.

Baum, D. A., and L. C. Hileman. 2006. A developmental genetic model for the origin of the flower. In C. Ainsworth (ed.), *Flowering and Its Manipulation. Annual Plant Reviews* 20: 3–27. Oxford: Blackwell.

Becker, A., M. Bey, T. R. Burglin, H. Saedler, and G. Theissen. 2002. Ancestry and diversity of BEL1-like homeobox genes revealed by gymnosperm (*Gnetum gnemon*) homologs. *Development Genes and Evolution* 212: 452–457.

Benlloch, R., A. Berbel, A. Serrano-Mislata, and F. Madueño. 2007. Floral initiation and inflorescence architecture: a comparative view. *Annals of Botany* 100: 659–676.

Boss, P. K., R. M. Bastow, J. S. Mylne, and C. Dean. 2004. Multiple pathways in the decision to flower: enabling, promoting, and resetting. *Plant Cell* 16 (Suppl.): S18–S31.

Bowman, J. L., and D. R. Smyth. 1999. CRABS CLAW, a gene that regulates carpel and nectary development in *Arabidopsis*, encodes a novel protein with zinc finger and helix–loop–helix domains. *Development* 126: 2387–2396.

Brenner, G. J. 1996. Evidence for the earliest stage of angiosperm pollen evolution: a paleoequatorial section from Israel. In D. W. Taylor and L. J. Hickey (eds.), *Flowering Plant Origin, Evolution and Phylogeny*, pp. 91–115. New York: Chapman and Hall.

Brenner, E. D., M. S. Katari, D. W. Stevenson, S. A. Rudd, A. W. Douglas, W. N. Moss, R. W. Twigg, S. J. Runko, G. M. Stellari, W. R. McCombie, and G. M. Coruzzi. 2005. EST analysis in *Ginkgo biloba*: an assessment of conserved developmental regulators and gymnosperm specific genes. *BMC Genomics* 6: 143.

Buzgo, M., D. E. Soltis, P. S. Soltis, and H. Ma. 2004. Towards a comprehensive integration of morphological and genetic studies of floral development. *Trends in Plant Science* 9: 164–173.

Buzgo, M., P. S. Soltis, S. Kim, and D. E. Soltis. 2005. The making of the flower. *The Biologist* 52: 149–154.

Carles, C. C., K. Lertpiriyapong, K. Reville, and J. C. Fletcher. 2004. The ULTRAPETALA1 gene functions early in *Arabidopsis* development to restrict shoot apical meristem activity and acts through WUSCHEL to regulate floral meristem determinacy. *Genetics* 167: 1893–1903.

Chanvivattana, Y., A. Bishopp, D. Schubert, C. Stock, Y.-H. Moon, Z. R. Sung, and J. Goodrich. 2004. Interaction of polycomb-group proteins controlling flowering in *Arabidopsis*. *Development* 131: 5263–5276.

Chevalier, D., P. Sieber, and K. Schneitz. 2002. The genetic and molecular control of ovule development. In S. D. O'Neill and J. A. Roberts (eds.), *Plant Reproduction: Annual Plant Reviews* 6: 61–85. Boca Raton, Fla.: CRC Press.

Coen, E. S., and E. M. Meyerowitz. 1991. The war of the whorls: genetic interactions controlling flower development. *Nature* 353: 31–37.

Colombo, L., J. Franken, E. Koetje, J. van Went, H. J. Dons, G. C. Angenent, and A. J. van Tunen. 1995. The petunia MADS box gene FBP11 determines ovule identity. *Plant Cell* 7: 1859–1868.

Costa, M. M. R., S. Fox, A. I. Hanna, C. Baxter, and E. Coen. 2005. Evolution of regulatory interactions controlling floral asymmetry. *Development* 132: 5093–5101.

Crane, P. R., K. R. Pedersen, E. M. Friis, and A. N. Drinnan. 1993. Early Cretaceous (early to middle Albian) platanoid inflorescences associated with *Sapindopsis* leaves from the Potomac Group of eastern North America. *Systematic Botany* 18: 328–344.

Crepet, W. L. 2008. The fossil record of angiosperms: requiem or renaissance? *Annals of the Missouri Botanical Garden* 95: 3–33.

Cubas, P. 2004. Floral zygomorphy, the recurring evolution of a successful trait. *BioEssays* 26: 1175–1184.

Devlin, P. 2007. Photocontrol of flowering. In G. C. Whitelam and K. J. Halliday (eds.), *Light and Plant Development: Annual Plant Reviews* 30: 185–210. Oxford: Blackwell.

Dilcher, D. L., G. Sun, Q. Ji, and H. Li. 2007. An early infructescence *Hyrcantha decussata* (comb. nov.) from the Yixian Formation in northeastern China. *Proceedings of the National Academy of Sciences* 104: 9370–9374.

Dinneny, J. R., R. Yadegari, R. L. Fischer, M. F. Yanofsky, and D. Weigel. 2004. The role of JAGGED in shaping lateral organs. *Development* 131: 1101–1110.

Dinneny, J. R., D. Weigel, and M. F. Yanofsky. 2006. NUBBIN and JAGGED define stamen and carpel shape in *Arabidopsis*. *Development* 133: 1645–1655.

Drinnan, A. N., P. R. Crane, E. M. Friis, and K. R. Pedersen. 1991. Angiosperm flowers and tricolpate pollen of buxaceous affinity from the mid-Cretaceous of eastern North America. *American Journal of Botany* 78: 153–176.

Egea-Cortines, M., H. Saedler, and H. Sommer. 1999. Ternary complex formation between the MADS-box proteins SQUAMOSA, DEFICIENS and GLOBOSA is involved in the control of floral architecture in *Antirrhinum majus*. *The EMBO Journal* 18: 5370–5379.

Endress, P. K. 1986. Reproductive structures and phylogenetic significance of extant primitive angiosperms. *Plant Systematics and Evolution* 152: 1–28.

———. 2006. Angiosperm floral evolution: morphological developmental framework. *Advances in Botanical Research* 44: 1–61.

Endress, P. K., and J. A. Doyle. 2009. Recontructing the ancestral angiosperm flower and its initial specializations. *American Journal of Botany* 96: 22–66.

Endress, P. K., and L. D. Hufford. 1989. The diversity of stamen structures and dehiscence patterns among Magnoliidae. *Botanical Journal of the Linnean Society of London* 100: 45–85.

Espinosa-Soto, C., P. Padilla-Longoria, and E. R. Alvarez-Buylla. 2004. A gene regulatory network model for cell-fate determination during *Arabidopsis thaliana* flower development that is robust and recovers experimental gene expression profiles. *Plant Cell* 16: 2923–2939.

Fletcher, J. C. 2002. Shoot and floral meristem maintenance in *Arabidopsis. Annual Review of Plant Biology* 53: 45–66.

Floyd, S. K., and J. L. Bowman. 2007. The ancestral developmental tool kit of land plants. *International Journal of Plant Sciences* 168: 1–35.

Fourquin, C., M. Vinauger-Douard, B. Fogliani, C. Dumas, and C. P. Scutt. 2005. Evidence that CRABS CLAW and TOUSLED have conserved their roles in carpel development since the ancestor of the extant angiosperms. *Proceedings of the National Academy of Sciences U.S.A.* 102: 4649–4654.

Fourquin, C., M. Vinauger-Douard, P. Chambrier, A. Berne-Dedieu, and C. P. Scutt. 2007. Functional conservation between CRABS CLAW orthologues from widely diverged angiosperms. *Annals of Botany* 100: 651–657.

Friedman, W. E., R. C. Moore, and M. D. Purugganan. 2004. The evolution of plant development. *American Journal of Botany* 91: 1726–1741.

Friis, E. M., H. Eklund, K. R. Pedersen, and P. R. Crane. 1994. *Virginianthus calycanthoides* gen. et sp. nov. — a calycanthaceous flower from the Potomac Group (Early Cretaceous) of Eastern North America. *International Journal of Plant Sciences* 155: 772–785.

Friis, E. M., J. A. Doyle, P. K. Endress, and Q. Leng. 2003. *Archaefructus* — angiosperm precursor or specialized early angiosperm? *Trends in Plant Science* 8: 369–373.

———. 2000a. Fossil floral structures of a basal angiosperm with monocolpate, reticulate–acolumellate pollen from the Early Cretaceous of Portugal. *Grana* 39: 226–239.

———. 2000b. Reproductive structure and organization of basal angiosperms from the Early Cretaceous (Barremian or Aptian) of western Portugal. *International Journal of Plant Science* 161 (Suppl. 6): S169–S182.

———. 2001. Fossil evidence of water lilies (Nymphaeales) in the Early Cretaceous. *Nature* 410 (6826): 357–60.

———. 2004. Araceae from the early Cretaceous of Portugal: evidence on the emergence of monocotyledons. *Proceedings of the National Academy of Sciences U.S.A.* 101: 16565–16570.

———. 2006. Cretaceous angiosperm flowers: innovation and evolution in plant reproduction. *Palaeogeography, Palaeoclimatology, and Palaeoecology* 232: 251–293.

Frohlich, M. W. 2001. A detailed scenario and possible tests of the Mostly Male theory of flower evolutionary origins. In M. L. Zelditch (ed.), *Beyond Heterochrony: The Evolution of Development*, pp. 59–104. New York: Wiley-Liss.

———. 2002. The Mostly Male theory of flower origins: summary and update regarding the Jurassic pteridosperm *Pteroma*. In Q. C. B. Cronk, R. M. Bateman, and J. A. Hawkins (eds.), *Developmental Genetics and Plant Evolution*. Systematics Association special volume series 65: 85–108. London: Taylor and Francis.

———. 2003. An evolutionary scenario for the origin of flowers. *Nature Reviews Genetics* 4, 559–566.

———. 2006a. Recent developments regarding the evolutionary origin of flowers. *Advances in Botanical Research* 44: 63–127.

———. 2006b. Recommendations and goals for evo-devo research: scenarios, genetic constraint, and developmental homeostasis. In J. T. Columbus, E. A. Friar, J. M. Porter, L. M. Prince, and M. G. Simpson (eds.), *Monocots: Comparative Biology and Evolution*, 2 vols., pp. 172–187. Claremont, Calif.: Rancho Santa Ana Botanic Garden.

Frohlich, M. W., and M. W. Chase. 2007. After a dozen years of progress the origin of angiosperms is still a great mystery. *Nature* 450 (7173): 1184–1189.

Frohlich, M. W., and D. S. Parker. 2000. The Mostly Male theory of flower evolutionary origins: from genes to fossils. *Systematic Botany* 25: 155–170.

Fulton, L., M. Batoux, R. K. Yadav, and K. Schneitz. 2006. The genetic control of flower size and shape. In C. Ainsworth (ed.), *Flowering and Its Manipulation: Annual Plant Reviews* 20: 71–97. Oxford: Blackwell.

Gambin, A., S. Lasota, and M. Rutkowski. 2006. Analyzing stationary states of gene regulatory network using petri nets. *In Silico Biology* 6: 93–109.

Gómez-Mena, C., S. de Folter, M. M. R. Costa, G. C. Angenent, and R. Sablowski. 2005. Transcriptional program controlled by the floral homeotic gene AGAMOUS during early organogenesis. *Development* 132: 429–438.

Gregis, V., A. Sessa, L. Colombo, and M. M. Kater. 2006. AGL24, SHORT VEGETATIVE PHASE, and APETALA1 redundantly control AGAMOUS during early stages of flower development in *Arabidopsis. Plant Cell* 18: 1373–1382.

Guillet-Claude, C., N. Isabel, B. Pelgas, and J. Bousquet. 2004. The evolutionary implications of *knox-I* gene duplications in conifers: correlated evidence from phylogeny, gene mapping, and analysis of functional divergence. *Molecular Biology and Evolution* 21: 2232–2245.

Hake, S., H. M. S. Smith, H. Holtan, E. Magnani, G. Mele, and J. Ramirez. 2004. The role of *knox* genes in plant development. *Annual Review of Cell and Developmental Biology* 20: 125–151.

Hedman H., Källman T., and Lagercrantz U. 2009. Early evolution of the MFT-like gene family in plants. *Plant Molecular Biology* 70:359–369.

Hepworth, S. R., Y. Zhang, S. McKim, X. Li, and G. W. Haughn. 2005. BLADE-ON-PETIOLE-dependent signaling controls leaf and floral patterning in *Arabidopsis. Plant Cell* 17: 1434–1448.

Hernández-Hernández, T., L. P. Martinez-Castilla, and E. R. Alvarez-Buylla. 2007. Functional diversification of B MADS-box homeotic regulators of flower development: adaptive evolution in protein–protein interaction domains after major gene duplication events. *Molecular Biology and Evolution* 24: 465–481.

Hickey, L. J., and D. W. Taylor. 1996. Origin of the angiosperm flower. In D. W. Taylor and L. J. Hickey (eds.), *Flowering Plant Origin, Evolution, and Phylogeny*, pp. 176–231. New York: Chapman and Hall.

Hileman, L. C., and V. F. Irish. 2009. More is better: the uses of developmental genetic data to reconstruct perianth evolution. *American Journal of Botany* 96: 83–95.

Hill, C. R. 1996. A plant with flower-like organs from the Wealden of the Weald (Lower Cretaceous), southern England. *Cretaceous Research* 17: 27–38.

Irish, V. 2006. Duplication, diversification and comparative genetics of angiosperm MADS-box genes. *Advances in Botanical Research* 44: 129–161.

Ito, T., F. Wellmer, H. Yu, P. Das, N. Ito, M. Alves-Ferreira, J. L. Riechmann, and E. M. Meyerowitz. 2004. The homeotic protein AGAMOUS controls microsporogenesis by regulation of SPOROCYTELESS. *Nature* 430: 356–360.

Ito, T., K.-H. Ng, T.-S. Lim, H. Yu, and E. M. Meyerowitz. 2007. The homeotic protein AGAMOUS controls late stamen development by regulating a Jasmonate biosynthetic gene in *Arabidopsis*. *Plant Cell* 19: 3516–3529

Ji, Q., H. Li, L. M. Bowe, Y. Liu, and D. W. Taylor. 2004. Early Cretaceous *Archaefructus eoflora* sp. nov. with bisexual flowers from Beipiao, Western Liaoning, China. *Acta Geologica Sinica* 78: 883–896.

Katz, A., M. Oliva, A. Mosquna, O. Hakim, and N. Ohad. 2004. FIE and CURLY LEAF polycomb proteins interact in the regulation of homeobox gene expression during sporophyte development. *Plant Journal* 37: 707–719.

Kim, S., M.-J. Yoo, V. A. Albert, J. S. Farris, P. S. Soltis, and D. E. Soltis. 2004. Phylogeny and diversification of B-function MADS-box genes in angiosperms: evolutionary and functional implications of a 260-million-year-old duplication. *American Journal of Botany* 91: 2102–2118.

Kim, S., J. Koh, M.-J. Yoo, H. Kong, Y. Hu, H. Ma, P. S. Soltis, and D. E. Soltis. 2005. Expression of floral MADS-box genes in basal angiosperms: implications for the evolution of floral regulators. *Plant Journal* 43: 724–744.

Kim, S., P. S. Soltis, K. Wall, and D. E. Soltis. 2006. Phylogeny and domain evolution in the APETALA2-like gene family. *Molecular Biology and Evolution* 23: 107–120.

Komeda, Y. 2004. Genetic regulation of time to flower in *Arabidopsis thaliana*. *Annual Review of Plant Biology* 55: 521–535.

Kramer, E. M. 2006. Floral patterning and control of floral organ formation. In C. Ainsworth (ed.), *Flowering and Its Manipulation*. Annual Plant Reviews 20: 49–70. Oxford: Blackwell.

Kramer, E. M., and E. A. Zimmer. 2006. Gene duplication and floral developmental genetics of basal eudicots. *Advances in Botanical Research* 44: 353–384.

Kramer, E. M., M. A. Jaramillo, and V. S. Di Stilio. 2004. Patterns of gene duplication and functional evolution during the diversification of the AGAMOUS subfamily of MADS box genes in angiosperms. *Genetics* 166: 1011–1023.

Krassilov, V. A. 1984. New paleobotanical data on origin and early evolution of angiospermy. *Annals of the Missouri Botanical Garden* 71: 557–592.

———. 1997. *Angiosperm Origins: Morphological and Ecological Aspects*. Sofia: Pensoft.

Krassilov, V. A., P. V. Shilin, and V. A. Vachrameev. 1983. Cretaceous flowers from Kazakhstan. *Review of Palaeobotany and Palynology* 40: 91–113.

Krizek, B. A., and J. C. Fletcher. 2005. Molecular mechanisms of flower development: an armchair guide. *Nature Reviews Genetics* 6: 688–698.

Leibfried, A., J. P. To, W. Busch, S. Stehling, A. Kehle, M. Demar, J. J. Kieber, and J. U. Lohmann. 2005. WUSCHEL controls meristem function by direct regulation of cytokinin-inducible response regulators. *Nature* 438 (7071): 1172–1175.

Leng, Q., and E. M. Friis. 2003. *Sinocarpus decussatus* gen. et sp. nov., a new angiosperm with basally syncarpous fruits from the Yixian Formation of Northeast China. *Plant Systematics and Evolution* 241: 77–88.

———. 2006. Angiosperm leaves associated with *Sinocarpus* infructescences from the Yixian Formation (mid-Early Cretaceous) of NE China. *Plant Systematics and Evolution* 262: 173–187.

Lenhard, M., G. Jurgens, and T. Laux. 2002. The WUSCHEL and SHOOTMERISTEMLESS genes fulfill complementary roles in *Arabidopsis* shoot meristem regulation. *Development* 129: 3195–3206.

Litt, A., and V. F. Irish. 2003. Duplication and diversification in the APETALA1/FRUITFULL floral homeotic gene lineage: implications for the evolution of floral development. *Genetics* 165: 821–833.

Liu, Z., R. G. Franks, and V. P. Klink. 2000. Regulation of gynoecium marginal tissue formation by LEUNIG and AINTEGUMENTA. *Plant Cell* 12: 1879–1892.

Lohmann, J., R. Hong, M. Hobe, M. Busch, F. Parcy, R. Simon, and D. Weigel. 2001. A molecular link between stem cell regulation and floral patterning in *Arabidopsis*. *Cell* 105: 793–803.

Maizel, A., M. A. Busch, T. Tanahashi, J. Perkovic, M. Kato, M. Hasebe, and D. Weigel. 2005. The floral regulator LEAFY evolves by substitutions in the DNA binding domain. *Science* 308: 260–263.

Maizel, A., and D. Weigel. 2004. Temporally and spatially controlled induction of gene expression in *Arabidopsis thaliana*. *Plant Journal* 38: 164–171.

Masiero, S., M.-A. Li, I. Will, U. Hartmann, H. Saedler, P. Huijser, Z. Schwarz-Sommer, and H. Sommer. 2004. INCOMPOSITA: a MADS-box gene controlling prophyll development and floral meristem identity in *Antirrhinum*. *Development* 131: 5981–5990.

Mathews, S., J. G. Burleigh, and M. J. Donoghue. 2003. Adaptive evolution in the photosensory domain of phytochrome A in early angiosperms. *Molecular Biology and Evolution* 20: 1087–1097.

Melzer, R., K. Kaufmann, and G. Theissen. 2006. Missing links: DNA-binding and target gene specificity of floral homeotic proteins. *Advances in Botanical Research* 44: 209–236.

Mohr, B. A. R., and M. E. C. Bernardes-de-Oliveira. 2004. *Endressinia brasiliana*, a magnolialean angiosperm from the Lower Cretaceous Crato Formation (Brazil). *International Journal of Plant Sciences* 165: 1121–1133.

Mohr, B. A. R., and H. Eklund. 2003. *Araripia florifera*, a magnoliid angiosperm from the Lower Cretaceous Crato Formation (Brazil). *Review of Palaeobotany and Palynology* 126: 279–292.

Mohr, B. A. R., and E. M. Friis. 2000. Early angiosperms from the Lower Cretaceous Crato Formation (Brazil): a preliminary report. *International Journal of Plant Science* 161 (Suppl. 6): S155–S167.

Mohr, B. A. R., M. E. C. Bernardes-de-Oliveira, and D. W. Taylor. 2008. *Pluricarpellatia*, a nymphaealean angiosperm from the Lower Cretaceous of northern Gondwana (Crato Formation, Brazil). *Taxon* 57: 1157–1148.

Moon, J., S.-S. Suh, H. Lee, K.-R. Choi, C. B. Hong, N.-C. Paek, S.-G. Kim, and I. Lee. 2003. The SOC1 MADS-box gene integrates vernalization and gibberellin signals for flowering in *Arabidopsis*. *Plant Journal* 35: 613–623.

Nagasaki, H., M. Matsuoka, and Y. Sato. 2005. Members of TALE and WUS subfamilies of homeodomain proteins with potentially important functions in development form dimers within each subfamily in rice. *Genes and Genetic Systems* 80: 261–267.

Nam, J., C. W. de Pamphilis, H. Ma, and M. Nei. 2003. Antiquity and evolution of the MADS box gene family controlling flower development in plants. *Molecular Biology and Evolution* 20:1435–1447.

Ni, W., D. Xie, L. Hobbie, B. Feng, D. Zhao, J. Akkara, and H. Ma. 2004. Regulation of flower development in *Arabidopsis* by SCF complexes. *Plant Physiology* 134: 1574–1585.

Niklas, K. J. 1997. *The Evolutionary Biology of Plants*. Chicago: University of Chicago Press.

Parcy, F., K. Bomblies, and D. Weigel. 2002. Interaction of LEAFY, AGAMOUS and TERMINAL FLOWER1 in maintaining floral meristem identity in *Arabidopsis*. *Development* 129: 2519–2527.

Park, S. O., Z. Zheng, D. G. Oppenheimer, and B. A. Hauser. 2005. The PRETTY FEW SEEDS2 gene encodes an *Arabidopsis* homeodomain protein that regulates ovule development. *Development* 132: 841–849.

Pedersen, K. R., M. von Balthazar, P. R. Crane, and E. M. Friis. 2007. Early
    Cretaceous floral structures and in situ tricolpate–striate pollen: new early
    eudicots from Portugal. *Grana* 4: 176–196.

Prusinkiewicz, P., Y. Erasmus, B. Lane, L. D. Harder, and E. Coen. 2007.
    Evolution and development of inflorescence architectures. *Science* 316:
    1452–1456.

Rasmussen, D. A., E. M. Kramer, and E. A. Zimmer. 2009. One size fits all?
    Molecular evidence for a commonly inherited petal identity program in
    Ranunculales. *American Journal of Botany* 96: 96–109.

Ratcliffe, O. J., I. Amaya, C. A. Vincent, S. Rothstein, R. Carpenter, E. S. Coen,
    and D. J. Bradley. 1998. A common mechanism controls the life cycle and
    architecture of plants. *Development* 125: 1609–1615.

Ratcliffe, O. J., D. J. Bradley, and E. S. Coen. 1999. Separation of shoot and floral
    identity in *Arabidopsis*. *Development* 126: 1109–1120.

Ronse de Craene, L. P. 2008. Homology and evolution of petals in the core
    eudicots. *Systematic Botany* 33: 301–325.

Rudall, P. J., M. V. Remizowa, G. Prenner, C. J. Prychid, R. E. Tuckett, and
    D. D. Sokoloff. 2009. Nonflowers near the base of extant angiosperms?
    Spatiotemporal arrangement of organs in reproductive units of
    Hydatellaceae and its bearing on the origin of the flower. *American Journal
    of Botany* 96: 67–82.

Schubert, D., O. Clarenz, and J. Goodrich 2005. Epigenetic control of plant
    development by polycomb-group proteins. *Current Opinion in Plant Biology*
    8: 553–561.

Scott, R. J., M. Spielman, and H. G. Dickinson. 2004. Stamen structure and
    function. *Plant Cell* 16: S46–S60.

Scutt, C. P., M. Vinauger-Douard, C. Fourquin, C. Finet, and C. Dumas. 2006.
    An evolutionary perspective on the regulation of carpel development.
    Journal of Experimental Botany 57: 2143–2152.

Scutt, C. P., G. Theissen, and C. Ferrandiz. 2007. The evolution of plant
    development: past, present, and future. *Annals of Botany* 100: 599–601.

Sieber, P., J. Gheyselinck, R. Gross-Hardt, T. Laux, U. Grossniklaus, and
    K. Schneitz. 2004. Pattern formation during early ovule development in
    Arabidopsis thaliana. *Developmental Biology* 273: 321–334.

Siegfried, K. R., Y. Eshed, S. F. Baum, D. Otsuga, G. N. Drews, and J. L.
    Bowman. 1999. Members of the YABBY gene family specify abaxial cell fate
    in *Arabidopsis*. *Development* 126: 4117–4128.

Singer, S. R. 2006. Inflorescence architecture—moving beyond description to
    development, genes and evolution. In C. Ainsworth (ed.), *Flowering and Its
    Manipulation*. *Annual Plant Reviews* 20: 98–113. Oxford: Blackwell.

Skinner, D. J., T. A. Hill, and C. S. Gasser. 2004. Regulation of ovule
    development. *Plant Cell* 16: S32–S45.

Sokoloff, D., P. J. Rudall, and M. Remizowa. 2006. Flower-like terminal
    structures in racemose inflorescenses: a tool in morphogenetic and
    evolutionary research. *Journal of Experimental Botany* 57: 3517–3530.

Soltis, D. E., V. A. Albert, S. Kim, M.-J. Yoo, P. S. Soltis, M. J. Frohlich,
    J. Leebens-Mack, H. Kong, K. Wall, C. de Pamphilis, and H. Ma. 2005a.
    Evolution of the flower. In R. J. Henry (ed.), *Plant Diversity and Evolution:
    Genotypic and Phenotypic Variation in Higher Plants*, pp. 165–200.
    Cambridge, Mass.: CABI.

Soltis, D. E., P. S. Soltis, P. K. Endress, and M. W. Chase. 2005b. *Phylogeny and
    Evolution of Angiosperms*. Sunderland, Mass.: Sinauer Associates.

Soltis, P. S., D. E. Soltis, S. Kim, A. Chanderbali, and M. Buzgo. 2006.
    Expression of floral regulators in basal angiosperms and the origin and
    evolution of ABC-function. *Advances in Botanical Research* 44: 483–506.

Soltis, P.S., S. F. Brockington, M.-J. Yoo, A. Piedrahita, M. Latvis, M. J. Moore, A. S. Chanderbali, and D. E. Soltis. 2009. Floral variation and floral genetics in basal angiosperms. *American Journal of Botany* 96: 110–128.

Stewart, W. N., and G. W. Rothwell. 1993. *Paleobotany and the Evolution of Plants*, 2nd ed. Cambridge: Cambridge University Press.

Stuessy, T. F. 2004. A transitional-combinational theory for the origin of angiosperms. *Taxon* 53: 3–16.

Sun, G., and D. L. Dilcher. 1997. Discovery of the oldest known angiosperm inflorescence in the world from Lower Cretaceous of Jixi, China. *Acta Palaeontologica Sinica* 36: 135–142

———. 2002. Early angiosperms from the Lower Cretaceous of Jixi, eastern Heilongjiang, China. *Review of Palaeobotany and Palynology* 121: 91–112.

Sun, G., D. L. Dilcher, S. Zheng, and Z. Zhou. 1998. In search of the first flower: a Jurassic angiosperm, *Archaefructus*, from northeast China. *Science* 282: 1692–1695.

Sun, G., S. Zheng, D. L. Dilcher, Y. Wang, and S. Mei. 2001. *Early Angiosperms and Their Associated Plants from Western Liaoning, China.* Shanghai: Shanghai Scientific and Technological Education Publishing House. [In Chinese and English.]

Sun, G., Q. Ji, D. L. Dilcher, S. Zheng, K. C. Nixon, and X. Wang. 2002. Archaefructaceae, a new basal angiosperm family. *Science* 296: 899–904.

Szécsi, J., C. Joly, K. Bordji, E. Varaud, J. M. Cock, C. Dumas, and M. Bendahmane. 2006. BIGPETALp, a bHLH transcription factor is involved in the control of *Arabidopsis* petal size. *The EMBO Journal* 25: 3912–3920.

Takeda, S., N. Matsumoto, and K. Okada. 2003. RABBIT EARS, encoding a SUPERMAN-like zinc finger protein, regulates petal development in *Arabidopsis thaliana*. *Development* 131: 425–434.

Taylor, D. W. 1991. Angiosperm ovules and carpels: their characters and polarities, distribution in basal clades and structural evolution. *Postilla* 208: 1–40.

Taylor, D. W., and L. J. Hickey. 1990. An Aptian plant with attached leaves and flowers: implications for angiosperm origin. *Science* 247: 702–704.

———. 1992. Phylogenetic evidence for the herbaceous origin of angiosperms. *Plant Systematics and Evolution* 180: 137–156.

———. 1996. Evidence for and implications of an herbaceous origin for angiosperms. In D. W. Taylor and L. J. Hickey (eds.), *Flowering Plant Origin, Evolution, and Phylogeny*, pp. 232–266. New York: Chapman and Hall.

Taylor, D. W., and G. Kirchner. 1996. The origin and evolution of the angiosperm carpel. In D. W. Taylor and L. J. Hickey (eds.), *Flowering Plant Origin, Evolution, and Phylogeny*, pp. 116–140. New York: Chapman and Hall

Taylor, D. W., H. Li, J. Dahl, F. J. Fago, D. Zinniker, and J. M. Moldowan. 2006. Biogeochemical evidence for the presence of the angiosperm molecular fossil oleanane in Paleozoic and Mesozoic nonangiospermous fossils. *Paleobiology* 32: 179–190.

Taylor, D. W., G. J. Brenner, and S. H. Basha. 2008. *Scutifolium jordanicum* gen. et sp. nov. (Cabombaceae), an aquatic fossil plant from the Lower Cretaceous of Jordan, and the relationships of related leaf fossils to living genera. *American Journal of Botany* 95: 340–352.

Taylor, T. N., E. L. Taylor, and M. Krings. 2009. *Paleobotany: The Biology and Evolution of Fossil Plants*, 2nd ed. San Diego: Academic Press.

Theissen, G., and A. Becker. 2004. The ABCs of flower development in *Arabidopsis* and rice. *Progress in Botany* 65: 193–218.

Theissen, G., and R. Melzer. 2007. Molecular mechanisms underlying origin and diversification of the angiosperm flower. *Annals of Botany* 100: 603–619.

Theissen, G., A. Becker, K.-U. Winter, T. Munster, C. Kirchner, and H. Saedler. 2002. How the land plants learned their floral ABCs: the role of MADS-box genes in the evolutionary origin of flowers. In C. B. Quentin, R. M. Cronk, and J. A. Bateman (eds.), *Developmental Genetics and Plant Evolution*, pp. 173–205. London: Taylor and Francis.

Tremblay, R., and J. Colasanti. 2006. Floral induction. In C. Ainsworth (ed.), *Flowering and Its Manipulation. Annual Plant Reviews* 20: 28–48. Oxford: Blackwell.

Twell, D. 2002. The developmental biology of pollen. In S. D. O'Neill and J. A. Roberts (eds.), *Plant Reproduction. Annual Plant Reviews* 6: 86–153. Boca Raton, Fla.: CRC Press.

von Balthazar, M., K. R. Pedersen, and E. M. Friis. 2005. *Teixeiria lusitanica*, a new fossil flower from the Early Cretaceous of Portugal with affinities to Ranunculales. *Plant Systematics and Evolution* 255: 55–75.

von Balthazar, M., K. R. Pedersen, P. R. Crane, M. Stampanoni, and E. M. Friis. 2007. *Potomacanthus lobatus* gen. et sp. nov., a new flower of probable Lauraceae from the Early Cretaceous (Early to Middle Albian) of eastern North America. *American Journal of Botany* 94: 2041–2053.

Warner, K. A., P. J. Rudall, and M. W. Frohlich. 2009. Environmental control of sepalness and petalness in perianth organs of waterlilies: a new Mosaic Theory for the evolutionary origin of a differentiated perianth. *Journal of Experimental Botany* 60: 3559–3574.

Weiss, J., L. Delgado-Benarroch, and M. Egea-Cortines. 2005. Genetic control of floral size and proportions. *International Journal of Developmental Biology* 49: 513–525.

Wellmer, F., J. L. Riechmann, M. Alves-Ferreira, and E. M. Meyerowitz. 2004. Genome-wide analysis of spatial gene expression in *Arabidopsis* flowers. *Plant Cell* 16: 1314–1326.

William, D. A., Y. Su, M. R. Smith, M. Lu, D. A. Baldwin, and D. Wagner. 2004. Genomic identification of direct target genes of LEAFY. *Proceedings of the National Academy of Sciences U.S.A.* 101: 1775–1780.

Yu, H., T. Ito, F. Wellmer, and E. M. Meyerowitz. 2004. Repression of AGAMOUS-LIKE 24 is a crucial step in promoting flower development. *Nature Genetics* 36: 157–161.

Zahn, L. M., H. Kong, J. H. Leebens-Mack, S. Kim, P. S. Soltis, L. L. Landherr, D. E. Soltis, C. W. de Pamphilis, and H. Ma. 2005a. The evolution of the SEPALLATA subfamily of MADS-box genes: a preangiosperm origin with multiple duplications throughout angiosperm history. *Genetics* 169: 2209–2223.

Zahn, L. M., J. Leebens-Mack, C. W. de Pamphilis, H. Ma, and G. Theissen. 2005b. To B or not to B a flower: the role of DEFICIENS and GLOBOSA orthologs in the evolution of the angiosperms. *Journal of Heredity* 96: 225–240.

Zahn, L. M., B. Feng, and H. Ma. 2006a. Beyond the ABC-model: regulation of floral homeotic genes. *Advances in Botanical Research* 44: 163–207.

Zahn, L. M., J. H. Leebens-Mack, J. M. Arrington, Y. Hu, L. L. Landherr, C. W. de Pamphilis, A. Becker, G. Theissen, and H. Ma. 2006b. Conservation and divergence in the AGAMOUS subfamily of MADS-box genes: evidence of independent sub- and neofunctionalization events. *Evolution and Development* 8: 30–45.

Zhang, P., H. T. W. Tan, K.-H. Pwee, and P. P. Kumar. 2004. Conservation of class C function of floral organ development during 300 million years of evolution from gymnosperms to angiosperms. *Plant Journal* 37: 566–577.

Zik, M., and V. F. Irish. 2003. Flower development: initiation, differentiation, and diversification. *Annual Review of Cell and Developmental Biology* 19: 119–140.

PART 2

Phylogeny of Mesozoic Plants

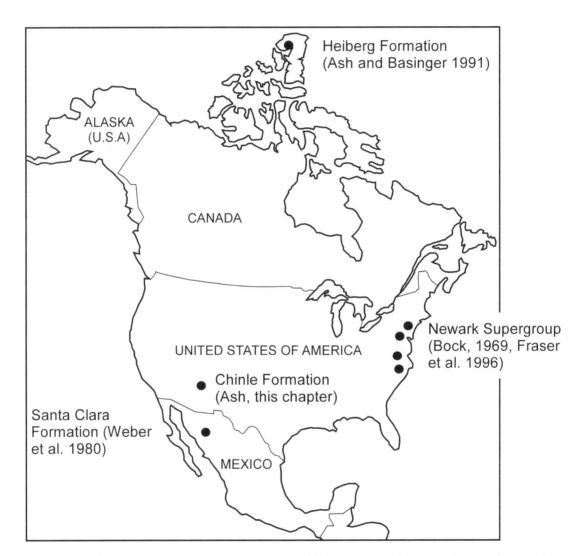

Fig. 8.1. Map of North America showing the areas that have yielded Late Triassic ginkgoaleans. Sketches of some of the species from each area are given in Figure 8.2. Details on each area are given in Table 8.1 and the text.

LATE TRIASSIC GINKGOALEANS OF NORTH AMERICA

8

Sidney R. Ash

Ginkgoaleans of Late Triassic age are uncommon in North America and are known from only three widely separated parts of the continent. The productive areas are situated along the mid-Atlantic seaboard, in the Arctic Archipelago of Canada, and in the southwestern United States and adjacent parts of northwestern Mexico. Fossils that have been attributed to the Ginkgoales range from unmistakable ginkgoalean leaves to some fossils that are virtually unidentifiable or clearly misidentified. In this chapter, all of these occurrences are briefly reviewed and a new species of *Ginkgoites*, *G. watsoniae*, is described from the Late Triassic Chinle Formation in Petrified Forest National Park, Arizona.

Although the records of most major groups of nonangiospermous land plants of Late Triassic age are well documented in North America, that of the Ginkgoales is not (Ash 1980). Those ginkgoalean fossils that have been described are remains of leaves that come from three widely spaced parts of the continent (Fig. 8.1). A survey of all published records (Table 8.1) indicates that only about 50 ginkgoalean leaves have been collected and examined in North America. The apparent scarcity of ginkgoalean fossils is surprising because moderately large Late Triassic floras have been collected and studied from Late Triassic strata at many localities on the continent (Ash 1989a), and ginkgoalean fossils are common components in Late Triassic floras in many other parts of the world, including the southern continents (Anderson and Anderson 2003), China (Li 1995), and Eurasia (Dobruskina 1994).

The Late Triassic ginkgoalean fossils known from North America range from creditable ginkgoalean leaves to some specimens that are virtually unidentifiable or misidentified (Table 8.1). In this chapter, all of the reported occurrences are briefly reviewed, and a new species of *Ginkgoites* is described from the Chinle Formation of Late Triassic age in Petrified Forest National Park, Arizona (PEFO).

Introduction

In terms of number of specimens, most of the Late Triassic ginkgoalean leaves presently known from North America were found in the Newark Supergroup in the eastern United States (Figs. 8.1, 8.2A, E) where they were first recognized by Emmons (1857) in North Carolina and described by him as *Baiera gracilis* and *Noeggerathia striata*. Subsequently, the two

Survey of North American Ginkgoaleans

Table 8.1. Summary of Ginkgoaleans Reported from the Upper Triassic Strata in North America, Compiled from Original Sources

Area	Stratigraphic Unit	Original Identifications and Author	Current Identifications and Author	Number of Specimens
Borden Clay Products Pit Gulf, North Carolina	Pekin Formation of the Newark Supergroup	*Noeggerathia striata* Emmons 1857	*Sphenobaiera striata* (Emmons) Bock 1969	6 (e)
Borden Clay Products Pit Gulf, North Carolina	Pekin Formation of the Newark Supergroup	*Baiera gracilis* Emmons 1857	*Sphenobaiera striata* Bock 1969	4 (e)
Winterpock and Clover Hill, Virginia	Tuckahoe Formation of the Newark Supergroup	*Baiera multifida* Fontaine 1883	*Sphenobaiera striata* Bock 1969	5 (e)
Solite Quarry, Cascade, Virginia	Cow Branch Formation of the Newark Supergroup	*Sphenobaiera* sp. in Fraser et al. 1996	No change	1 (e)
Milford, New Jersey	Brunswick Formation of the Newark Supergroup	*Ginkgoites milfordensis* Bock 1952	No change	2
Eureka Quarry, Bucks County, Pennsylvania	Lockatong Formation of the Newark Supergroup	*Sphenobaiera striata* Bock 1969	Unidentifiable	0
Carver's Quarry, Carversville, Pennsylvania	Lockatong Formation of the Newark Supergroup	*Eoginkgoites sectoralis* Bock 1952 and *E. gigantean* Bock 1969	No change except that the genus has been transferred to the Bennettitales; see Ash 1976	0
Ellesmere Island, Arctic Archipelago, Canada	Fosheim Member of the Heiberg Formation	*Ginkgo sibirica* Heer 1876 in Ash and Basinger 1991	No change	6
Ellesmere Island, Arctic Archipelago, Canada	Fosheim Member of Heiberg Formation	*Sphenobaiera spectabilis* (Nathorst) Florin 1936 in Ash and Basinger 1991	No change	5
San Javier, Sonora, Mexico	Santa Clara Formation	*Jeanpaulia radiata* Newberry 1876	*Baiera radiata* (Newberry) Weber et al. 1980	1
La Cuesta-Buenavista, Sonora, Mexico	Santa Clara Formation	*Sphenobaiera* sp. Weber et al. 1980	*Sphenobaiera spectabilis* (Nathorst) Florin 1936; see this chapter	2
Fort Wingate, New Mexico	Monitor Butte Member	*Baiera* sp. Ash 1967	Uncertain; see this chapter	6

Note. The number of specimens of each species is based on the actual number of specimens known to me or are estimates based on the literature (e = estimated).

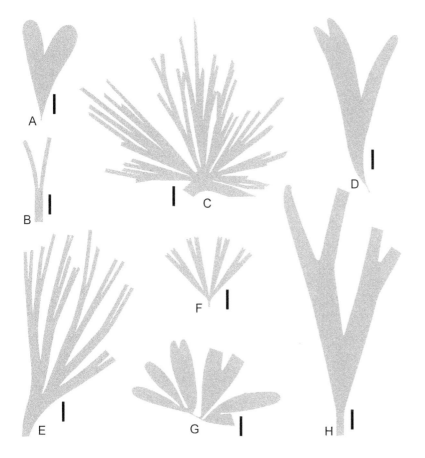

species were attributed to *B. munsteriana* and *B. multifida*, respectively, by Fontaine (1883) and later by Ward (1905). Fontaine (1883) also reported the discovery of more ginkgoalean leaves in the Newark Supergroup at two localities in Virginia (Fig. 8.2E). He assigned them to *Baiera multifida*, but many years later, they were attributed to *Sphenobaiera striata* by Bock (1969). In 1952, Bock described two small fan-shaped leaves from the Newark Supergroup in New Jersey as *Ginkgoites milfordensis* (Fig. 8.2A) and some unusually large pinnate leaves composed of several large wedge-shaped lobes from a locality in Pennsylvania as a new genus and species, *Eoginkgoites sectoralis* Bock. A few years later, Bock (1969) assigned another new species from the same locality in Pennsylvania to *E. gigantea*. Cuticular studies of the same types of leaves from other Late Triassic localities in the southwestern United States have clearly demonstrated that the fossils Bock (1952, 1969) referred to *Eoginkgoites* are actually unusual bennettitalean leaves (Ash 1976). In 1969, Bock combined *Baiera gracilis* Emmons and *Noeggerathia striata* Emmons and redescribed them together with some scrappy material he had collected from the Newark Supergroup in Virginia and Pennsylvania as a new combination, *Sphenobaiera striata* (Emmons) Bock. Unfortunately, some of the new specimens are so poorly preserved and/or incomplete that they cannot be attributed to any species, much less classified as a ginkgoalean.

An unidentified species of *Sphenobaiera* has been reported more recently by Fraser et al. (1996) from the Cow Branch Formation of the Newark Supergroup in southern Virginia. It is associated with a generally typical Newark flora that includes *Metreophyllum* Anderson et Anderson (1983), a rather large linear leaf that has been tentatively assigned to the Ginkgoales. Previously it was known only from the Permian of South Africa. These fossils are not discussed further here because neither has been illustrated or clearly described.

Probably the best-documented and best-preserved Late Triassic ginkgoalean leaves from North America occur in the Canadian High Arctic on Ellesmere Island and were described by Ash and Basinger (1991) (Fig. 8.1). However, the age of these fossils is uncertain because they occur in the Fosheim Member of the Heiberg Formation, which straddles the Triassic–Jurassic boundary. Nevertheless, they are included here because the associated flora is Late Triassic in aspect. These ginkgoalean leaves, which show the venation clearly and have cuticles, were attributed to *Ginkgo sibirica* Heer 1876 (Fig. 8.2G) and to *S. spectabilis* (Nathorst) Florin 1936 (Fig. 8.2H) by Ash and Basinger (1991).

Well-preserved impressions of a few ginkgoalean leaves have been described from two localities in the Late Triassic Santa Clara Formation in the state of Sonora in northwestern Mexico. They include a large, finely divided, wedge-shaped leaf that Newberry (1876) attributed to *Jeanpaulia radiata* Newberry 1876 but which Bock (1969) referred to *S. striata* and Weber et al. (1980) later assigned to *Baiera radiata* (Fig. 8.2C). Two additional impressions of ginkgoalean leaves have been described from the same general area in Mexico by Weber et al. (1980) as *Sphenobaiera* sp. In my opinion, because these fossils seem to closely resemble the common ginkgoalean leaf *S. spectabilis* (Nathorst) Florin (Fig. 8.2D), they should be referred to this species.

The first fossil attributed to the Ginkgoales from the Late Triassic of the southwestern United States was described by Daugherty in 1941 from the Chinle Formation in Petrified Forest National Park, Arizona (Figs. 8.1, 8.3) from the mudstone facies of the Newspaper Rock Bed of Parker (2006). Unfortunately, the fossil that Daugherty named *Baiera arizonica* is poorly preserved and incomplete, and it lacks key diagnostic features of any known ginkgoalean. A recent reexamination of the fossil indicates that it is probably the remains of the enigmatic leaf *Sanmiguelia* Brown (1956), as was suggested to me some time ago by Chester Arnold (personal communication, 1974). Now that specimens of undoubted *S. lewisii* leaves have been identified at closely adjacent sites in the Petrified Forest, such an identification is more attractive than when it was suggested by Arnold. Fragments of a deeply dissected, fan-shaped leaf of clear ginkgoalean affinity were discovered in the Chinle Formation a few years ago, but its description was delayed in hopes that more material would be found. Because that has not happened, the fossil is described here as a new species, *Ginkgoites watsoniae* (Fig. 8.2F). A petrified log found in the city of Holbrook, Arizona, a few miles west of the Petrified

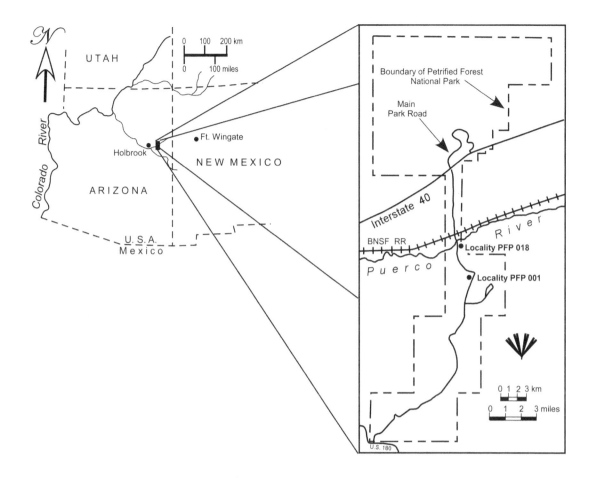

Forest (Fig. 8.3), in the Shinarump Member of the Chinle Formation was attributed to the order Ginkgoales by Savidge (2006). He named it *Ginkgoxylpropinquus hewardii* on the basis of its anatomy. A possible ginkgoalean leaf was identified from the Chinle Formation in the Fort Wingate area of New Mexico (Figs. 8.1, 8.2B) as *Baiera* sp. by Ash (1967, 1989b). This species is represented by several linear leaves with a long, narrow apical notch. However, the leaf base and apex of the specimens are not preserved, and these fossils may well represent linear leaves that accidentally split before burial

Fig. 8.3. Index map of a part of the Four Corners area in the southwestern United States showing the location of Petrified Forest National Park in east-central Arizona and a detailed map showing the localities in the central part of the park that are discussed in this chapter.

Locality

The fossils attributed to the new *Ginkgoites* were collected from a locality in the same bed of massive greenish mudstone in the Blue Mesa Member of the Chinle Formation (Fig. 8.4) that has yielded most of the plant compressions that have been described from Petrified Forest National Park in the past (e.g., Daugherty 1941; Ash 1970, 1980, 2005). This mudstone unit is now acknowledged by most workers to be the lateral equivalent of a prominent layer of hard, brown-weathering sandstone that had been called the Newspaper Sandstone Bed for many years (Cooley 1959). Both

The New Species of *Ginkgoites*

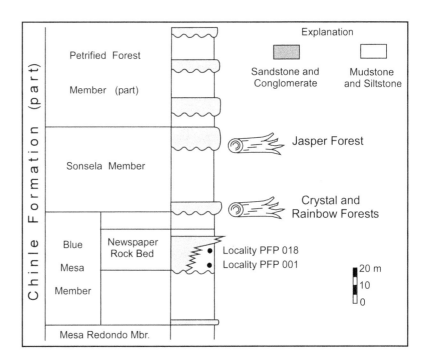

Fig. 8.4. Composite stratigraphic section of the Chinle Formation in the southern part of Petrified Forest National Park, Arizona. The relative position of localities PFP 001 and PFP 018 and the major concentrations of petrified wood (= petrified forests) in this part of the park are shown.

lithologic units measure up to about 16 m in thickness and are exposed on both sides of the main road south of the Puerco River in the central part of the park (Fig. 8.3). Stratigraphic studies by Demko (1995), Woody (2006), and Parker (2006) indicated that the bed of brown-weathering sandstone represents a shallow, meandering stream and that the fossil-bearing mudstone unit is an overbank deposit derived from that stream. These two lithologic units intertongue with each other and with a thinner overlying bed of reddish pedogenic siltstone about 5 m thick. Because the three units are genetically related, Parker (2006) proposes treating them as a single unit that he calls the Newspaper Rock Bed. Furthermore, he recommends that the individual lithologic units be considered facies of the Newspaper Rock Bed.

The many plant fossils found in the mudstone facies of the Newspaper Rock Bed are mainly the remains of the riparian flora that inhabited the banks of the stream that deposited the sandstone facies of the Newspaper Rock Bed. Probably the most important leaf locality in Petrified Forest National Park is locality PFP 001, which contained most of the plant remains described by Daugherty in 1941 and revealed that the Chinle Formation contained a large and distinctive flora. This locality is near the southern end of the outcrops of the mudstone facies (Fig. 8.3). Since the discovery of Daugherty's locality, more than a dozen more have been found in the mudstone facies. The locality with the new *Ginkgoites* is near the northern extent of exposures of the unit, just south of the Puerco River, and has been assigned the number PFP 018 in the records of Petrified Forest National Park.

A small compression flora is associated with the new *Ginkgoites* at locality PFP 018. This flora is dominated by the ferns and fern allies

and includes the stems of *Equisetites bradyi* Daugherty, leaves of the ferns *Cynepteris lasiophora* Ash, *Phlebopteris smithii* Daugherty emend. Arnold, and an undescribed filmy fern. It also contains the leaves of the probable pteridosperm *Sphenopteris arizonica* Daugherty emend. Ash, the leaves of several gymnosperms including *Zamites powellii* Fontaine (in Fontaine and Knowlton), the leafy shoot of *Czekanowskia* sp., the fronds of *Marcouia neuropteroides* (Daugherty) Ash, and *Podozamites* sp., as well as the leafy shoot, samara, and pollen cone of *Dinophyton spinosus* Ash. A large number of narrow, unidentifiable broken stems and fragments of linear leaves of various sizes bearing parallel venation also occur at this locality. These fossil leaves may represent *Pelourdea* and/or *Pseudotorellia*. The strata also contain a large palynoflora, which has been investigated by Litwin et al. (1992), who demonstrate that the Blue Mesa Member of the Chinle Formation is Late Triassic (Carnian stage) in age.

Fig. 8.5. *Ginkgoites watsoniae* sp. nov. (A) The largest and most complete specimen. The apical region of the segment marked with an "x" is shown at a higher magnification in Figure 8.5B. The remains of a second specimen of the new species are present at the lower right. Scale bar = 10 mm. PEFO 34353. (B) An enlargement of the segment marked with an "x" in (A) showing the venation in the ultimate lobes. PEFO 34353. Scale bar = 5 mm.

Materials and Methods

The new species described here is represented by one nearly complete dissected fan-shaped leaf (Fig. 8.5A) and several fragments of segments. These fossils are preserved as compressions on the surface of a soft, greenish mudstone. The compressions consist of a thin translucent layer of organic matter that is light gold in color and presumably represents the lower cuticle because it bears the incomplete remains of stomata in places. Scattered long simple hairs, small resin bodies, and veins are also visible at places in this layer. The underlying impressions of the leaf on the rock surface are too faint to clearly show any features.

Initially, the fossils were studied in reflected light with a dissecting microscope. Later, a few small fragments of the rock were removed with dissecting needles and made into cellulose acetate transfers using the techniques previously applied so successfully to some of the fern leaves and other fossils in the Chinle flora by Ash (1970, 1975). The resulting preparations were mounted in glycerine jelly on standard glass microscope slides and examined with a compound microscope. These preparations allowed for a more careful study and accurate measurement of the structures in the organic layer than would have been otherwise possible.

Systematic Paleobotany

Division Ginkgophyta Bessey 1907
Order Ginkgoales Engler in Engler and Prantl 1897
Genus *Ginkgoites* Seward emend. Watson et al. 1999

The nomenclature of fossil ginkgoalean leaves has fluctuated widely during the last century as scientific understanding of the Mesozoic vegetation has improved. This history has been summarized recently by Watson et al. (1999), who recommended a "return to the use of *Ginkgoites* Seward" for sparse and/or fragmentary vegetative material, such as that described herein, that is clearly of ginkgoalean affinity.

GINKGOITES WATSONIAE SP. NOV.
FIGS. 8.5, 8.6

Holotype. Specimen PEFO 34353 is proposed here as the holotype (Fig. 8.5A, B).

Paratypes. Specimens PEFO 34354 and 34355 are proposed here as paratypes.

Disposition of specimens. All specimens are deposited in the collections of Petrified Forest National Park, Arizona, U.S.A.

Locality. PFP 018.

Stratigraphy. Mudstone facies of the Newspaper Rock Bed in the Blue Mesa Member of the Chinle Formation.

Age. Late Triassic (Late Carnian Stage).

Etymology. The specific epithet honors Professor Joan Watson of The University, Manchester, U.K., for her careful work on fossil Ginkgoales and other gymnosperms in the English Wealden floras (Watson et al. 1999, 2001).

Diagnosis. Entire lamina fan shaped, radius about 40 mm, basal angle about 80 degrees, more or less equally and deeply divided into four primary lobes, central division deeper than others. Primary lobes wedge shaped (width increasing distally), basal angles 12 to 20 degrees, lateral margins smooth, entire, divided once into apically notched secondary lobes by shallow notches. Primary lobes about 3 mm wide near base, gradually widening to a maximum width of nearly 5 mm just below bifurcation. Secondary lobes about 2.5 mm wide just above bifurcation, widening gradually to 3.0 mm immediately below apical notch and narrowing slightly near apex. Apical notches short, ranging from 1.5 to 4.0 mm deep, about 1.0 mm wide at top, sides gradually narrowing to rounded floor,

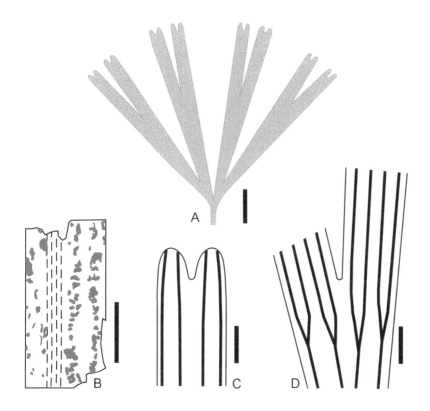

Fig. 8.6. *Ginkgoites watsoniae* sp. nov. (A) Reconstruction based on the most complete specimen. The leaf base and petiole are hypothetical. PEFO 34353. Scale bar = 10 mm. (B) Fragment of a naturally macerated pinna showing the course of a vein (dashes) and resin bodies (irregular black figures) on either side of the vein. The particles are more common on the right side of the vein than on the left side. PEFO 34354. Scale bar = 200 μm. (C) Reconstruction of the venation and morphology of the apical area of a ultimate segment of the leaf based on the most complete specimen. PEFO 34353. Scale bar = 2 mm. (D) Reconstruction of the venation and morphology of the area where two ultimate segments separate based on the most complete specimen. PEFO 34353. Scale bar = 2 mm.

ultimate lobes on either side of notch about 1.2 mm wide at base, apices rounded. Veins inconspicuous, narrow (about 80–100 μm wide), straight, typically 400 to 600 μm apart, about two per millimeter, converging slightly near apex of ultimate lobes, ending in apical margins of ultimate lobes on either side of apical notch, typically four veins per primary and secondary lobes except where veins divide, vein dichotomies occurring typically just below bifurcations of primary and secondary lobes, usually two veins in ultimate lobes. Resin bodies numerous, oval to irregularly rectangular, ranging up to about 25 μm wide and 35 μm long, sporadically distributed between veins in longitudinal files.

Discussion. The fossil described here is noteworthy because, as reported above, ginkgoalean fossils are comparatively rare in Late Triassic rocks of North America, especially in the southwestern United States. The presence of this leaf and the log of *Ginkgoxylpropinquus hewardii* Savidge in the Chinle Formation confirms the presence of the Ginkgoales in the southwestern United States during the Late Triassic, although they appear to have constituted only a minor part of the land flora at that time. However, it is possible that their scarcity in this flora is partially the result of the plants inhabiting mesic areas some distance from watercourses, which would mean a lesser chance of fossilization in comparison to the horsetails and ferns that are so common in the flora.

Because the basal region of the fan-shaped leaf described here is missing, it is not possible to determine whether the leaf was apetiolate, as in *Sphenobaiera*, or petiolate, as in other ginkgoaleans such as *Ginkgoites* and *Baiera*. In the absence of this knowledge, a search of the literature was made to determine which previously described fossil species might have other features in common with the new species. It was found that the Chinle leaf is smaller and more delicate than most leaves assigned to *Sphenobaiera* and that the ultimate segments in *G. watsoniae* have two veins, while those in *Sphenobaiera* usually have only one vein. Although some of the leaves previously assigned to *Baiera* are delicate and have narrow lobes, they contain only a maximum of four veins (see Harris and Millington 1974), in contrast to those of *G. watsoniae*, which have as many as eight (Fig. 8.6D) where the veins divide. This leaf falls within the general size range of many of the fossil leaves assigned to *Ginkgoites*, and the lobes in some of them are similarly narrow and delicate. It is especially significant that the penultimate lobes on *Ginkgoites* often have a narrow shallow notch similar to the notches on the penultimate lobes of the new species. Also, the ultimate lobes in the two genera contain a pair of veins. Because the fossil described here seems to compare more closely with some species of *Ginkgo* and *Ginkgoites* than any other genera, it is assigned here to the latter genus.

Comparisons. Although the base and petiole are missing in the ginkgoalean leaf described here, its small size and other features differentiate the fossil from most species in this order. For example, there are no leaves in the Chinle Formation or in any of the other Triassic units in the southwestern United States that come close to resembling *G. watsoniae*. The only fan-shaped leaves described from the Newark that might be confused with *G. watsoniae* are the specimens described as *Baiera multifida* by Fontaine (1883) and later transferred to *Sphenobaiera striata* by Bock (1952, 1969). These fossils do have quite narrow lobes, but they are at least four times as long as the present specimen. None of the other ginkgoalean leaves described from the Newark Supergroup in the eastern United States, the Santa Clara Formations of northern Mexico, or the Heiberg Formation in Canada is at all similar to the Chinle form described here.

To an extent, this species does compare with the Late Triassic–Early Jurassic form *Baiera muensteriana* (Presl) Heer 1876, as emended by Barnard 1967. Both have rather small, delicate, fan-shaped leaves with a dichotomously dissected lamina. However, the basal segments in *B. muensteriana* are spread at a wider angle (120 degrees to 180 degrees) than in the new species (approximately 80 degrees), and also the lamina is unevenly divided in *B. muensteriana*, in contrast to the evenly divided lamina in *G. watsoniae*. Furthermore, the ultimate lobes of the segments in *B. muensteriana* typically contain only a single vein, in contrast to those of the new species, which contain two. Some specimens of *G. minuta* (Nathorst) Harris (1935) from the Early Jurassic of east Greenland are similar in size but the lobes are more than twice as wide, and although the secondary lobes maybe notched as in the present fossil, the resulting ultimate lobes generally contain only one vein each. In contrast, the secondary lobes in *Sphenobaiera pecten* Harris and

Millington (1974) from the Middle Jurassic of Yorkshire and a few other species have similarly narrow lobes, but they are much longer and their ultimate lobes contain only a single vein. At least some of the ultimate lobes in *Ginkgoites taeniata* (Braun) Harris (1935) from the Triassic of east Greenland have apical notches and the adjacent lobes contain two veins, as in the present species. Several species in the Triassic of the southern continents have narrow lobes throughout, but they are three to four times as large as *G. watsoniae*. They include *Sphenobaiera pontifolia* Anderson et Anderson (1989) and *S. schenkii* (Feistmantal) Florin (1936). Another similar leaf from the Triassic of Australia and South Africa is *S. africana* Baldoni (1980), which also has narrow lobes; however, they are more numerous and much longer in each leaf, and only an occasional ultimate lobe is notched.

Conclusions

A review of the Late Triassic ginkgoalean fossils described from North America shows that they are relatively rare and occur only in small numbers along the mid-Atlantic seaboard, in the Arctic Archipelago of Canada, and in the southwestern United States and adjacent parts of northwestern Mexico. Most of the fossils are the remains of leaves, and although many are not very well preserved, it is evident that they represent *Ginkgoites*, *Sphenobaiera*, and *Baiera*. A petrified log attributed to the Ginkgoales has recently been described from the southwestern United States as *Ginkgoxylpropinquus hewardii*. The first unequivocal ginkgoalean leaf found in the southwestern United States is described herein as *Ginkgoites watsoniae* sp. nov. from the Late Triassic Chinle Formation in Petrified Forest National Park, Arizona.

Acknowledgments

I am grateful to the late Michele Helickson, Superintendent of Petrified Forest National Park, for authorizing me to collect the fossil described here and others found in the area under her jurisdiction and to her staff for assisting me with my work. This study was supported in part by a grant from the Petrified Forest Museum Association.

References Cited

Anderson, J. M., and H. M. Anderson. 1983. Vascular plants from the Devonian to Lower Cretaceous in Southern Africa. *Bothalia* 14: 337–344.
———. 1989. *Palaeoflora of southern Africa. Molteno Formation (Triassic). Volume 2, Gymnosperms (excluding* Dicroidium). Rotterdam: A. A. Balkema.
———. 2003. Heyday of the gymnosperms: systematics and biodiversity of the Late Triassic Molteno fructifications. *Strelitzia* 15: 1–398.
Ash, S. R. 1967. The Chinle (Upper Triassic) megaflora, Zuni Mountains, New Mexico. *New Mexico Geological Society Guidebook, 18th Annual Field Conference, Defiance-Zuni-Mount Taylor Region, Guidebook*, pp. 125–131.
———. 1970. Ferns from the Chinle Formation (Upper Triassic) in the Fort Wingate area, New Mexico. *U.S. Geological Survey Professional Paper* 613D.
———. 1975. *Zamites powellii* and its distribution in the Upper Triassic of North America. *Palaeontographica* 149, Abt. B, 139–152.

———. 1976. The systematic position of *Eoginkgoites*. *American Journal of Botany* 63: 1327–1331.

———. 1980. Upper Triassic floral zones of North America. In D. L. Dilcher and T. N. Taylor (eds.), *Biostratigraphy of Fossil Plants*, pp. 153–170. Stroudsburg: Dowden, Hutchinson and Ross.

———. 1989a. A catalog of Upper Triassic plant megafossils of the western United States through 1988. In S. G. Lucas, and A. P. Hunt (eds.), *Dawn of the Age of Dinosaurs in the American Southwest*, pp. 189–222. Albuquerque: New Mexico Museum of Natural History.

———. 1989b. The Upper Triassic Chinle flora of the Zuni Mountains, New Mexico. *New Mexico Geological Society Guidebook, 40th Field Conference, Southeastern Colorado Plateau*, pp. 225–230.

———. 2005. Synopsis of the Upper Triassic flora of Petrified Forest National Park and vicinity. In S. J. Nesbitt, W. G. Parker, and R. B. Irmis (eds.), Guidebook to the Triassic formations of the Colorado Plateau in northern Arizona. *Mesa Southwest Museum Bulletin* 9: 53–62.

Ash, S. R., and J. H. Basinger. 1991. A high latitude Upper Triassic flora from the Heiberg Formation, Sverdrup Basin, Arctic Archipelago. *Geological Survey of Canada Bulletin* 412: 101–131.

Baldoni, A. M. 1980. *Baiera africana*, una nueva especie de ginkgoal del Triasico de Sudafrica. *Ameghiniana* 17 (2): 156–162.

Barnard, P. D. W. 1967. Flora of the Shemshak formation. Part 2. Liassic plants from Shemshak and Astar. *Rivista Italiana di Palaeontologia* 73: 539–588.

Bessey, C. E. 1907. A synopsis of plant phyla. *University of Nebraska Studies* 7: 321.

Bock, W. 1952. New eastern Triassic ginkgos. *Bulletin of the Wagner Free Institute of Science* 27: 9–14.

———. 1969. The American Triassic flora and global correlations. *Geological Center Research Series* 3/4: 1–340.

Brown, R. W. 1956. Palmlike plants from the Dolores Formation (Triassic) southwestern Colorado. *U.S. Geological Survey Professional Paper* 274-H: 205–209.

Cooley, M. 1959. Triassic stratigraphy in the state line region of west-central New Mexico and east-central Arizona. *New Mexico Geological Society Guidebook 10th Field Conference, West-Central New Mexico*, pp. 66–73.

Daugherty, L. H. 1941. The Upper Triassic flora of Arizona. *Carnegie Institution of Washington Publication* 526: 1–108.

Demko, T. M. 1995. Taphonomy of fossil plants in Petrified Forest National Park, Arizona. *Fossils of Arizona* 3. *Mesa Southwest Museum* 37–51.

Dobruskina, I. A. 1994. Triassic floras of Eurasia. *Österreichische Akademie der wissenschaften Schriftenreiche der Erdwissenschaftlichen Kommissionen* 10: 1–422.

Emmons, E. 1857. *American Geology*, pt. 5. Albany: Sprague.

Engler, A., and K. Prantl. 1897. *Die natürlichen Pflanzenfamilien, Nachtrag zu Teilen* 2–4.

Florin, R. 1936. Die fossilen Ginkgophyten von Franz-Joseph-Land nebst Erörterungen über vermeintliche Cordaitales mesozoischen Alters. *Palaeontographica*, 81, 82, Abt. B, 1–173.

Fontaine, W. M. 1883. Contributions to the knowledge of the older Mesozoic flora of Virginia. *U.S. Geological Survey Monograph* 6: 1–144.

Fraser, N. C., D. A. Grimaldi, P. E. Olsen, P. E, and B. Axsmith. 1996. A Triassic lagerstätte from eastern North America. *Nature* 380: 615–619.

Harris, T. M. 1935. The fossil flora of Scoresby Sound, east Greenland. Part 4: Ginkgoales, Coniferales, Lycopodiales and isolated fructifications. *Meddelelser om Grønland* 112 (1): 1–176.

Harris, T. M., and W. Millington. 1974. Ginkgoales. In T. M. Harris, W. Millington, and José Miller. *The Yorkshire Jurassic Flora, IV*, pp. 1–150. London: British Museum (Natural History).

Heer, O. 1876. Beiträge zur Jura-Flora Ostsibiriens und des Amurlandes. *Mémoires de l'Académie Imperial Sciences de St.-Pétersbourg*, ser. 7, 25 (6): 1–122.

Li, X., ed. 1995. *Fossil Floras of China Through the Geologic Ages.* Guangzhou: Guangdong Science and Technology Press.

Litwin, R. J., A. Traverse, and S. R. Ash. 1992. Preliminary palynological zonation of the Chinle Formation, southwestern U.S.A., and its correlation with the Newark Supergroup (eastern U.S.A.). *Review of Palaeobotany and Palynology* 68: 269–287.

Newberry, J. S. 1876. Geological report. In J. N. Macomb, *Report of the Exploring Expedition from Santa Fe, New Mexico, to the Junction of the Grand and Green Rivers of the Great Colorado of the West, in 1859, under the command of Captain J. N. Macomb, Corps of Topographical Engineers*, pp. 9–118. Washington, D.C.: U.S. Army Engineers.

Parker, W. G. 2006. The stratigraphic distribution of major fossil localities in Petrified Forest National Park, Arizona. In W. G. Parker, S. R. Ash, and R. B. Irmis (eds.), *A Century of Research at Petrified Forest National Park, 1906–2006, Geology and Paleontology.* Museum of Northern Arizona Bulletin 62: 46–61.

Savidge, R. A. 2006. Xylotomic evidence for two new conifers and a ginkgo within the Late Triassic Chinle Formation of Petrified Forest National Park, Arizona, U.S.A. In W. G. Parker, S. R. Ash, and R. B. Irmis (eds.), A Century of Research at Petrified Forest National Park, 1906–2006, Geology and Paleontology. *Museum of Northern Arizona Bulletin* 62: 147–149.

Ward, L. F. 1905. Status of the Mesozoic floras of the United States (2nd paper). *U.S. Geological Survey Monograph* 48, pt. 1: 1–616.

Watson, J., S. J. Lyndon, and N. A. Harrison. 1999. Considerations of the genus *Ginkgoites* Seward and a redescription of two species from the Lower Cretaceous of Germany. *Cretaceous Research* 20: 719–734.

———. 2001. A revision of the English Wealden Flora, III: Czekanowskiales, Ginkgoales and allied conifers. *Bulletin of the Natural History Museum of London (Geology)* 57: 29–82.

Weber, R., A. Zambrano-García, and F. Amozurrutia-Silva. 1980. Nuevas contribuciones al conocimiento de la Tafoflora de la Formación Santa Clara (Triásico Tardío) de Sonora. *UNAM Revista Mexicana de Ciencias Geologia* 4: 125–137.

Woody, D. T. 2006. Revised stratigraphy of the Lower Chinle Formation (Upper Triassic) of Petrified Forest National Park, Arizona. In W. G. Parker, S. R. Ash, and R. B. Irmis (eds.), A Century of Research at Petrified Forest National Park, 1906–2006, Geology and Paleontology. *Museum of Northern Arizona Bulletin* 62: 17–45.

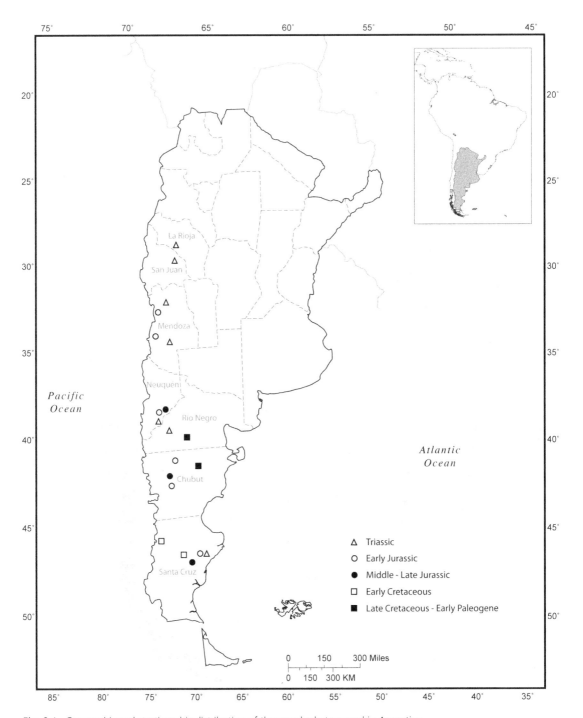

Fig. 9.1. Geographic and stratigraphic distribution of the cycadophyte record in Argentina.

REVIEW OF THE CYCADS AND BENNETTITALEANS FROM THE MESOZOIC OF ARGENTINA

9

N. Rubén Cúneo, Ignacio Escapa, Liliana Villar de Seoane, Analía Artabe, and Silvia Gnaedinger

Cycads and bennettitaleans from Argentina are thoroughly reviewed in light of their fossil record and recent findings from a geographic and stratigraphic perspective. The history of these groups is traced from their first appearance in the Triassic to the last record of cycads in the early Paleogene. Selected cycadophyte taxa have been particularly useful for interpreting their phylogenetic context and possible relationships with modern representatives. Biochronostratigraphic and paleoecological aspects are also considered that suggest that the biogeographic distribution of the cycadophytes was much farther poleward than previously thought, owing to greenhouse conditions prevailing during their time of occurrence in southern South America.

Introduction

Bennettitaleans and cycads are separate groups of seed plants that can be traced as far back as the Triassic for the bennettitaleans (Taylor et al. 2009) and probably even farther into the Late Paleozoic for the cycads (Mamay 1969; Hermsen et al. 2006). In Argentina, where living cycads are absent, both groups can definitely be recognized as early as the Triassic, throughout the Mesozoic, and into the early Paleogene. This record consists mostly of compressions and/or impressions, and, less commonly, permineralizations and mummifications. As elsewhere in the world, it is clear that these plant groups in South America were much more diverse in the past, particularly during the Jurassic and Cretaceous. As a result of the effect of greenhouse conditions during the Mesozoic and early Cenozoic, they had a large biogeographic range that extended toward the poles in both hemispheres, a distribution that strongly contrasts with the more limited range of the cycads in the tropics and subtropics today. This is the case for South America, especially Patagonia, which was located at middle to high latitudes during the late Mesozoic.

Although bennettitaleans (also known as cycadeoids) and cycads are currently considered a paraphyletic group of seed plants (Rothwell and Serbet 1994), they are commonly lumped in a systematically informal group called the cycadophytes in the paleobotanical literature (Taylor et al. 2009). We will continue to use this practical name when referring to the two groups here.

In this chapter, we offer an up-to-date review of these two groups of gymnosperms in Argentina, from the Triassic to the early Paleogene. Besides this survey of the complete fossil cycadophyte record, implications will be discussed regarding their phylogenetic significance, past distribution and diversity, and subsequent biostratigraphical importance. We will also examine their ecological niche in the Mesozoic plant communities as well as the paleoclimatological conditions that governed their presence in southern South America.

Triassic

Triassic floras in Argentina are mainly recorded from the western Cuyo region (Uspallata, Sorocayense, and Rincón Blanco Groups) and the Patagonian region (Tronquimalal Group and Vera/Paso Flores Formations in northern Patagonia, and the El Tranquilo Group in southern Patagonia) (Fig. 9.1). These floras characterize a Southwest Gondwana phytogeographic province (Artabe et al. 2003). Here, Triassic deposits are interpreted as fluvial and lacustrine sedimentary facies with a highly diverse vegetation dominated by ferns and corystosperms, but including cycads and bennettitaleans that amount to approximately 25% of the species diversity (Artabe et al. 2007). This represents a pattern that apparently extended to most of the Western Gondwana supercontinent (Anderson and Anderson 1989).

Because the preservational mode of the Triassic cycadophytes in Argentina is comprised almost exclusively of leaf compressions and impressions, with the exception of one permineralized trunk (see below), their true botanical affinities are uncertain. As a result, the use of form genera, such as *Nilssonia* Brongniart, *Pseudoctenis* Seward, and *Ctenis* Lindley et Hutton for cycads, or *Anomozamites* Schimper and *Pterophyllum* Brongniart for bennettitaleans, is common (Artabe and Stevenson 1999; Artabe et al. 2001; Leppe and Moisan 2003). As a result of a lack of sufficient diagnostic characters, Gnaedinger (1999) and Herbst and Troncoso (2000) have suggested a more informal inclusion of these taxa into the "Cycadopsida."

The genus *Pseudoctenis* (Fig. 9.2A) is morphologically the most variable taxon, with approximately 15 species (Artabe et al. 2007), while *Ctenis* and *Nilssonia* have each two species. *Taeniopteris* Brongniart (Fig. 9.2B) is another form genus that occurs profusely in Triassic floras of Argentina as impressions of at least 15 morphospecies described as "Cycadophyta incertae sedis" (Artabe et al. 2007) or "Pteridophylla incertae sedis" (Gnaedinger and Herbst 1998, 2004).

The genus *Kurtziana* Frenguelli (Fig. 9.2C), known from Late Triassic beds in Argentina, Chile, South Africa, and Australia, is a bipinnate frond of possible cycadophyte affinity (Artabe et al. 1991; Spalletti et al. 1999), although Herbst and Gnaedinger (2002) have suggest a possible pteridosperm origin. *Yabeiella* Oishi is another form genus of cycadophytes, with five species in Argentina.

Fig. 9.2. Triassic cycadophytes from Argentina. (A) *Pseudoctenis spatulata* Du Toit. (B) *Taeniopteris vittata* Brongt. (C) *Kurtziana*

Plate 1. Reconstruction of a Late Triassic forest in Greenland, showing the tufted appearance of Mesozoic forest tree crowns. *Image courtesy of Marlene Hill Donnelly and Jennifer McElwain. © by Marlene Hill Donnelly.*

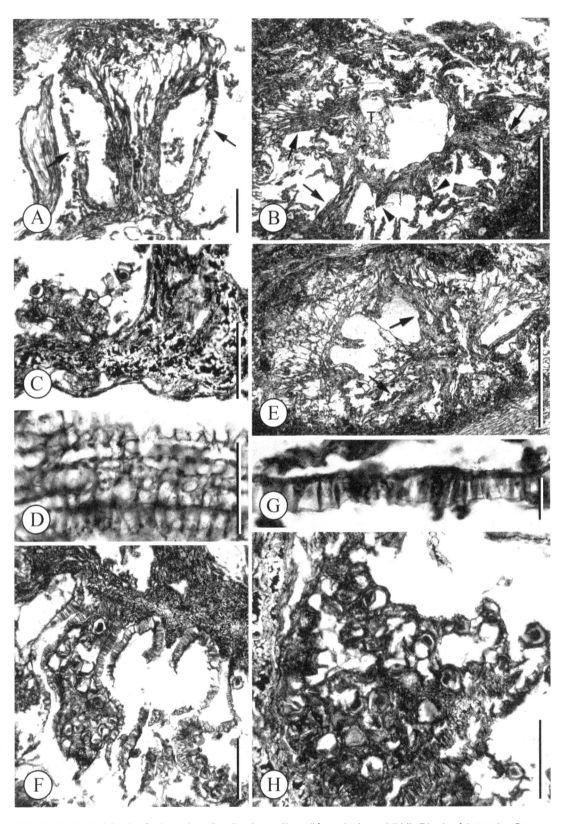

Plate 2. Anatomical details of sphenophyte *Spaciinodum collinsonii* from the lower Middle Triassic of Antarctica. For more information, please refer to the caption for Figure 2.2.

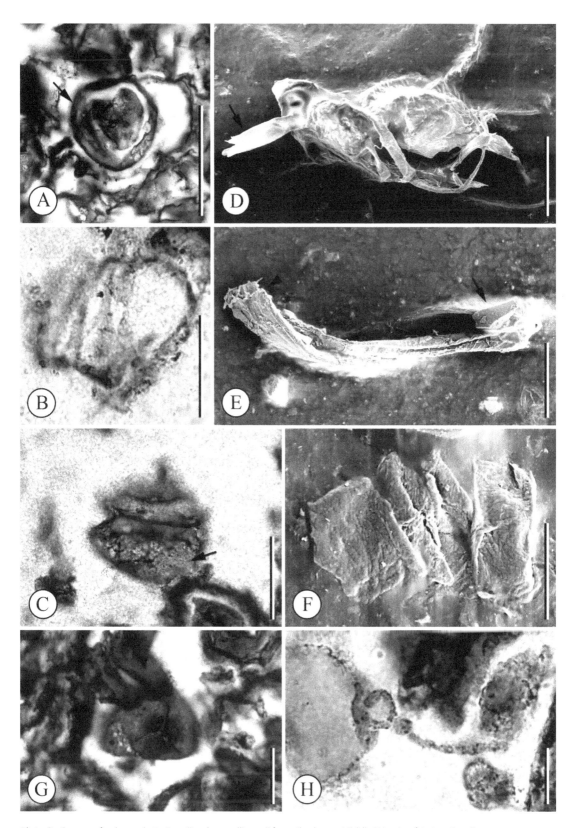

Plate 3. Spores of sphenophyte *Spaciinodum collinsonii* from the lower Middle Triassic of Antarctica. For more information, please refer to the caption for Figure 2.3.

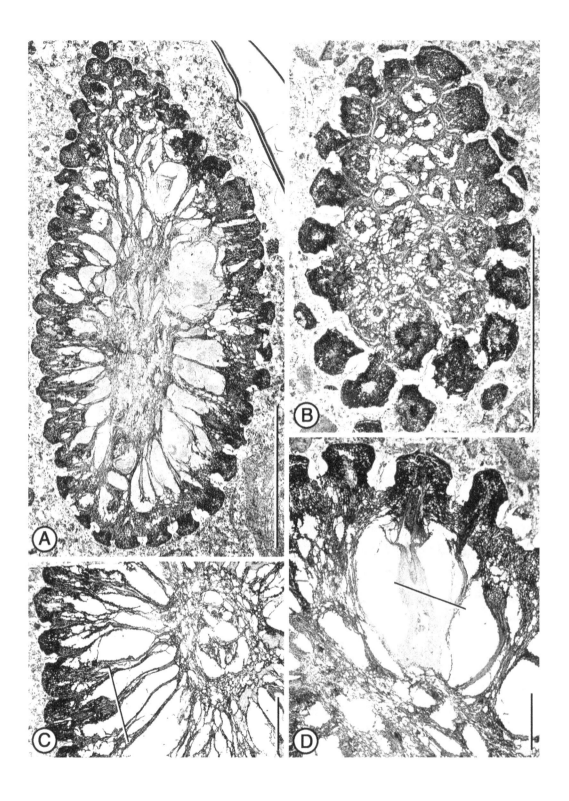

Plate 4. Sections through the ovulate cone of bennettitalean *Foxeoidea connatum* gen. et sp. nov. Holotype. For more information, please refer to the caption for Figure 4.2.

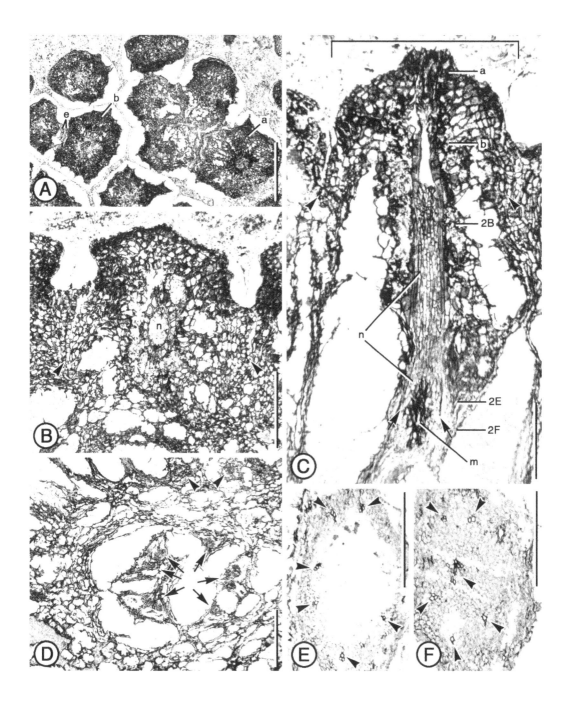

Plate 5. Additional anatomical details of the ovulate cone of *Foxeoidea connatum* gen. et sp. nov. Holotype. For more information, please refer to the caption for Figure 4.3.

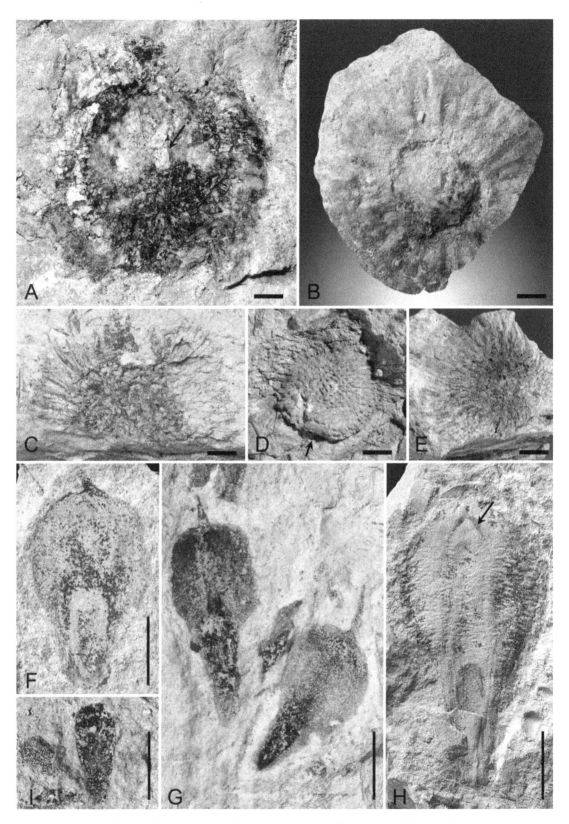

Plate 6. Compression fossils of the ovulate reproductive parts of the conifer *Araucaria delevoryasii* Gee sp. nov. For more details, please refer to the caption for Figure 5.3.

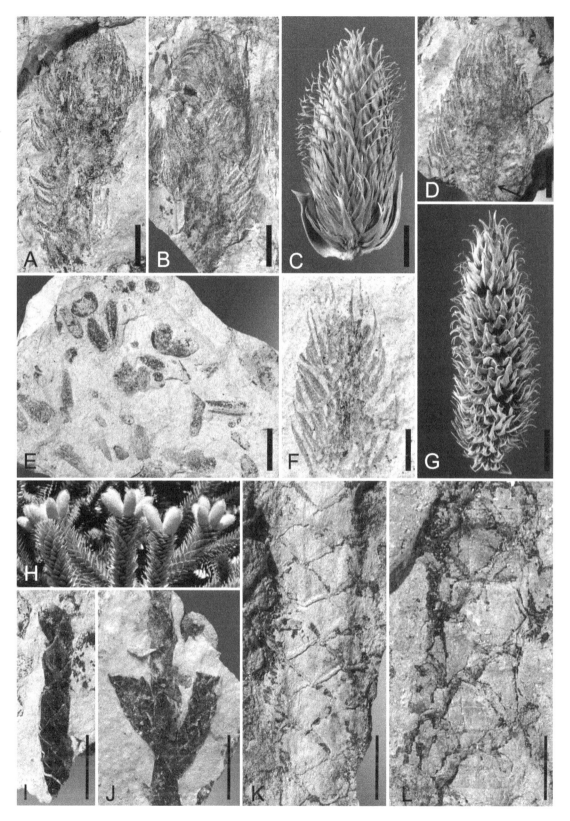

Plate 7. Compression fossils of the other plant organs of *Araucaria delevoryasii* Gee sp. nov., with living *Araucaria araucana* (Mol.) K. Koch for comparison. For more details, please refer to the caption for Figure 5.4.

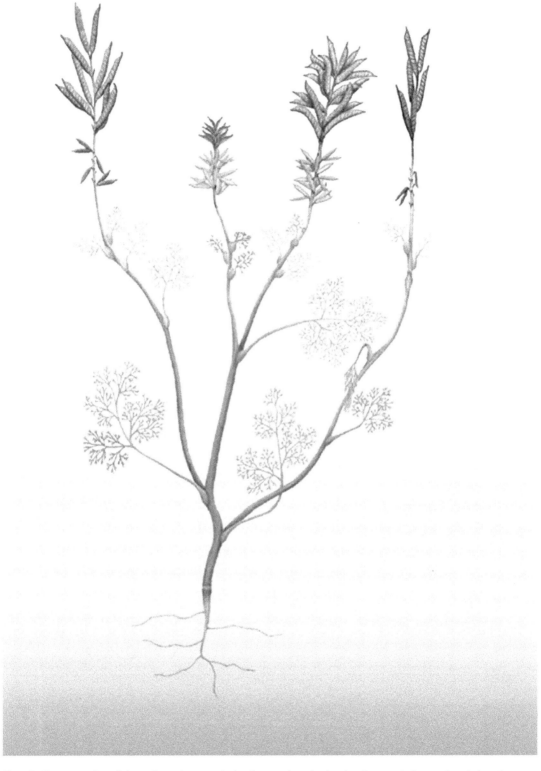

Plate 8. Reconstruction of the early angiosperm *Archaefructus sinensis,* showing the vegetative portions below the water surface and the reproductive organs just at or slightly above the water surface. *Modified by D. L. Dilcher from an illustration by K. Simons and D. Dilcher.*

Plate 9. A sampler of angiosperm flowers and fruits. For more details, please refer to the caption for Figure 6.12.

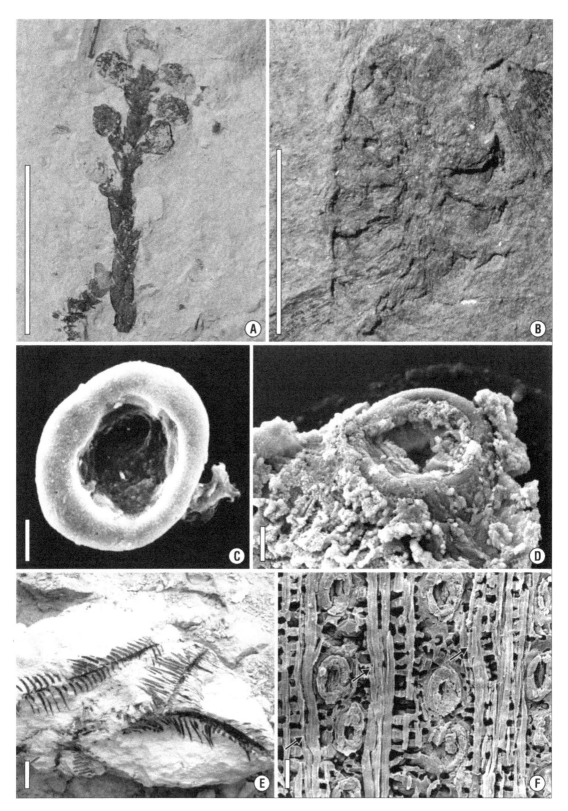

Plate 10. Pollen cones, pollen, leafy branches, and leaf cuticle of endemic Early Cretaceous conifers in Western Gondwana. For more information, please refer to the caption for Figure 11.2.

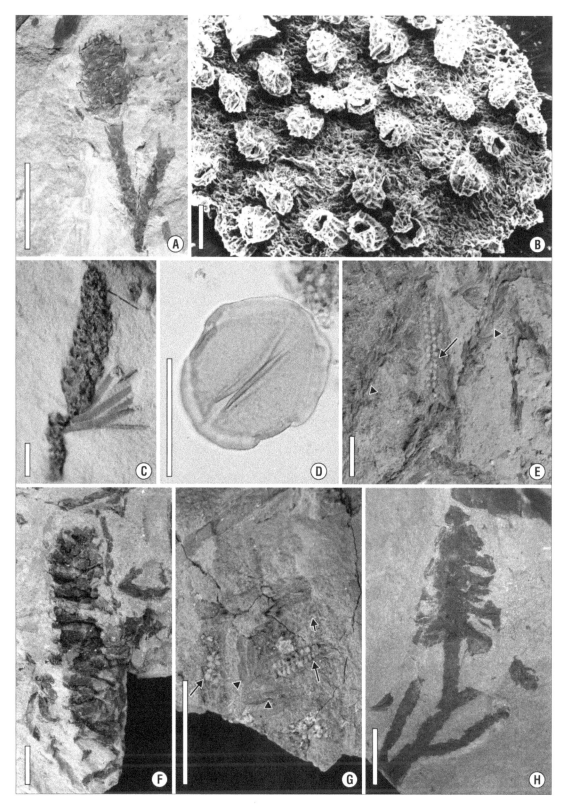

Plate 11. Seed and pollen cones, leaf cuticle, and foliage of endemic Early Cretaceous conifers in Western Gondwana. For more information, please refer to the caption for Figure 11.3.

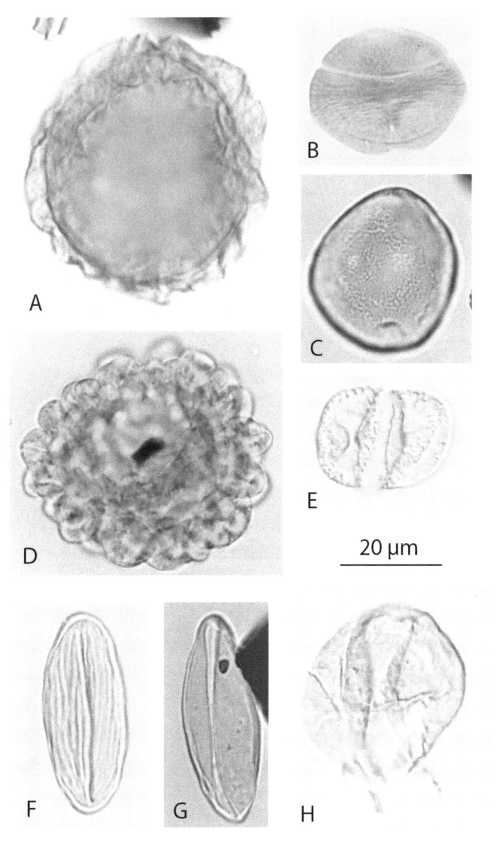

20 µm

Plate 12. Light micrographs of common palynomorphs in the Late Jurassic Morrison Formation, U.S.A. For more details, please refer to the caption for Figure 13.2.

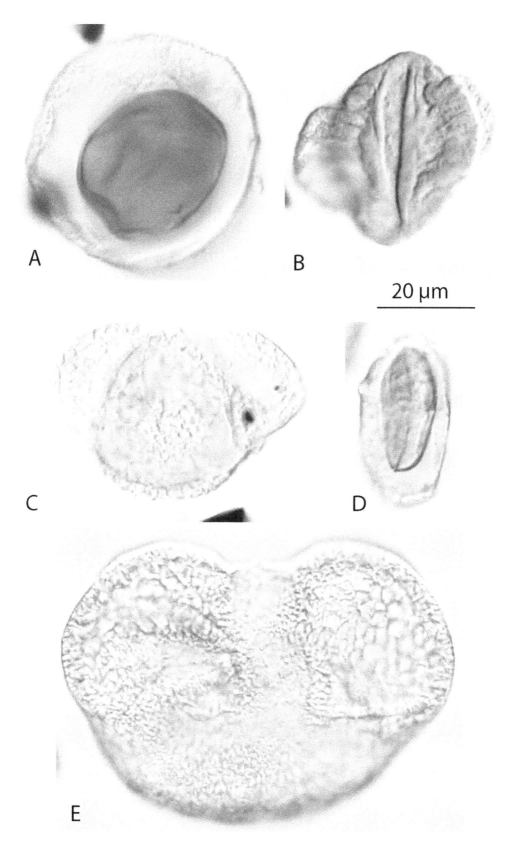

A

B

20 μm

C

D

E

Plate 13. Light micrographs of common saccate pollen in the Late Jurassic Morrison Formation, U.S.A. For more details, please refer to the caption for Figure 13.3.

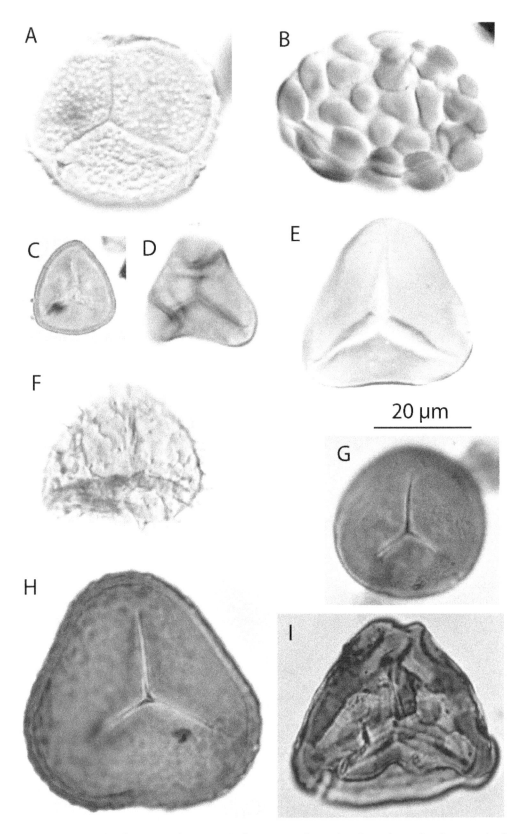

Plate 14. Light micrographs of common trilete spores in the Late Jurassic Morrison Formation, U.S.A. For more details, please refer to the caption for Figure 13.4.

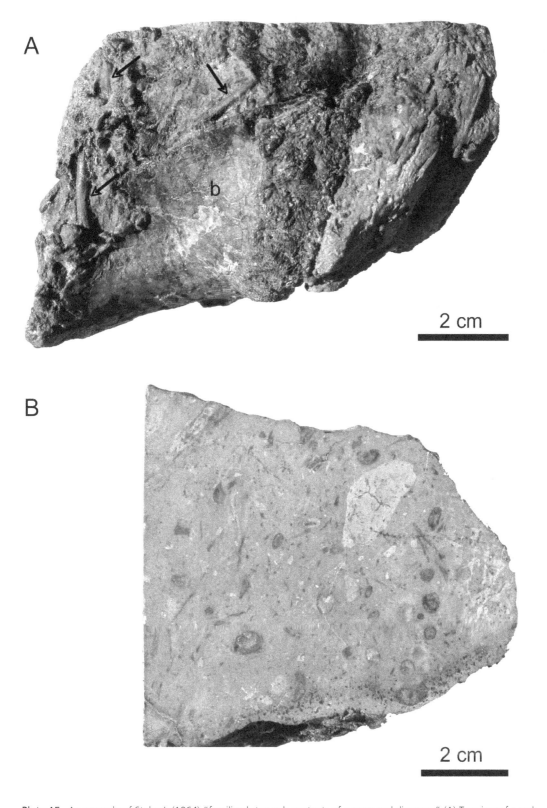

Plate 15. An example of Stokes's (1964) "fossilized stomach contents of a sauropod dinosaur." (A) Top view of a rock sample from the fossilerous deposit in Figure 14.5. Note the slender bones (arrows) and larger, flatter bone (b), all presumably of sauropod origin, on the left. (B) Lateral view of cut and polished section of the rock sample in Figure 14.6A. Note the inclusion of numerous multisized clasts in this matrix-supported calcareous mass. *Photo courtesy of S. R. Ash.*

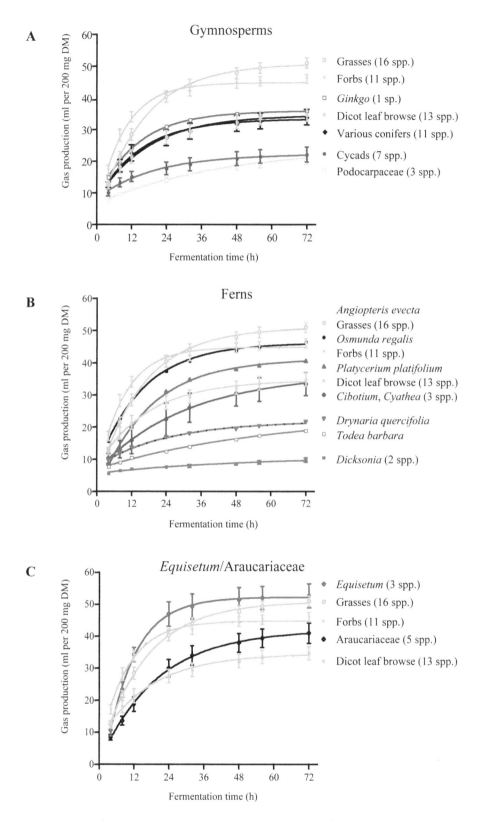

Plate 16. Fermentative behavior of potential dinosaur food plants compared to that of angiosperms (plotted in a light blue color). Gas production in the Hohenheim gas test is plotted versus fermentation time. DM = dry matter. The mean and standard error of the mean (SEM) of each data point are indicated. For more details, please refer to the caption for Figure 14.8.

Possible bennettitaleans are only represented by the leaf impressions *Pterophyllum*, with four species, and *Anomozamites*, with two species (Bonetti 1968, 1972; Artabe et al. 2007).

Finally, it is interesting to note the occurrence of permineralized cycad trunks in the Triassic of northwestern Argentina. *Michelilloa waltonii* Archangelsky et Brett (1963) is a slender stem (up to 10 cm in diameter)

brandmayri Frenguelli. (D) *Michelilloa waltonii* Archangelsky et Brett; magnification, 10×. (E, F) *Michelilloa waltonii* Archangelsky et Brett, magnification, 20×. Scale bars (A–C) = 1 cm.

without persistent foliar bases or a well-developed periderm (Fig. 9.2D, E, F). It has a wide pith, a monoxylic vascular cylinder divided by leaf gaps into wedges, and a parenchymatous cortex. The genus also has masses of long filamentous hairs on its leaf bases, which is considered a typical cycad feature.

Jurassic

Jurassic cycadophytes in Argentina are apparently represented by only bennettitaleans because true cycads have remained unrecognized or have been doubtfully recorded. Compressions and impressions of reproductive and vegetative organs tentatively assigned to the bennettitaleans have been recovered mostly from Patagonia (Fig. 9.1). Interestingly, there is a great difference in diversity between the Early and the Middle–Late Jurassic record. In the former, the bennettitaleans have a relatively broad geographic distribution, whereas in Middle–Late Jurassic floras, they are highly reduced in both abundance and diversity.

Early Jurassic floras in Patagonia occur in terrestrial or shoreline sedimentary deposits. Here, the most commonly found plant organ is the leaf genus *Otozamites* Braun in Münster, with approximately 10 species, which led Herbst (1968) and Arrondo and Petriella (1980) to call them the classical representatives of Liassic floras in Argentina. In this regard, the following taxa were recognized by Herbst (1966a) from the Piedra Pintada Formation in northwest Patagonia: *O. ameghinoi* Kurtz, *O. barthianus* Kurtz, *O. bunburyanus* var. *major* Kurtz (Fig. 9.3A), *O. simonatoi* Orlando, and two species of *Ptilophyllum* Morris, *P. cutchense* Morris and *P. acutifolium* Morris.

The bennettitalean seed cone *Williamsonia* cf. *gigas* (Bonetti 1964; Herbst and Anzotegui 1968) (Fig. 9.3B) was reported from the Early Jurassic Taquetrén locality in Chubut Province.

From the Nestares Formation in northwestern Patagonia, Arrondo and Petriella (1980) reported *O. albosaxatilis* Herbst, *O. ameghinoi*, *O. bechei* Brongniart, and *O. hislopii* (Oldham) Feist. A similar flora was recovered outside Patagonia from the El Freno Formation in Mendoza, which also included *Taeniopteris* sp., *Ptilophyllum acutifolium*, *Kurtziana brandmayri* Frenguelli, and *Williamsonia* sp. (Spalletti et al. 2007). From the nearby Los Patos Formation, Herbst (1980) described *Otozamites volkheimeri* Herbst.

The genus *Alicurana* Herbst et Gnaedinger (which formerly included some species of *Kurtziana*; Artabe et al. 1991) from the Early Jurassic Nestares Formation in northwestern Patagonia is a bipinnate frond with preserved epidermis that suggests a possible cycadalean affinity (Artabe et al. 1991; Herbst and Gnaedinger 2002).

From the Pampa de Agnia Formation in the central Patagonian Chubut province, Herbst (1966b) described *Otozamites albosaxatilis* (Fig. 9.3C), *O. hislopii*, *O.* cf. *oldhamii*, *O. chubutensis* Herbst (Fig. 9.3D), and *O. sueroi* Herbst. He noted the presence of *O. albosaxatilis* in the Roca Blanca Formation in southern Patagonia as well (Herbst 1965).

Fig. 9.3. Early Jurassic cycadophytes from Argentina. (A) *Otozamites*

Bonetti (1964) reported *Zamites* cf. *gigas* (Fig. 9.3E) and *Otozamites sanctae-crucis* (Feruglio) Archangelsky from the Early Jurassic Taquetrén flora. Herbst and Anzotegui (1968) suggested that the latter species should be assigned to *Zamites pusillus* Halle (Fig. 9.3F).

By the Middle–Late Jurassic, cycadophytes had nearly vanished from Argentina. The only genus noted from this time period is the bennettitalean form genus *Otozamites*, as represented by *O. sanctae-crucis* and

cf. *bunburyanus* De Zigno. (B) *Williamsonia* cf. *gigas* Lindley et Hutton (Carr.). (C) *O.* cf. *albosaxatilis* Herbst. (D) *Otozamites* cf. *chubutensis* Herbst. (E) *Zamites* cf. *gigas* Lindley et Hutton. (F) *Zamites pusillus* Halle. Scale bars = 1 cm.

Fig. 9.4. Early Cretaceous cycads from Argentina. (A) *Mesodescolea* cf. *obtusa* Archangelsky. Scale bar = 1 cm. (B) Cuticle of *Mesosingeria parva* Villar de Seoane. Scale bar = 50 μm. (C) *Pseudoctenis ornata* Archangelsky, Andreis, Archangelsky et Artabe. Scale bar = 1 cm. (D) *Ticoa lamellata* Archangelsky. Scale bar = 1 cm. (E) Stoma of *Ticoa lanceolata* Villar de Seoane. Scale bar = 20 μm. (F) Cuticle of *Ticoa lanceolata* Villar de Seoane. Scale bar = 50 μm.

Otozamites sp. (Baldoni 1977) from the Middle–Late Jurassic La Matilde Formation in southern Patagonia. Conversely, the species *O. traversoi* Baldoni, *O. linearis* Halle, *O. sanctae-crucis*, and *O. simonatoi* Orlando were reported from northwestern Patagonia (Baldoni 1978, 1981).

The genus *Ptilophyllum*, mentioned earlier in this chapter in association with Late Jurassic floras in Patagonia, probably does not truly occur here. *Ptilophyllum antarcticum* (Seward) Halle was reported from the Lago Argentino locality, thought to be of Late Jurassic age (Bonetti 1974). However, because the age of this flora was based on its close resemblance to the Hope Bay flora from Antarctica (Gee 1989), whose age has recently been reassigned to either the Early Jurassic (Rees and Cleal 2004) or the Middle Jurassic (Hunter 2005), the real presence of the genus in Late Jurassic floras of Argentina is therefore questionable.

In summary, it is definitely clear that, for some reason, the cycadophytes experienced a marked retraction in their range, diversity, and abundance during the Middle–Late Jurassic in Argentina. A good example is the genus *Otozamites*. Nonetheless, the group was able to recover during the Early Cretaceous.

Cretaceous

Cretaceous cycads and bennettitaleans from Argentina are known only from Patagonia (Fig. 9.1). Two different floras can be envisaged: an Early Cretaceous flora in southern Patagonia, and a Late Cretaceous flora in northern Patagonia. These two floras have completely different preservational modes, as compressions with cuticle and impressions occur in the former and permineralizations in the latter.

Early Cretaceous Cycads and Bennettitaleans in Southern Patagonia

Both groups of cycadophytes have been consistently recovered from the Central Deseado Massif intracratonic basin and the marginal Austral/Magellanean basin (Fig. 9.1). The main record comes from the Barremian–Aptian Baqueró Group, which probably preserves the richest Early Cretaceous flora in the Southern Hemisphere. Cycads and bennettitaleans, along with conifers, represent the most diverse and abundant plant groups in these sediments. In many cases, even mummified cuticles and epidermal features have been found.

Cycads in the Baqueró Group have been assigned to five genera of pinnate leaves (Figs. 9.4, 9.5), i.e., *Almargemia* Florin, *Mesodescolea* Archangelsky (Fig. 9.4A), *Mesosingeria* Archangelsky (Fig. 9.4B), *Pseudoctenis* (Fig. 9.4C), and *Ticoa* Archangelsky (Fig. 9.4D), as well as to two genera with lanceolate and entire laminae, *Nilssonia* and *Sueria* Menéndez. These leaf genera share xeric traits such as trichomes on both leaf surfaces, sunken stomata protected by large or small epistomatal chambers (Fig. 9.4E), and thick cuticular membranes. Similar features are also observed in the microsporophyll cuticles of the cycad

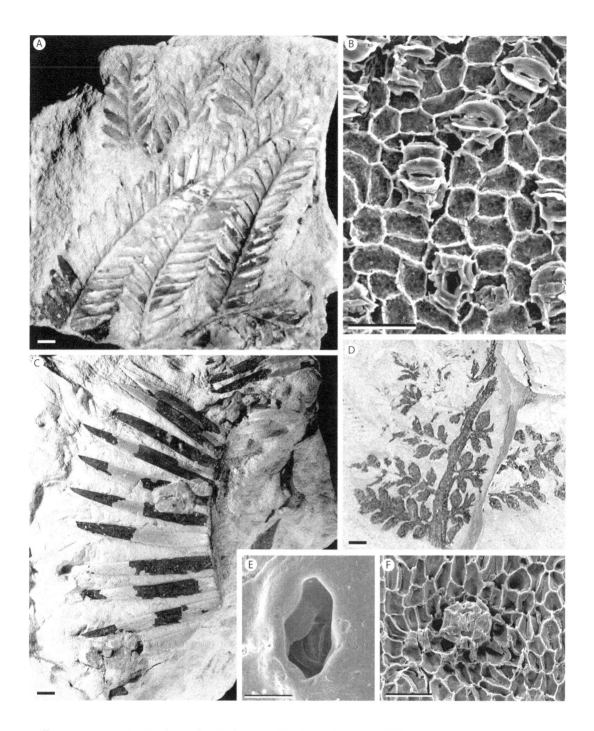

pollen organ species *Androstrobus* Schimper (Archangelsky and Villar de Seoane 2004).

Pseudoctenis and *Ticoa* have the most ornamented epidermis, with *P. crassa* Archangelsky et Baldoni bearing the most varied types of trichomes. *Mesosingeria oblonga* Villar de Seoane has the deepest stomata, with epistomatal chambers up to 18 μm. *Almargemia incrassata* Archangelsky has monocyclic to imperfectly dicyclic abaxial stomata and sparse

Fig. 9.5. Early Cretaceous
cycads and bennettitale-
ans from Argentina. (A)
Pseudoctenis giganteus
Archangelsky and *Ptilophyllum* sp. Scale bar = 10 cm.
(B) *Williamsonia bulbiformis*
Menéndez. Scale bar = 1
cm. (C) *Cycadolepis oblonga*
Menéndez. Scale bar = 1
cm. (D) Stoma of *Otozamites
ornatus* Villar de Seoane.
Scale bar = 10 µm. (E) Cuticle of *O. ornatus*. Scale bar
= 1 mm.

papillae on both leaf surfaces (Archangelsky 1966), while *Mesodescolea plicata* Archangelsky has monocyclic stomata and trichome bases on its abaxial epidermis (Archangelsky 1963a).

The endemic genus *Mesosingeria* contains seven species: *M. coriacea* Archangelsky, *M. herbstii* Archangelsky, *M. mucronata* Archangelsky, *M. obtusa* Archangelsky, *M. striata* Archangelsky, *M. parva* Villar de Seoane, and *M. oblonga* Villar de Seoane (Archangelsky 1963b; Villar de Seoane 1997, 2005). They differentiate from one another by their cuticular characters.

Mesodescolea is a leaf genus that compares with the living cycad *Stangeria* T. Moore in its cuticular features (Artabe and Archangelsky 1992)

Pseudoctenis is another common genus of pinnate leaves with three species defined by their epidermal features: *P. crassa*, *P. dentata* Archangelsky et Baldoni (Archangelsky and Baldoni 1972; Artabe 1994), and *P. ornata* A. Archangelsky, Andreis, S. Archangelsky et Artabe (Fig. 9.4C) (Archangelsky et al. 1995). *Pseudoctenis giganteus* A. Archangelsky (Fig. 9.5A) has been assigned to impressions of huge leaves (Archangelsky 1997), while *P. ensiformis* Halle is known from the Kachaike Formation (Longobucco et al. 1985).

Ticoa is another endemic genus with at least five species that are principally distinguished by cuticular characters, such as stomata distribution and density, and trichome types (Fig. 9.4E, F). *Ticoa harrisii* Archangelsky, *T. magnipinnulata* Archangelsky, *T. lamellata* Archangelsky, and *T. lanceolata* Archangelsky were described from the lower Baqueró Group (Archangelsky 1963a, 1966, 1976; Villar de Seoane 2005), while *T. magallanica* Archangelsky is known from the Springhill Formation, Austral basin.

Entire leaves of *Nilssonia and Sueria* were also described from the Baqueró Group. *Nilssonia clarkii* Berry is a large, coriaceous leaf (Berry 1924). *Sueria rectinervis* Menéndez has a cuticle with hairs and fewer adaxial stomata but more abundant abaxial ones (Menéndez 1965; Artabe 1994), while *Sueria elegans* Villar de Seoane has some hairs and abundant abaxial stomata (Villar de Seoane 1997). According to Hermsen et al. (2007), the presence of perforated external cells and accessory cell coronas in *Sueria* is unique to the Cycadales.

Ultrastructural studies of the cuticle of five of the seven genera of cycad leaves from the Early Cretaceous Baqueró flora show that the cutinized layer is the thickest part of the epidermal wall (Villar de Seoane 2005). This cuticle is divided into an outer layer without cellulose and an inner layer with cellulose and pectinaceous microchannels. Transmission electronic microscopic studies have shown the following similarities: *Mesodescolea*, *Mesosingeria*, and *Ticoa* have a granulate outer layer, whereas in *Pseudoctenis* it is lamellate and in *Sueria* compact. *Mesosingeria*, *Sueria*, and *Ticoa* have an alveolate inner layer, whereas it is granulate in *Mesodescolea* and *Pseudoctenis*. Interestingly, the layers differ in thickness, with *Sueria elegans* having the thinnest cuticular membrane (2.5 µm) and *Mesosingeria oblonga* the thickest (13.5 µm).

Three different species of *Androstrobus* (cycad microsporophylls with in situ pollen grains) were described from the Anfiteatro de Ticó Formation of the Lower Baqueró Group (Archangelsky and Villar de Seoane 2004). These microsporophylls are rhomboidal and distally truncate, with a thick cuticle in which the typical (actinocytic) stomatal apparatus are embedded. Microsporangia bear pollen masses stuck to their internal walls. *Androstrobus munku* Archangelsky et Villar de Seoane has monocolpate pollen, which is ellipsoidal to fusiform in polar view, 39 µm long and 28 µm wide, and possesses a smooth to scabrate exine. *Androstrobus patagonicus* Archangelsky et Villar de Seoane has monocolpate pollen grains, which are 34 µm long and 28 µm wide, are circular to subcircular in polar view, and have a smooth exine with a basal laminated layer, an alveolate ectexine, and abundant orbicules. *Androstrobus rayen* Archangelsky et Villar de Seoane has monocolpate pollen, which is ellipsoidal to circular in outline in polar view, 25 µm long and 20 µm wide, and has a scabrate exine, an alveolate–tectate ectexine, and abundant orbicules.

Bennettitaleans in the Early Cretaceous of Argentina are represented by five leaf genera (*Dictyozamites* Medlicott et Blanford, *Otozamites*, *Pterophyllum*, *Ptilophyllum* [Fig. 9.5A], and *Zamites* Brongniart), one genus of scale leaves (*Cycadolepis* Saporta), and one genus of seed cones (*Williamsonia* Carruthers, Fig. 9.5B, C), all of which are recorded from both the Baqueró Group, and the Springhill and Kachaike formations.

There is a large variety of leaf species within each genus that are mostly differentiated from one another by the size of their pinnae, the different protection structures associated with their sunken stomata, and the trichome and papillae types present on their abaxial epidermis (Fig. 9.5D, E). They have in common a falcate shape to their pinnae and paracytic stomata on the leaf epidermis.

Dictyozamites is the only genus with reticulate venation. On the basis of cuticular features, four species were described: *D. areolatus* Archangelsky et Baldoni, *D. crassinervis* Menéndez, *D. latifolius* Menéndez, and *D. minusculus* Menéndez (Archangelsky and Baldoni 1972; Menéndez 1966).

Otozamites is the most diverse leaf genus with seven species: *O. archangelskyi* Baldoni et Taylor (1983), *O. grandis* Menéndez, *O. patagonicus* Villar de Seoane (1995), *O. parvus* Villar de Seoane (2001), *O. parviauriculata* Menéndez (1966), *O. ornatus* Villar de Seoane (1999), and *O. waltonii* Archangelsky et Baldoni (1972). *Pterophyllum* is only represented by two species, *P. trichomatosum* (Archangelsky and Baldoni 1972) and *Pterophyllum* sp. (Menéndez 1966).

Ptilophyllum is represented by two impressions—*Ptilophyllum antarcticum* and *P. hislopii* (Oldham) Seward (Menéndez 1966) (Fig. 9.5A)—and four compressions—*P. longipinnatum* Menéndez (1966), *P. valvatum* Villar de Seoane (1995), *P. angustus* Baldoni et Taylor (1983), *P. ghiense* Baldoni (1977), and *P. acutifolium* (cf. Passalia 2007). *Zamites*, on the other hand, is another genus scarcely represented with only two species: *Z. decurrens* Menéndez (1966) and *Z. grandis* Archangelsky et Baldoni (1972).

Reproductive structures have been referred to two species of *Williamsonia*, *W. bulbiformis* Menéndez (Fig. 9.5B) and *W. umbonata* Menéndez (1966), and several species of *Cycadolepis* (Menéndez 1966; Baldoni 1974) (Fig. 9.5C).

Ultrastructurally, a cuticle proper, an outer layer, and an inner layer with microchannels or pathways form the external epidermal wall of bennettitalean organs, except the receptacle of *Williamsonia*. On this basis, three different groups of taxa can be delineated. The first group contains *Dictyozamites*, *Otozamites*, and *Zamites* with an outer lamellate layer, and an inner reticulate layer only in the case of *Dictyozamites*. The second contains *Otozamites* and *Zamites* with an inner alveolate layer. The third contains *Ptilophyllum* and *Pterophyllum* with an inner lamellate–reticulate layer, the former having an outer alveolate layer and the latter an outer reticulate one. The receptacle of *Williamsonia* has a cuticular membrane with a fibrous aspect, while its bracts have an outer reticulate layer and an inner lamellate layer, which is similar to *Cycadolepis* (Villar de Seoane 2003).

The Late Cretaceous to Earliest Paleogene Record

During this time span, cycads are the only cycadophytes in Argentina and are represented by 5 of the 14 genera of permineralized cycad stems known worldwide at this time. They are *Chamberlainia* Artabe, Zamuner et Stevenson, *Worsdellia* Artabe, Zamuner et Stevenson, and *Brunoa* Artabe, Zamuner et Stevenson from the Late Cretaceous (Artabe et al. 2004, 2005), *Bororoa* Petriella (1972, 1978) from the latest Cretaceous (?)–Paleocene, and *Menucoa* Petriella (1969) from the Paleocene, all recovered from northern Patagonia (Allen and Bororo Formations) (Artabe et al. 2004, 2005).

Chamberlainia pteridospermoidea Artabe, Zamuner et Stevenson (Fig. 9.6A, B) is another multilacunar polyxylic (centripetal/centrifugal) stem, which is covered by persistent rhomboidal foliar bases and smaller cataphylls in cycles. The central parenchymatous pith contains idioblasts, mucilage canals, and arcs of medullary vascular bundles. It has a vascular system with three rings of centrifugal secondary xylem and phloem, endarch or mesarch primary xylem, and centripetal secondary bundles. The parenchymatous cortex possesses extrafascicular concentric bundles. A few bulbils are present along the stem.

Worsdellia bonettii Artabe, Zamuner et Stevenson (Fig. 9.6C) is also a multilacunar polyxylic stem covered by alternating cycles of cataphylls and persistent rhomboidal bases of small dimensions. The parenchymatous pith is wide and contains mucilage canals and peripheral medullary bundles. It has a manoxylic vascular cylinder with two concentric rings of centrifugal secondary xylem and phloem and centripetal vascular bundles (inverted) with respect to the inner ring. The leaf traces run irregularly and produce a girdling configuration, while the parenchymatous cortex has extrafascicular concentric bundles.

Brunoa santarrosensis Artabe, Zamuner et Stevenson (Fig. 9.6D, E) is a multilacunar polyxylic stem covered by cataphylls and persistent rhomboidal foliar bases in alternating cycles. It has mucilage cavities in

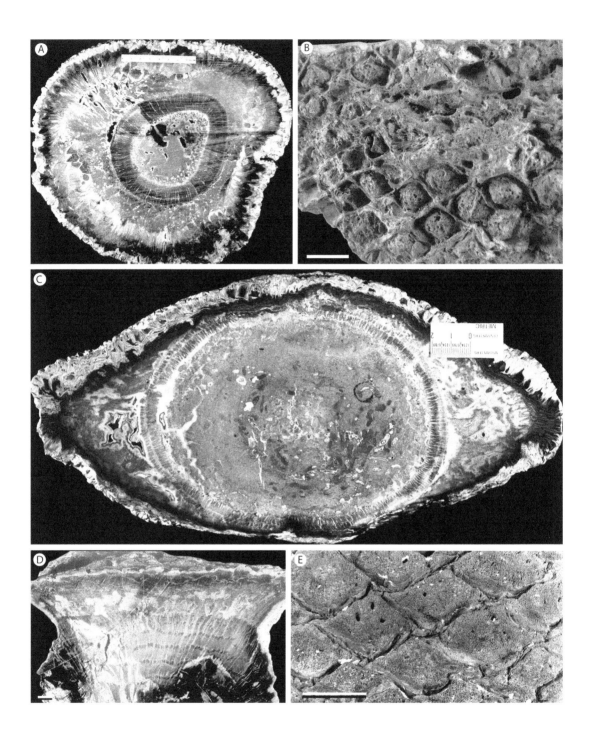

its parenchymatous cortex and medullary vascular bundles that form rings related to cone domes. The wood parenchyma in *B. santarrosensis* is scanty outside cortical steles. The leaf traces run irregularly and produce a girdling configuration. Its vascular cylinder is composed of six rings of collateral vascular bundles with phloem toward the outside and divided into fanlike sectors by multiseriate, parenchymatous rays.

Bororoa (Fig. 9.7A–C) was originally established by Petriella (1972), who described two species, *B. anzulovichii* and *B. andreisii*. The generic concept of *Bororoa* refers to a polyxylic stem with a persistent armor of leaf bases. The parenchymatous pith has medullary vascular bundles forming rings related to cone domes and mucilage channels. Wide primary rays divide these vascular rings. The tracheids of its secondary xylem have alternate, multiseriate, and areolate pittings. The parenchymatous cortex is differentiated into an inner zone with leaf traces running obliquely and an outer zone with girdling traces. *Bororoa anzulovichii* differs from *B. andreisii* in its larger stem diameter, the presence of stone cells in the pith, uniseriate secondary rays, and a cortex without periderm layers. *B. anzulovichii* also has leaf bases lacking periderm layers and but containing hypodermic fibers.

The monospecific genus *Menucoa* (*M. cazaui* Petriella) was erected by Petriella (1969) for polyxylic stems with a persistent armor of leaf bases (Fig. 9.7D). The parenchymatous pith in this stem has collateral bundles and mucilage channels, while wide primary rays divide vascular rings. The parenchymatous cortex has mucilage ducts and collateral traces running obliquely in the inner zone and showing girdling in the outer one. The leaf bases in this genus have several collateral bundles arranged in no apparent order.

Taxonomic and Phylogenetic Considerations

Cycads and bennetittaleans, as well as pteridosperms, have traditionally been placed into the informal taxon of the "cycadophytes" (Taylor et al. 2009), which clearly represents a paraphyletic group within the spermatophyte clade (Crane 1985; Doyle and Donoghue 1992; Rothwell and Serbet 1994). However, the phylogenetic position of each group is not well defined, especially that of the fossil order Bennettitales.

Living cycads consist of 10 to 12 genera and almost 300 species that have been treated differently in various classification systems (Stevenson 1981, 1990, 1992; Jones 2002). Depending on the author, modern cycads can be grouped into three or five families. The most commonly accepted classification (Stevenson 1992) recognizes the suborder Cycadineae with the family Cycadaceae (*Cycas*), and the suborder Zamiineae with the families Stangeriaceae (*Stangeria, Bowenia*) and Zamiaceae (*Dioon, Encephalartos, Macrozamia, Lepidozamia, Ceratozamia, Microcycas, Zamia,* and *Chigua*). The monophyletic character of these families is supported by both morphological (Crane 1988; Stevenson 1990; Brenner et al. 2003; Hermsen et al. 2006) and molecular phylogenetic analyses (Rai et al. 2003). However, among the latter, the monophyly of the Stangeriaceae and Zamiaceae has been questioned (Rai et al. 2003; Chaw et al. 2005).

Cycads show a unique combination of plesiomorphic and apomorphic characters (Brenner et al. 2003) and have been interpreted as the

Fig. 9.6. Late Cretaceous cycads from Argentina. (A, B) *Chamberlainia pteridospermoidea* Artabe, Zamuner et Stevenson. (c) *Worsdellia bonettii* Artabe, Zamuner et Stevenson. (D, E) *Brunoa santarrosensis* Artabe, Zamuner et Stevenson. Scale bars (A) = 8 cm, (B, D, E) = 1 cm, (C) = 2 cm.

Discussion

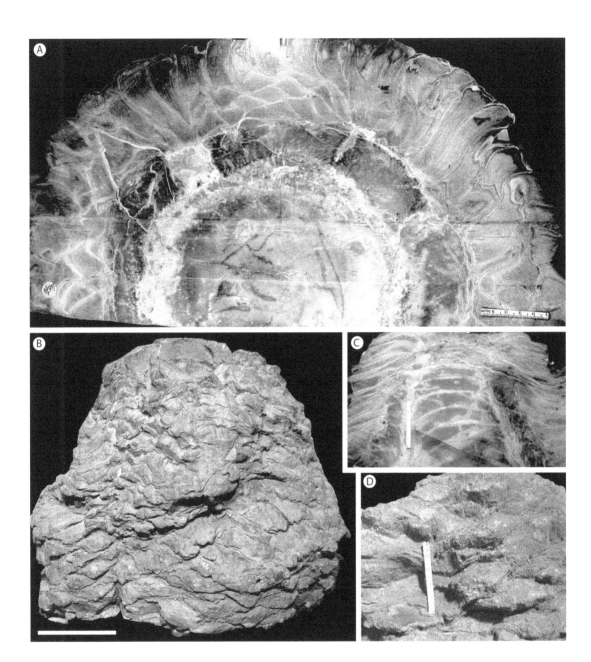

Fig. 9.7. Paleocene cycads from Argentina. (A–C) *Bororoa anzulovichii* Petriella. (D) *Menucoa cazaui* Petriella. Scale bars (A, C, D) = 5 cm, (B) = 10 cm.

group basal to the remaining modern gymnosperms (Ginkgoales, Coniferales, and Gnetales). This means they are sister to all living seed plants (Crane 1988; Stevenson 1990; Chaw et al. 2000), or sister to *Ginkgo* (Chaw et al. 1997). However, when fossils—the ancient remains of once-living plants—are included, new relationships must be considered. For instance, it has long been known that the cycads and seed ferns are related to one another (Worsdell 1906; Delevoryas 1955; Stewart and Rothwell 1993), and some studies postulate that Paleozoic medullosas can either be included in the same clade (Crane 1985) or are basal to the cycads (Rothwell and Serbet 1994). Furthermore, Doyle (1996) and Bateman et al. (2006) simply count the cycads as part of a pteridosperm clade.

In order to define phylogenetic relationships among the cycads, several cladistic analyses have been performed with living genera (Stevenson 1990; De Laubenfeld 1999; Chaw et al. 2005) or with both living and fossil genera (Brenner et al. 2003; Hermsen et al. 2006). The latter included morphological and anatomical characters that allowed for the inclusion of fossil morphotaxa of vegetative organs. Taxa such as the stems of *Michelilloa* (Triassic), *Menucoa* (Paleocene), and *Bororoa* (Cretaceous?–Paleocene) and the leaves of *Kurtziana* (Triassic–Early Jurassic), *Mesodescolea*, *Ticoa*, *Ctenis*, *Almargemia*, and *Sueria* (Early Cretaceous) were run in the analyses as representative cycads from Argentina. As a result *Mesodescolea* (formerly assigned to Stangeriaceae; Archangelsky and Petriella 1971), *Sueria*, and *Ctenis* occupy positions clearly basal to the Stangeriaceae. On the other hand, Triassic–Early Jurassic *Kurtziana* (*Alicurana*) and Early Cretaceous *Ticoa* are both basal to a clade that includes all extant Zamiaceae and related fossil genera. This is supported by synapomorphies related to epidermal characters such as wall thickness and occurrences of perforations in the epidermal cell walls. *Almargemia*, in turn, is basal to the extant subfamily Zamiodeae (sensu Stevenson 1992), a position that is supported by cuticular characters as well.

For this study, we have run a new cladistic analysis incorporating the genera *Brunoa*, *Chamberlainia*, and *Worsdellia* that were not included in the original matrixes (Tables 9.1, 9.2). Cladistic analysis of the resulting matrix (69 characters and 35 taxa) was performed by TNT (Goloboff et al. 2009); characters were treated as nonordered following Hermsen et al. (2006). A heuristic search (1000 replicates starting from a random Wagner tree followed by tree bisection and reconnection, or TBR) was conducted under equal weights. A single most parsimonious tree of 155 steps was obtained (Fig. 9.8) without changing the topologies initially obtained by Brenner et al. (2003) and Hermsen et al. (2006).

As expected from their morphological features, the Triassic genera *Michelilloa* and *Antarcticycas* occur outside the clade, with *Michelilloa* representing the basal form of the clade that includes all extant and fossil cycads. This position is supported by two synapomorphies, the presence of multilacunar nodes and girdling leaf traces, both of which are considered a prerequisite for the order Cycadales (Loconte and Stevenson 1990). On the other hand, *Brunoa*, *Worsdellia*, and *Chamberlainia* sort out into the clade that represents the family Zamiaceae. The last two genera occupy a position basal to the clade exclusively conformed by the extant genera of the Encephalartoideae tribe (*Lepidozamia*, *Macrozamia*, and *Encephalartos*); this position is mainly supported by the presence of cortical steles. On the other hand, *Brunoa* occurs as sister to *Fascisvarioxylon* in a clade placed sister to the group conformed by *Chamberlainia*, *Worsdellia*, and the extant tribe Encephalartoideae. Finally, the Cretaceous–Paleogene genera *Menucoa* and *Bororoa* form a trichotomy together with the clade involving the Encephalartoideae tribe and the fossil genera *Brunoa*, *Chamberlainia*, *Worsdellia*, and *Fascisvarioxylon*. On the basis of their phylogenetic position, all these taxa

Table 9.1. List of Characters
and States Used for the
Cladistic Analysis

*Source. From Hermsen et al.
(2006).*

Character	State
0 Coralloid roots	0, absent; 1, present
1 Root buds	0, absent; 1, present
2 Primary thickening meristem	0, internal derivatives; 1, external derivatives
3 Cone domes	0, absent; 1, present
4 Medullary bundles	0, absent; 1, present
5 Cortical steles	0, absent; 1, present
6 Secondary polyxyly	0, monoxylic; 1, polyxylic
7 Wood parenchyma	0, abundant; 1, scanty
8 Nodal anatomy	0, unilacunar; 1, multilacunar
9 Axillary buds	0, absent; 1, present
10 Leaf traces	0, radial; 1, girdling
11 Petiole bundle pattern	0, C-shaped; 1, omega; 2, obscure omega
12 Prickles	0, absent; 1, present
13 Petiole spines	0, absent; 1, transitional; 2, abrupt
14 Leaf base	0, vascularized stipules; 1, vestigial stipules; 2, estipulate
15 Ptyxis	0, circinate; 1, noncircinate
16 Pinna ptyxis	0, circinate; 1, adplicate; 2, conduplicate; 3, involute
17 Acroscopic basal callus	0, absent; 1, present
18 Pinna traces	0, two or more; 1, one
19 Trichomes	0, colored; 1, transparent
20 Curved trichomes	0, absent; 1, present
21 Equally branched trichomes	0, absent; 1, present
22 Cataphylls	0, absent; 1, present; 2, irregular
23 Peduncular cataphylls	0, absent; 1, present
24 Sporophyll pubescence	0, present; 1, absent
25 Sporophyll vasculature	0, planar; 1, three-dimensional; 2, *Macrozamia* type
26 Megasporophyll	0, absent; 1, pinnatifid; 2, simple
27 Megasporophyll spines	0, absent; 1, single; 2, two
28 Megasporophyll lobing	0, absent; 1, lateral
29 Megasporophyll shape	0, flat; 1, peltate; 2, adaxially thickened
30 Megasporophyll skirt	0, absent; 1, above; 2, below
31 Ovule position	0, lateral; 1, below; 2, above; 3, terminal
32 Ovule number	0, more than two; 1, two
33 Ovule vasculature	0, simple; 1, medullosan; 2, cycadean
34 Integument vasculature	0, single; 1, double
35 Micropyle	0, distal; 1, proximal
36 Endospermic jacket	0, absent; 1, present
37 Buffer cells	0, absent; 1, present
38 Seed	0, absent; 1, radiospermic; 2, platyspermic
39 Coronula	0, absent; 1, distinct; 2, indistinct
40 Cotyledon bundle	0, collateral; 1, concentric
41 Pollen shape	0, oblong; 1, elliptic
42 Proximal shape	0, convex; 1, concave
43 Pollen exine	0, psilate; 1, fossulate; 2, foveolate
44 Sperm number	0, many; 1, 4–8; 2, 2
45 Leaf bases	0, persistent; 1, ephemeral

Character	State
46 Inverse polyxyly	0, absent; 1, continuous ring; 2, random
47 Mucilage	0, canals; 1, absent; 2, cavities
48 Accessory cell corona	0, absent; 1, present
49 Guard cell corona	0, absent; 1, present
50 Number of accessory cell layers	0, zero; 1, one; 2, two; 3, more than two
51 Overarching accessory cells	0, absent; 1, present
52 Epidermal cells	0, without perforations; 1, with perforations; 2, corners only
53 Anticlinal pegs	0, absent; 1, present
54 Cuticular lamellae	0, absent; 1, present
55 Stomate position	0, hypostomatic; 1, amphistomatic
56 Epidermal cells	0, thin walled; 1, thick and thin walled
57 Lamina position	0, lateral; 1, adaxial
58 Leaves	0, bipinnate; 1, pinnate; 2, simple
59 Terminal pinna	0, all stages; 1, seedlings; 2, absent
60 Pinna attachment	0, decurrent; 1, articulate
61 Midrib	0, *Marattia* type; 1, absent; 2, *Cycas* type; 3, *Chigua* type
62 Veins	0, free; 1, anastomosing
63 Stomates	0, flush; 1, sunken
64 Guard cell orientation	0, irregular; 1, longitudinal
65 Stomatal shape	0, oblong; 1, circular
66 Stomatal type	0, haplocheilic; 1, syndetocheilic
67 Microsporangia	0, absent; 1, free; 2, clustered with some fused; 3, clustered with all fused
68 Pollen wall	0, spongy; 1, alveolar

could be considered as fossil form genera that could pertain to the tribe Encephalartoideae. Even though our cladogram shows a high degree of resolution, the phylogenetic hypotheses show low support values (e.g., all the nodes show Bremmer support values of 1), which suggests that if additional data (characters and taxa) were to be added, they could lead to changes in the postulated relationships.

The extinct bennettitaleans have previously been placed within the "anthophyte" clade (Crane 1985; Doyle and Donoghue 1992; Rothwell and Serbet 1994; Doyle 1996), which also includes flowering plants and the gnetophytes. However, some molecular studies reject the existence of an "anthophyte" clade and postulate that the Gnetales are more closely related to the conifers than to the angiosperms (Chaw et al. 1997). It is clear that relationships among extant seed plants (i.e., *Ginkgo*, conifers, cycads, gnetophytes, and angiosperms) do not yet show a consensus. This could be explained by many reasons, such as the clash between morphological and molecular studies, or the alternative homology hypothesis for certain morphological structures. In addition, extant seed plants represent just a small portion of the complete diversity of this clade, and the inclusion of new fossil information could readily change the relationships among major groups of seed plants (Donoghue et al. 1989).

Table 9.2. Full Coding for *Brunoa, Worsdellia,* and *Chamberlania*

Brunoa
???110111?1?????????????1????????????????????????012??????????? ??????????
Worsdellia
???011101?1?????????????1????????????????????????010??????????? ??????????
Chamberlainia
???011101?1?????????????1????????????????????????010??????????? ??????????

Bennettitaleans are widely accepted as a monophyletic group, which is supported by synapomorphies such as the cutinization of the guard cells, bivalvate synangia, interseminal scales, and numerous ovules (Crane 1985). Although internal relationships within the Bennettitales have not yet been resolved, two families are traditionally recognized, the Williamsoniaceae and the Cycadeoidaceae (see Watson and Sincock 1992; Taylor et al. 2009). However, these families are contradicted in the morphological phylogenetic analysis conducted by Crane (1985), although some pivotal taxa (especially from the Triassic) were not included in Crane's analysis. The current fossil record in Argentina indicates that only the Williamsoniaceae was ever present.

In summary, the internal relationships within the Bennettitales and phylogenetic position of this order within the context of the seed plants remain controversial (see also Crepet and Stevenson 2010; Rothwell and Stockey 2010). Nonetheless, it is clear that a more refined character discussion, the reconstruction of "whole plants," and their inclusion in broader phylogenetic studies could lead to the better understanding of the evolution of this particular group. Unfortunately, the bennettitalean record in Argentina does not contribute much to this matter because only compressions and impressions of reproductive organs have been preserved, and only few of these, such as the diverse bracts of *Cycadolepis*, have retained cuticular characters. Thus, it would be valuable to undertake more detailed study of the structure and ultrastructure of extremely well-preserved cutinized material to gain additional morphological characters useful for a better understanding of the phylogenetic relationships of Argentinian bennettitaleans.

Bio and Chronostratigraphic Aspects

Both cycads and bennettitaleans from Argentina have been used successfully for biostratigraphic and chronostratigraphic purposes. Stipanicic (2002) suggested that Triassic cycadophytes in Argentina include some morphospecies that show variable stratigraphic ranges; some are restricted to the Triassic (in particular Neotriassic species of *Pseudoctenis* and *Taeniopteris*), while others have longer time ranges that reach the Jurassic. This is the case for the species of *Anomozamites* and *Ctenis* and for some

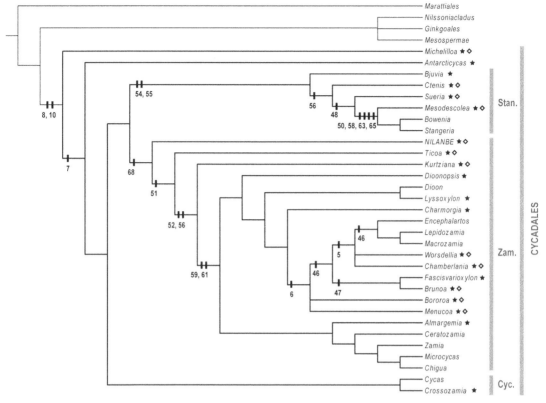

◇ Genera recorded from Argentina

★ Fossil genera

species of *Pseudoctenis, Nilssonia,* and *Taeniopteris.* In their bio- and chronostratigraphic characterization of the Argentinian Triassic, Spalletti et al. (1999) do not consider cycadophytes as time markers, with the exception of a few species of the possible cycadophyte genera *Yabeiella* and *Kurtziana.*

During the Jurassic, several species of the bennettitalean genus *Otozamites* are typical of the Early Jurassic plant assemblages from a biochronostratigraphic perspective ("the *Otozamites* flora"), but not much can be said for the Middle–Late Jurassic.

A different situation occurs in the Early Cretaceous, a time when both cycads and bennettitaleans become major components of the vegetation in Patagonia, reversing the strangely absent role that they played during most of the Jurassic. Archangelsky (2003), for example, was able to define a *Ptilophyllum* biozone on the basis of the presence of highly diversified bennettitaleans and cycads in the Deseado Massif of southern Patagonia. Interestingly, bennettitaleans disappear in the younger *Gleichenites* biozone and only the cycads remain. Thus, the Bennettitales become completely extinct by the beginning of the Late Cretaceous, while cycads, although diminished, remain until the early Paleogene, after which they finally disappear from the Patagonian region. Nevertheless, it is quite possible that the cycads in particular could have persisted

Fig. 9.8. Most parsimonious tree obtained in the cladistic analysis by TNT, based on the morphological matrix of Brenner et al. (2003) and Hermsen et al. (2006). *Brunoa, Worsdellia,* and *Chamberlainia* are included for the first time. Cyc = Cycadaceae; Zam = Zamiaceae; Stan = Stangeriaceae.

in the northern part of the Argentinian territory until later in the Paleogene or early Neogene, at which point they definitely abandoned their midlatitudinal distribution in this part of South America.

Paleoecological and Paleoclimatic Implications

Middle–Late Triassic plant communities in Argentina have been extensively analyzed by Artabe et al. (2001), who showed that cycadophytes were likely minor components. In these floras, cycadophytes are interpreted as shrubby plants that, along with the peltasperms, were part of the understory vegetation in various forest types (evergreen, deciduous, or sclerophyllous) dominated by corystosperms, or occupied open areas as shrubs. These types of vegetation grew on the alluvial plains with lakes and rivers under subtropical, but seasonal, climatic conditions that were apparently more humid during the Middle and early Late Triassic than in the latest Triassic.

By the Jurassic, plant communities in this region were represented by extensive conifer forests mainly dominated by the Araucariaceae and Cupressaceae (Archangelsky and Del Fueyo 2010), which occupied terrestrial or littoral settings with constant volcanic activity. There, the bennettitaleans and the ferns seem to have represented the understory vegetation, particularly during the Early Jurassic, when they were probably favored by warm and humid climatic conditions. During the Middle–Late Jurassic, the role of the cycadophytes in the local plant communities became restricted, apparently as a result of some kind of paleoclimatic control (displacement of arid belts, for instance; Rees et al. 2000). This sort of gap in southern South America is in stark contrast to coeval floras in the rest of the world, which show a climax in the role of cycadophytes in terrestrial plant communities.

Later, during the Early Cretaceous, cycads and bennettitaleans began to recover their former prominence in plant communities in Patagonia. This situation has been analyzed extensively in the Baqueró Group, where the two groups occur in great abundance. Interestingly, both cycads and bennettitaleans show marked adaptations to dry conditions—for example, reduced leaves, sculptured surfaces with strongly ornamented epidermis (trichomes, papillae, and hairs), sunken and protected stomata, and thick cuticles. Apparently, these autoecological features were a response to the warm and strongly seasonally dry climates constantly influenced by volcanic activity, which especially increased during sedimentation of the upper Baqueró Group (Archangelsky et al. 1995; Archangelsky 2003).

Cycadophytes probably occupied various niches in different plant communities (Cúneo 2003). During the sedimentation of the lower Baqueró Group, bennettitaleans and cycads were major components of the local vegetation, which, along with the pteridosperms and ferns, formed a low stratum of a relatively closed community dominated by conifers (podocarps, araucarians, and cheirolepidiaceans). This scenario changes toward the upper part of the Baqueró Group, when the bennettitaleans disappear and cycads (*Pseudoctenis, Nilssonia,* and *Mesosingeria*) remain present as midsized plants associated with gleicheniaceous/dipteridaceous ferns and araucarian conifers in a more open community structure (Cladera and Cúneo 2002).

Late Cretaceous and early Paleogene plant communities show an increased role on the part of the dicots, although conifers, ferns, and cycads still occur in differing amounts in the flora. Cycads in particular have been intensively recorded in northern Patagonia, which documents their northward retreat away from their more southern distribution in the Early Cretaceous. On the basis of the paleoecological and sedimentological settings in which they have been found, cycads could have been part of low, open communities in coastal areas, where they normally occurred alongside palms (Petriella 1972). Paleoclimatic analyses of latest Cretaceous–Paleocene floras suggest warm and humid conditions (Petriella and Archangelsky 1975; Iglesias et al. 2007; Cúneo et al. 2007) favorable for plant communities with thermophilic plants such as cycads. In this regard, the dominance of the Encephalartineae cycads, with xeric traits like mucilage channels and persistent leaf bases, also supports seasonal dry conditions.

Acknowledgments

We extend our highest respect and friendship to our good friend Ted Delevoryas for an entire life dedicated to paleobotany. We congratulate Carole Gee on this volume and thank her for the kind invitation to participate in Ted's Festschrift. The TNT program is freely available thanks to a subsidy from the Willi Hennig Society.

References Cited

Anderson, J. M., and H. M. Anderson. 1989. *Paleoflora of Southern Africa, Molteno Formation (Triassic),* vol. 2, *Gymnosperms (Excluding* Dicroidium). *National Botanic Institute* 2: 1–167. Rotterdam: Balkema.

Archangelsky, A. 1997. *Pseudoctenis giganteus,* nueva cycadal de la Formación Baqueró, Cretácico inferior de Argentina. *Ameghiniana* 34: 387–391.

Archangelsky, A., R. R. Andreis, S. Archangelsky, and A. E. Artabe. 1995. Cuticular characters adapted to volcanic stress in a new Cretaceous cycad leaf from Patagonia, Argentina. Considerations on the stratigraphy and depositional history of the Baqueró Formation. *Review of Palaeobotany and Palynology* 89: 213–233.

Archangelsky, S. 1963a. A new Mesozoic flora from Ticó, Santa Cruz Province, Argentina. *Bulletin of the British Museum (Natural History), Geology* 8: 45–92.

———. 1963b. Notas sobre la flora fósil de la zona de Ticó, Provincia de Santa Cruz. 2. Tres nuevas especies de *Mesosingeria. Ameghiniana* 3: 113–122.

———. 1966. New gymnosperms from the Ticó flora, Santa Cruz Province, Argentina. *Bulletin of the British Museum (Natural History), Geology* 13: 261–295.

———. 1976. Vegetales fósiles de la Formación Springhill, Cretácico, en el subsuelo de la Cuenca Magallánica, Chile. *Ameghiniana* 13: 141–158.

————, ed. 2003. La flora cretácica del Grupo Baqueró, Santa Cruz, Argentina. *Monografías Museo Argentino Ciencias Naturales* 4 (CD-ROM). Buenos Aires.

Archangelsky, S., and A. Baldoni. 1972. Notas sobre la flora fósil de la Zona de Ticó, Santa Cruz. X. Dos nuevas especies de *Pseudoctenis* (Cycadales). *Ameghiniana* 9: 241–257.

Archangelsky, S., and D. Brett. 1963. Studies on Triassic fossil plants from Argentina. II. *Michellloa waltonii* nov. gen. et sp. from the Ischigualasto Formation. *Annals of Botany* 27: 147–154.

Archangelsky, S., and G. M. Del Fueyo. 2010. Endemism of Early Cretaceous conifers in Western Gondwana. In C. T. Gee (ed.), *Plants in Mesozoic Time: Morphological Innovations, Phylogeny, Ecosystems*, pp. 247–268 Bloomington: Indiana University Press.

Archangelsky, S., and B. Petriella. 1971. Notas sobre la flora fósil de la zona de Ticó, Santa Curz. IX. Nuevos datos acerca de la morfología foliar de *Mesodescolea plicata* Arch. (Cycadales, Stangeriaceae). *Boletin Sociedad Argentina Botanica* 14: 88–94.

Archangelsky, S., and L. Villar de Seoane. 2004. Cycadean diversity in the Cretaceous of Patagonia, Argentina. Three new *Androstrobus* species from the Baqueró Group. *Review of Palaeobotany and Palynology* 131: 1–28.

Arrondo, O. G., and B. Petriella 1980. Alicurá, nueva localidad plantífera liásica de la provincia de Neuquén, Argentina. *Ameghiniana* 17: 121–133.

Artabe, A. 1994. Estudio al microscopio electronico de barrido (MEB) de dos cycadópsidas fósiles de Argentina. *Pseudoctenis dentata* Archangelsky Baldoni 1972 y *Sueria rectinervis* Menendez 1965. *Ameghiniana* 31: 115–124.

Artabe, A., and S. Archangelsky. 1992. Estudio al microscopio electronico de barrido y transmisión de una Stangeriaceae (Cycadales) *Mesodescolea plicata* Archangelsky emend. Archangelsky et Petriella 1971. *Ameghiniana* 28 (1–2): 202.

Artabe, A., and D. W. Stevenson. 1999. Fossil Cycadales of Argentina. *Botanical Review* 65: 219–238.

Artabe, A., A. Zamuner, and S. Archangelsky. 1991. Estudios cuticulares en Cycadopsida fósiles. El género *Kurtziana* Frenguelli. *Ameghiniana* 28: 365–374.

Artabe, A., E. M. Morel, and L. A. Saplletti. 2001. Paleoecología de las Floras triásicas argentines. In A. E. Artabe, E. M., Morel, and A. B. Zamuner (eds.), *El Sistema Triásico en la Argentina*, pp. 199–225. La Plata, Argentina: Fundación Museo La Plata.

————. 2003. Caracterización de las provincias fitogeográficas triásicas del Gondwana. *Ameghiniana* 40: 387–405.

Artabe, A., A. B. Zamuner, and D. W. Stevenson. 2004. Two new petrified cycad stems *Brunoa* gen. nov. and *Worsdellia* gen. nov., from the Cretaceous of Patagonia (Bajo de Santa Rosa, Río Negro province), Argentina. *Botanical Review* 70: 121–133.

————. 2005. Fossil cycad stems (Zamiaceae–Encephalartoideae) from the Late Cretaceous of Argentina, with reappraisal of the known forms. *Alcheringa* 29: 87–100.

Artabe, A., E. M. Morel, and D. G. Ganuza. 2007. Las floras triásicas de la Argentina. *Ameghiniana* 50° *Aniversario, Publicación Especial* 11: 75–86.

Baldoni, A. 1974. Revisión de las Bennettitales de la Formación Baqueró (Cretácico Inferior) Provincia de Santa Cruz. II. Brácteas. *Ameghiniana* 11: 328–356.

————. 1977. *Ptilophyllum ghiense* n. sp., una nueva Bennettital de Paso Roballos, Provincia de Santa Cruz. *Ameghiniana* 14: 53–58.

———. 1978. Plantas fósiles jurásicas del subsuelo de Plaza Huincul, provincia de Neuquén. *Boletin Asociación Latinoamericana Paleobotánica y Palinología* 5: 1–12.

———. 1981. Tafofloras Jurásicas y Eocretácicas de América del Sur. In W. Volkheimer and E. Mussachio (eds.), *Cuencas Sedimentarias del Jurásico y Cretácico de América del Sur* 2: 359–391.

Baldoni, A., and T. N. Taylor. 1983. Plant remains from a new Cretaceous site in Santa Cruz, Argentina. *Review of Palaeobotany and Palynology* 39: 301–311.

Bateman, R., J. Hilton, and P. J. Rudall. 2006. Morphological and molecular phylogenetic context of the angiosperms: contrasting the "top-down" and "bottom-up" approaches used to infer the likely characteristics of the first flowers. *Journal of Experimental Botany* 57: 3471–3503.

Berry, E. W. 1924. Mesozoic plants from Patagonia. *American Journal of Science*, ser. 5, 7: 473–482.

Bonetti, M. 1964. Flórula mesojurásica de la zona de Taquetrén (Cañadón del Zaino), Chubut. *Revista del Museo Argentino Ciencias Naturales* 1: 1–43.

———. 1968. Las especies del género *Pseudoctenis* en la flora triásica de Barreal (San Juan). *Ameghiniana* 5: 433–446.

———. 1972. Las Bennettitales de la flora triásica de Barreal (Prov. de San Juan). *Revista Museo Argentino de Ciencias Naturales "B. Rivadavia"* 1: 307–322.

———. 1974. Flórula jurásica del Lago Argentino (Santa Cruz). *I Congreso Argentino Paleontología y Bioestratigrafía, Resúmenes,* Tucumán, p. 9.

Brenner, E. D., D. Stevenson, and R. W. Twigg. 2003. Cycads: evolutionary innovations and the role of plant-derived neurotoxins. *Trends in Plant Science* 8: 446–452.

Chaw, S. M., A. Zharkikh, H. M. Sung, T. K. Lau, and W. H. Li. 1997. Molecular phylogeny of extant gymnosperms and seed plant evolution, analyses of 18S rRNA sequences. *Molecular Biology and Evolution* 14: 56–68.

Chaw, S. M., C. Parkinson, Y. Cheng, T. M. Vincent, and J. D. Palmer. 2000. Seed plant phylogeny inferred from all three plant genomes: monophyly of gymnosperms and sistehood of Gnetales and Pinaceae. *Proceedings of the National Academy of Sciences U.S.A.* 97: 4086–4091.

Chaw, S. M., T. W. Walter, C. C. Chang, S. H. Hu, and S. Chen. 2005. A phylogeny of cycads (Cycadales) inferred from chloroplast *matK* gene, *trnK* intron, and nuclear rDNA ITS region. *Molecular Phylogenetics and Evolution* 37: 214–234.

Cladera, G., and R. Cúneo. 2002. Fossil plants buried by volcanic ash in the Lower Cretaceous of Patagonia. In M. De Renzi, M. V. Pardo Alonso, M. Belinchón E. Peñalver, P. Montoya, and A. Márquez-Aliaga (eds.), *Current Topics on Taphonomy and Fossilization, Third Encounter Taphos 2002,* pp. 399–403. Valencia: Ayuntamiento de Valencia.

Crane, P. 1985. Phylogenetic analysis of seed plants and the origin of angiosperms. *Annals of the Missouri Botanical Garden* 72: 716–793.

———. 1988. Major clades and relationships in "higher" gymnosperms. In C. B. Beck (ed.), *Origin and Evolution of Gymnosperms,* pp. 218–272. New York: Columbia University Press.

Crepet, W. L., and D. W. Stevenson. 2010. The Bennettitales (Cycadeoidales): a preliminary perspective on this arguably enigmatic group. In C. T. Gee (ed.), *Plants in Mesozoic Time: Morphological Innovations, Phylogeny, Ecosystems,* pp. 215–244. Bloomington: Indiana University Press.

Cúneo, N. R. 2003. Early Cretaceous terrestrial ecosystems from Patagonia: the Baquero Group, a case study. Terrestrial paleobiology of South America, Cretaceous through Neogene (Paleont. Soc.). *Geological Society of America Annual Meeting-Exposition,* Seattle 35 (6): 58.

Cúneo, N. R., K. Johnson, P. Wilf, R. Scasso, A. Gandolfo, and A. Iglesias. 2007. A preliminary report on the diversity of latest Cretaceous floras from northern Patagonia, Argentina. *Geological Society of America Annual Meeting, Abstracts with Programs, Denver* 39: 584.

De Laubenfeld, D. J. 1999. The families of the Cycadaceae. *Encephalartos* 59: 7–9.

Delevoryas, T. 1955. The Medullosae—structure and relationships. *Palaeontographica*, Abt. B, 97: 114–167.

Donoghue, M., J. Doyle, J. Gauthier, A. G. Kluge, and T. Rowe. 1989. The importance of fossils in phylogeny reconstructions. *Annual Review of Ecology and Systematics* 20: 431–460.

Doyle, J. 1996. Seed plant phylogeny and the relationships of Gnetales. *International Journal of Plant Sciences* 157: 3–39.

Doyle, J., and M. Donoghue. 1992. Fossils and seed plant phylogeny reanalyzed. *Brittonia* 44: 89–106.

Gee, C. T. 1989. Revision of the Late Jurassic/Early Cretaceous fossil flora from Hope Bay, Antarctica. *Palaeontographica*, Abt. B, 213: 149–214.

Gnaedinger, S. 1999. La flora triásica del Grupo El Tranquilo, Provincia de Santa Cruz (Patagonia). Parte VII. Cycadophyta. In *Ameghiniana, Publicación Especial* 6: 27–32.

Gnaedinger, S., and R. Herbst. 1998. La flora triásica del Grupo El Tranquilo, provincia de Santa Cruz (Patagonia). Parte V. Pteridophylla. *Ameghiniana* 35: 53–65.

———. 2004. Pteridophylla del Triásico del Norte Chico de Chile. I. El género *Taeniopteris* Bgt. *Ameghiniana* 41: 91–110.

Goloboff, P., J. S. Farris, and K. Nixon. 2009. TNT, a free program for phylogenetic analysis. *Cladistics* 24: 774–786.

Herbst, R. 1965. La flora fósil de la Formación Roca Blanca, provincia de Santa Cruz, Patagonia. Con consideraciones geológicas y estratigráficas. *Opera Lilloana* 12: 7–101.

———. 1966a. Revisión de la flora liásica de Piedra Pintada, provincia de Neuquén, Argentina. *Revista del Museo de la La Plata*, n.s., 5: 27–53.

———. 1966b. La flora liásica del Grupo Pampa de Agnia, Chubut, Patagonia. *Ameghiniana* 4: 337–347.

———. 1968. Las floras liásica argentines con consideracioines estratigráficas. *III Jornadas Geológicas Argentinas* 1: 145–162.

———. 1980. Flórula fósil de la Fm. Los Patos (Sinemuriano) del río Los Patos (San Juan), Argentina. *II Congreso Argentino Paleontologia y Bioestratigrafia, I Congreso Latinoamericano Paleontología, Actas* 1: 175–189.

Herbst, R., and L. Anzotegui. 1968. Nuevas plantas de la flora del Jurásico medio (Matildense) de Taquetrén, provincia de Chubut. *Ameghiniana* 5: 183–190.

Herbst, R., and S. Gnaedinger. 2002. *Kurtziana* Frenguelli (Pteridospermae? incertae sedis) y *Alicurana* nov. gen. (Cycadopsida) del Triásico y Jurásico temprano de Argentina y Chile. *Ameghiniana* 39: 331–341.

Herbst, R., and A. Troncoso. 2000. Las Cycadophyta del Triásico de las Formaciones La Ternera y El Puquén (Chile). *Ameghiniana* 37: 283–292.

Hermsen, E. J., T. N. Taylor, E. L. Taylor, and D. W. Stevenson. 2006. Cataphylls of the Middle Triassic cycad *Antarcticycas schopfii* and new insights into cycad evolution. *American Journal of Botany* 93: 724–738.

———. 2007. Cycads from the Triassic of Antarctica: permineralized cycad leaves. *International Journal of Plant Sciences* 168: 1099–1112.

Hunter, M. A. 2005. Mid-Jurassic age for the Botany Bay Group: implications for the Weddell Sea Basin creation and southern hemisphere biostratigraphy. *Journal of the Geological Society* 162: 745–748.

Iglesias, A., P. Wilf, K. Johnson, A. Zamuner, R. Cúneo, S. Matheos, and B. Singer. 2007. A Paleocene lowland macroflora from Patagonia reveals significantly greater richness than North American analogs. *Geology* 35: 947–950.

Jones, D. L. 2002. *Cycads of the World: Ancient Plants in Today's Landscape,* 2nd ed. Washington, D.C.: Smithsonian Institution Press.

Leppe, M., and P. Moisan. 2003. Nuevos registros de Cycadales y Cycadeoidales del Triásico superior del río Biobío, Chile. *Revista Chilena Historia Natural* 76: 475–484.

Loconte, H., and D. Stevenson. 1990. Cladistics of the Spermatophyta. *Brittonia* 42: 197–211.

Longobucco, M. I., C. Azcuy, and B. Aguirre Urreta. 1985. Plantas fósiles de la Formación Kachaike, Cretácico de Santa Cruz. *Ameghiniana* 21: 305–315.

Mamay, S. 1969. Cycads: fossil evidence of Late Paleozoic origin. *Science* 164: 295–296.

Menéndez, C. A. 1965. *Sueria rectinervis* n. gen. et sp. de la flora fósil de Ticó, Provincia de Santa Cruz. *Ameghiniana* 4: 3–11.

———. 1966. Fossil Bennettitales from the Ticó flora, Santa Cruz Province, Argentina. *Bulletin of the British Museum (Natural History), Geology* 12: 1–42.

Passalia, M. 2007. Nuevos registros para la flora cretácica descripta por Halle (1913) en lago San Martín, Sancta Cruz, Argentina. *Ameghiniana* 44: 565–595.

Petriella, B. 1969. *Menucoa cazaui* nov gen. et sp., tronco petrificado de Cycadales, Provincia de Río Negro, Argentina. *Ameghiniana* 6: 291–302.

———. 1972. Estudio de las maderas petrificadas del Terciario inferior del área central de Chubut (Cerro Bororó). *Revista del Museo de La Plata,* n.s., *Paleontología* 6: 159–254.

———. 1978. Nuevos hallazgos de Cycadales fósiles en Patagonia. *Boletin de la Asociación Latinoamericana de Paleobotánica y Palinología* 5: 13–16.

Petriella, B., and S. Archangelsky. 1975. Vegetación y ambiente en el Paleoceno de Chubut. *I Congreso Argentino Paleontologia y Bioestratigrafia, Actas, Tucumán* 2: 257–270.

Rai, H. S., H. E. O'Brian, P. A. Reeves, R. G. Olmstead, and S. W. Graham. 2003. Inference of higher-order relationships in the cycads from a large chloroplast data set. *Molecular Phylogenetics and Evolution* 29: 350–359.

Rees, P. M., and C. J. Cleal. 2004. Lower Jurassic floras from Hope Bay and Botany Bay, Antarctica. *Special Papers in Palaeontology* 72: 1–90.

Rees, P. M., A. Ziegler, and P. Valdes. 2000. Jurassic phytogeography and climates: new data and model comparisons. In B. T. Huber, K. Macleod, and S. Wing (eds.), *Warm Climates in Earth History,* pp. 297–318. Cambridge: Cambridge University Press.

Rothwell, G., and R. Serbet. 1994. Lignophyte phylogeny and the evolution of the spermatophytes: a numerical cladistic analysis. *Systematic Botany* 19: 443–482.

Rothwell, G. W., and R. A. Stockey. 2010. Independent evolution of seed enclosure in the Bennettitales: evidence from the anatomically preserved cone *Foxeoidea connatum* gen. et sp. nov. In C. T. Gee (ed.), *Plants in Mesozoic Time: Morphological Innovations, Phylogeny, Ecosystems,* pp. 51–64. Bloomington: Indiana University Press.

Spalletti, L., A. Artabe, E. Morel, and M. Brea. 1999. Biozonación paleoflorística y cronoestratigrafía del Triásico argentino. *Ameghiniana* 36: 419–451.

Spalletti, L., E. Morel, J. Franzese, A. Artabe, D. Ganuza, and A. Zúñiga. 2007. Contribución al conocimiento sdimentológico y paleobotánico de la

Formación El Freno (Jurásico temprano) en el valle superior del río Atuel, Mendoza, Argentina. *Ameghiniana* 44: 367–386.

Stevenson, D. W. 1981. Observations on ptyxis, phenology, and trichomes in the Cycadales and their systematic implications. *American Journal of Botany* 68: 1104–1114.

———. 1990. Morphology and systematics of Cycadales. *Memoirs of the New York Botanical Garden* 57: 8–55.

———. 1992. A formal classification of the extant Cycadales. *Brittonia* 44: 220–223.

Stewart, W., and G. Rothwell. 1993. *Paleobotany and the Evolution of Plants*, 2nd ed. Cambridge: Cambridge University Press.

Stipanicic, P. N. 2002. Introducción. In P. N. Stipanicic and C. A. Marsicano (eds.), *Léxico Estratigráfico de la Argentina*, vol. 8, *Triásico. Asociación Geológica Argentina*, Serie "B" 26: 1–24.

Taylor, T. N., E. L. Taylor, and M. Krings. 2009. *Paleobotany: The Biology and Evolution of Fossil Plants*, 2nd ed. San Diego: Academic Press.

Villar de Seoane, L. 1995. Estudio cuticular de nuevas Bennettitales eocretácicas de Santa Cruz, Argentina. *VI Congreso Argentino Paleontología y Bioestratigrafía, Actas*, pp. 247–254.

———. 1997. Estudio cuticular comparado de nuevas Cycadales de la Formación Baqueró (Cretácico Inferior), Santa Cruz, Argentina. *Revista Española de Paleontología* 12: 129–140.

———. 1999. *Otozamites ornatus* sp. nov., a new bennettitalean leaf species from Patagonia, Argentina. *Cretaceous Research* 20: 499–506.

———. 2001. Cuticular study of Bennettitales from the Springhill Formation, Lower Cretaceous of Patagonia, Argentina. *Cretaceous Research* 22: 461–479.

———. 2003. Cuticle ultrastructure of the Bennettitales from the Anfiteatro de Ticó Formation (Early Aptian), Santa Cruz Province, Argentina. *Review of Palaeobotany and Palynology* 127: 59–76.

———. 2005. New cycadalean leaves from the Anfiteatro de Ticó Formation, Early Aptian of Patagonia, Argentina. *Cretaceous Research* 26: 540–550.

Watson, J., and C. A. Sincock. 1992. Bennettitales of the English Wealden. *Monograph of the Palaeontographical Society* 145 (588): 2–228.

Worsdell, W. C. 1906. The structure and origin of the Cycadaceae. *Annals of Botany* 20: 129–155.

THE BENNETTITALES (CYCADEOIDALES): A PRELIMINARY PERSPECTIVE ON THIS ARGUABLY ENIGMATIC GROUP

10

William L. Crepet and Dennis W. Stevenson

Most phylogenetic analyses of morphological and structural characters have placed the Bennettitales comfortably within an anthophyte clade that includes angiosperms and the Gnetales. The Bennettitales–Gnetales relationship has come into question because nucleic acid sequence–based phylogenies place the Gnetales with other gymnosperms outside of the anthophyte clade. Yet recent morphology-based phylogenetic analyses reinstate the Gnetales as part of the anthophytes, reuniting them with the Bennettitales. Questions about a possible gnetalean–bennettitalean relationship are nothing new and have persisted since the early twentieth century. Some aspects of this controversy are examined in light of certain characters that may be inadequately understood or misrepresented in most existing phylogenetic analyses of morphological characters. Phylogenetic analysis of a morphology-based data set that reflects new characters or revised character codings removes the Bennettitales from the anthophytes and places them in a latter-day cycadofilicalean clade. Although this analysis is preliminary and experimental, it draws attention to the need for additional studies to clarify the nature and distribution of certain characters, especially the ovulate receptacles, if we are to determine more precisely and less subjectively the phylogenetic position of the Bennettitales and their relationship to the Gnetales and other seed plants.

Although the phylogenetic position of the Bennettitales has a history of controversy (cf. Rothwell et al. 2009; Rothwell and Stockey 2010), recently, the Bennettitales have been securely nested in the anthophytes with the angiosperms and Gnetales (and other fossil taxa, depending on the analysis) in most phylogenetic analyses based on morphological and anatomical characters (e.g., Crane 1985; Doyle and Donoghue 1986, 1987, 1992; Nixon et al. 1994; Doyle 1996). In contrast, phylogenetic analyses of seed plants that are based on nucleic acid sequence data suggest that the Gnetales are variously positioned outside of the anthophytes closer to or nested with various gymnosperms (e.g., Chaw et al. 2000; Mathews 2009; Rothwell et al. 2009). Although some recent morphology-based analyses continue to support the anthophyte position of the Gnetales

Fig. 10.1. Habit of the cycad *Dioon edule. Redrawn from Wieland (1906).*

History and Phylogenetic Position of the Bennettitales

(Hilton and Bateman 2006), uncontested additional support for the anthophyte concept has not been forthcoming from the inclusion of new fossil taxa (Rothwell et al. 2009), and consensus trees of a recent analysis of seed plant relationships (Friis et al. 2007) do not resolve an anthophyte clade (cf. Rothwell et al. 2009). Moreover, the possibility that Gnetales are more securely allied with other gymnosperms as suggested by nucleic acid sequence–based analyses (e.g., Mathews 2009) would leave the angiosperms and Bennettitales as principal members of an "anthophyte" clade in some analyses. Such a juxtaposition draws attention to possible homologies among key angiosperm-defining organs and those of the Bennettitales, suggesting, as Arber and Parkin did long ago (1907), that both are related to an ancestor ("Hemiangiospermae") whose reproductive characteristics were compatible with evolutionary transformations in key characters into those characterizing the two disparate terminals; there is no uncontested support for this hypothesis from fossil evidence.

Complicating the matter is the fact that we cannot necessarily depend on the fossil record to support hypothesized transformations of such magnitude because underlying changes in homeotic genes, for example, need not be manifest in a sequence of extinct morphological intermediates leading from a hypothetical ancestor to known terminals because such transitional forms may have never existed. Although the phylogenetic position of the Bennettitales has not been directly challenged by contravening molecular analyses for obvious reasons (i.e., there is no available strictly bennettitalean DNA), the relationship between the Bennettitales and Gnetales has been placed into question, and at the same time such nucleic acid sequence–based analyses focus attention on possible angiosperm–bennettitalean homologies.

Long before the contrast between most phylogenetic analyses based on morphology and those based on nucleic acid sequence analyses was recognized, the possible relationship among the Bennettitales, Gnetales, and angiosperms was controversial. Morphological evidence was once invoked to support bennettitalean relationships with relatively basal seed plants (Cycadales, e.g., Chamberlain 1920; and sometimes, somehow, and strangely, also angiosperms, Wieland 1906). Such evidence was separately interpreted as suggesting bennettitalean affinities with the Gnetales (although based on at least some characters that are now and always have been in dispute; see Rothwell et al. 2009 for a historical summary). In contrasting assessments, evidence relating the Bennettitales to the Cycadales included not only cycadalean features, but also those thought to be held in common with both cycads and extinct seed ferns (see Wieland 1906, 1916, 1934; Arnold 1947; Delevoryas 1959, 1960, 1963, 1965, 1968a, 1968b; Bierhorst 1971; Gifford and Foster 1988; Stewart and Rothwell 1993; Norstog and Nicholls 1997; Taylor et al. 2009). Although implications of molecular-based phylogenetic analyses have indirectly complicated our understanding of bennettitalean relationships, still more recent morphology-based studies have in turn challenged molecular-based trees (Hilton and Bateman 2006), while other studies, building on Hilton and

Bateman's analysis, add support to the relationship among the Gnetales, Bennettitales, and angiosperms through inclusion of a new order, the Erdtmanithecales, which is a composite of dispersed organs known from different fossil localities and of different ages (Pedersen et al. 1989; Friis and Pedersen 1996; Kvaček and Pacltová 2001) and newly described charcoalified seeds of Cretaceous age purported to have characters that place them transitionally between the Gnetales and Bennettitales (Friis et al. 2007). The assessment of Rothwell et al. (2009) of seed plant relationships is characteristic of this dynamic taxonomic landscape inasmuch as the assertion that a Bennettitales–Erdtmanithecales–Gnetales (BEG) clade resolves upon phylogenetic analysis of morphology-based data sets has recently been refuted by contravening evidence based on the critical analysis of a number of data sets, including the original one of Friis et al. (2007). There is also some uncertainty about the exact nature of the Erdtmanithecales (Rothwell et al. 2009).

Thus, despite years of research focusing on the Bennettitales and the spectacular nature of some bennettitalean fossils, there is enduring discord on their probable phylogenetic relationships and thus on their significance in seed plant evolution. What is it specifically that makes these plants so interesting and yet so historically controversial with respect to affinities? As reflected in the historical literature, partly it is the contrast between their reproductive features, which are superficially similar to those of angiosperms and the Gnetales in at least some aspects, and their cycad or even seed fern–like morphology, which includes both vegetative and certain reproductive characters (e.g., Wieland 1906; Scott 1909; Chamberlain 1920; Delevoryas 1968a, 1968b; Norstog and Nicholls 1997). Attaining a consensus on bennettitalean phylogenetic relationships has been further complicated by the fact that some of their most vital characteristics are unknown (e.g., anatomical features of *Williamsoniella* or *Wielandiella*, and see discussion below). In addition, key organs are often not well enough preserved or often have not been sufficiently studied for various reasons contributing to the imperfect understanding of possible homologs (e.g., especially the anatomy of the ovulate receptacle, although there is recent progress in this area; e.g., Rothwell and Stockey 2002). Many of these issues and limitations also characterize our understanding of a number of other fossil groups resulting in dramatic differences in phylogenetic placement of such taxa from analysis to analysis (e.g., the Caytoniales, known from relatively few characters are quite mobile in phylogenetic studies, depending on the analysis; cf. Nixon et al. 1994; Rothwell et al. 2009). It is interesting in the above contexts to examine characters of the Bennettitales again and to explore their possible significance. Also, recent years have seen the development of a more thorough understanding of fossil cycads, including a cladistic analysis that encompasses both fossil and extant cycads (Artabe and Stevenson 1999; Artabe et al. 2004, 2005; Dower et al. 2004; Hermsen et al. 2006, 2007).

The Dispute Continues

Apropos to this volume, one of Ted Delevoryas's most interesting papers on the Cycadeoidales (Delevoryas 1968b) addresses the broader issue of bennettitalean homologies through his consideration of possible homologies of the cycadeoidalean cone. While precladistic, although based to some extent on phylogenetic principles, his cogent discussions are relevant today, over 40 years later, to bennettitalean relationships and homologies. These issues, emerging again in response to nucleic acid–based analyses of seed plant relationships and the uncertain results of particular morphology-based analyses, invite a revival of the questions involving gnetalean–bennettitalean relationships and possible angiosperm–bennettitalean homologies that have characterized botanical studies during the first half of the twentieth century.

In order to reexamine bennettitalean relationships, it is helpful to review comparisons among bennettitalean characters and those of other "anthophytes" with some attention to the cycadofilicalean ancestors once thought to be closely related to the Bennettitales. This has been done recently to a considerable extent by Rothwell et al. (2009), but the following discussion in this chapter goes further into the issue by considering several additional characters and their possible implications and by including the Cycadales.

Bennettitales/ Cycadales Character Comparison

Many of the characters of the Bennettitales are evocative of rather basal seed plants (Wieland 1906). They often have spirally arranged pinnate or pinnately veined leaves, often apparent manoxylic anatomy in specimens that are anatomically preserved, and persistent leaf bases on stems. The trunks branch sparsely or rarely, and there is some debate over their mode of branching, namely apical versus axillary, because there is no unambiguous evidence for axillary branching in those fossil specimens that are anatomically preserved (Rothwell et al. 2009), whereas in extant cycads, apical branching is known to occur, while axillary buds do not (Stevenson 1988). Their habit, pachycauly, is reminiscent of the habit of cycads (cf. Figs. 10.1 and 10.2) or even seed ferns, and there seems to have been a parallel evolution in habit among taxa of both the Bennettitales and Cycadales during the Mesozoic, with early representatives of each order having graceful stems that were either sparsely branched or apparently dichotomous (Delevoryas and Hope 1971; Crepet 1974), but apparently pachycaulous as in extant cycads with slender stems (Stevenson 1990). In keeping with the manoxylic nature of the two groups (obvious in most cases, less so in exceptions such as *Lyssoxylon grigsbyi*; Gould 1971), there is relatively sparse secondary wood, wide medullary rays, and a large pith and cortex as in extant cycads, independent of stem diameter and height (cf. Cúneo et al. 2010). Table 10.1 summarizes what are typically regarded as shared Bennettitales versus Cycadales characters. It is worth mentioning that contrasting characters between the Bennettitales and Cycadales in Table 10.2 distinguish the Bennettitales from the Cycadales, but not all are necessarily different from corresponding characters in seed ferns.

Fig. 10.2. Habit of the bennettitalean *Cycadeoidea. Redrawn from Delevoryas (1971) by Michael Rothman.*

For example, a nucellus free from the integuments is a characteristic of certain Paleozoic seed ferns (e.g., *Lyginopteris*), as are a granular pollen (or prepollen) wall and synangiate microsporangia on pinnate microsporophylls. Similarities to other older seed plants, while possibly retained plesiomorphies in a derived taxon, underscore the unusual character mix of the Bennettitales and their possible mosaic nature.

Although superficially similar in habit, the Cycadales and Bennettitales have usually been regarded as separated by a number of anatomical, cuticular, and reproductive characters (Table 10.2), and many of these are incorporated in most morphology-based phylogenetic analyses.

Cycadalean–Bennettitalean Character Contrasts

Upon close scrutiny, however, some of the characters traditionally thought to separate the Cycadales from the Bennettitales turn out to overlap or are somewhat ambiguous, and thus merit closer inspection. Such characters (a subset of those listed in Table 10.2) are listed in Table 10.3, with the actual distributions of states noted. Although some character distributions will be better understood and coded after more detailed studies of anatomy and morphology of living and fossil taxa are conducted, it is already clear that the distributions of some characters have not been accurately portrayed in many phylogenetic analyses, and that some characters that

Character	Bennettitales	Cycadales
Branching	Sparse[a]	Sparse
Axillary branching	?	Absent
Manoxylic wood	Present	Present
Leaves	Pinnate/pinnate venation	Pinnate
Persistent leaf bases	Present	Present
Omega-like petiole vascular bundle arrangement	Present	Present
Unisexual cones	Present	Predominant
Pollen aperture	Monosulcate	Monosulcate

have worked into the conventional wisdom of plant systematics appear to have been coded incorrectly as a result of inadequate sampling of the character states in the Cycadales, even though this information has been available in the literature. Nonetheless, significant distinctions remain between the Cycadales and the Bennettitales on the bases of morphology and anatomy (Table 10.2). Furthermore, it is clear from Table 10.4 that there are numerous new characters and character states now known for extant and fossil cycads that have yet to be examined in the Bennettitales. Although some characters relevant to reproduction seem unlikely to be preserved in fossils (e.g., alveolar megagametophyte development), the quality of preservation seen in the bennettitalean cones described by Rothwell and Stockey (2002) suggests otherwise, and in any case, there are other characters likely to be preserved, such as cell packets (Stevenson 1980) and epidermal cell features, which can be more extensively and intensively studied in the Bennettitales.

Bennettitales/Gnetales Characters

Table 10.5 summarizes the contrasting characters of the Bennettitales/Gnetales. Once again, there are some dramatic but possibly superficial similarities (analogs possibly, if the preponderance of molecular sequence–based estimates of seed plant relationships are correct), which have been traditionally regarded as synapomorphies. However, there are also significant differences in Table 10.5, and these have not always been included in morphology-based phylogenetic analyses.

The gnetalean–bennettitalean connection has been reviewed in some detail by Rothwell et al. (2009), and their review includes analyses of morphological data, in addition to an extensive discussion and review of comparative seed morphology, in response to reports of fossil charcoalified seeds alleged to link the Gnetales and Bennettitales (Friis et al. 2007). However, the analyses of Rothwell et al. (2009) suggest that Bennettitales are taxonomically distinct from the Gnetales (and not part of a BEG clade), and similar results (with respect to a BEG clade) have been obtained here in the analysis of an expanded and modified morphology matrix (analysis 3 of Rothwell et al. 2009). Such findings, in conjunction

Character	Bennettitales	Cycadales
Girdling leaf traces	Absent	Present[a]
Pitting	Scalariform	Circular bordered[a]
Stomata	Paracytic	Anomocytic
Indeterminate ovulate cones	Absent	In *Cycas*
Bisporangiate cones	Present	Absent
Perianth of bracts	Present	?Absent
Microsporangia	Adaxial	Abaxial
Microsporangia	Synangiate/encapsulated	Aggregated
Microsporophylls	Pinnate/paddlelike	Bladelike
Nucellus separates from integument	At base	In midregion
Pollen wall structure	Granular	Spongy[a]/alveolar
Micropyle	Tubular	Simple
Leaf epidermis	With sinuous anticlinal walls	Without sinuous anticlinal walls
Regularly oriented stomata	Present	Absent[a]
Accessory cells	Two	More than two

Table 10.2. Convention View of Contrasting Characters between the Bennettitales and Cycadales

Source. Based on Stevenson (1990) and Hermsen et al. (2006).
[a] *Plesiomorphic cycad states.*

with the results of numerous independent nucleic acid–based analyses (Chaw et al. 2000; Mathews 2009), are consistent with the morphological differences between these clades and imply that the similarities, although dramatic because they are unusual, may be analogous rather than homologous. This possibility is consistent with the opinions of some early twentieth-century proponents of a gnetalean–bennettitalean phylogenetic disjunction. Rothwell et al. (2009) quote W. P. Thompson (1912: 1099), as we do again here, to illustrate the conviction of certain researchers on this issue:

> It may be stated at once that on the anatomical side there is very little evidence for connecting the Bennettitales with *Ephedra*, although this genus, being the most primitive of the Gnetales, is the one where the evidence ought to be found. . . . Thus almost every tissue presents grave obstacles to this view.

To some degree, Thompson's reservations are applicable today, and upon analyses of various relevant data matrices, Rothwell et al. (2009) did not find support for a BEG clade or for a Bennettitales-Gnetales (BG) clade. Thus, depending on phylogenetic context, the Gnetales seem different enough from the Bennettitales in the details of their morphology and anatomy, including seed structure, so that nucleic acid sequence analyses may not misplace them with the rest of the gymnosperms, which is consistent with the fact that not all morphology sets place them unambiguously with anthophytes. It is interesting, too, that if one accepts the reality of a BEG clade (Rothwell et al. 2009), and if nucleic acid sequence analyses ultimately irrefutably place the Gnetales with other gymnosperms, then the phylogenetic position of the entire BEG clade would be affected and in turn also challenge the anthophyte concept.

Table 10.3. Actual Distribution of Certain Bennettitales Versus Cycadales Characters

Source. Based on Stevenson (1990) and Hermsen et al. (2006).
[a] Plesiomorphic cycad states.
[b] Data from Osborn and Taylor (1995) and unpublished data on *Williamsoniella* pollen from W.L.C.

Character	Bennettitales	Cycadales
Secondary tracheid pitting	Primarily scalariform	Circular bordered[a] or scalariform
Accessory cells	Two	Two but only in certain taxa
Stomatal orientation	Regular	Random[a] or regular
Anticlinal walls in leaf epidermis	Sinuous	Sinuous[a] or straight
Microsporangia borne	?Adaxially	Abaxially
Endexine[b]	Thin ?lamellated	Thick lamellated
Pinnate megasporophylls	?Absent	In *Cycas*
Girdling leaf traces	Absent	Unequivocally present in leaves in most taxa, but not in cataphylls or megasporophylls

Table 10.4. Unexplored Comparisons between Cycads (Fossil and Extant) and Bennettitales

Source. Based on Hermsen et al. (2006).

Character	Bennettitales	Gnetales
Ovule vasculature	?	Present
Embryo feeder	All?	Absent
Pollen chamber	Absent	Present
Alveolar megagametophyte	?	Present
Branched pollen tube	?	Present
Branched trichomes	?	Present
Epidermal cells with simple pitting	?	Present
Epidermal cell perforations	?	Present
Buffer cells surrounding archegonium	?	Present
Pith cell packets	?	Present
Primary thickening meristem	?	Present
Leaf ptyxis	?	Erect
Pinna ptyxis	?	Involute
Pinna traces	?	2
Cataphylls	Present	Present
Mucilage canals	?	Present
Leaf bases	Persistent	Persistent
Epidermal cell wall thickenings	?	Present
Leaf trace protoxylem transition	?	Present
Spongy ovular tapetum	?	Present
Endospermic jacket	?	Present

Bennettitales–Angiosperm Characters

Character differences notwithstanding (Table 10.6), angiosperms and the Bennettitales remain linked in most phylogenetic analyses that incorporate conventional coding for cycadalean characters and that also include the order "Bennettitales" rather than discrete bennettitalean taxa. Even the analysis of Rothwell et al. (2009), which incorporated certain character state changes reflective of their actual distribution in modern Cycadales, resulted in an anthophyte clade that included the Bennettitales (Rothwell et al. 2009, fig. 30). Yet there are no fossil or extant intermediates that form a morphocline linking the Bennettitales to angiosperms, and the transition to a carpel with bitegmic ovules would require a complex transformation with a number of

Table 10.5. Contrasting
Characters of the Bennet-
titales and Gnetales

Character	Bennettitales	Gnetales
Unequivocal axillary branching	?	Present
Pitting of tracheary elements in secondary wood	Primarily scalariform	Primarily circular bordered pits
Vessels in wood	Absent	Present
Leaf arrangement	Helical	Opposite/whorled
Leaf form	Pinnate	Nonpinnate (simple)
Dichotomous leaf venation	Absent	Present (*Gnetum*)
Sheathing leaf bases	Absent	Present
Origin of leaf trace bundles	Variable (one to several)	Two
Rachis vascular configuration	Omega pattern	Non-omega pattern
Cone complexity	Simple	Compound
Ovules borne on	Modified sporophyll(s)	Stem tip
Ovules enclosed by bracteoles	Absent	Present
Nucellus free from integument	From chalaza	From midregion (variable)
Pollen chamber	Absent	Present
Biseriate tubular micropyle	Absent	Present
Micropyle closure	By nucellar plug	By ingrowth of cells

intermediates stages that for now exist only in hypothetical form (e.g., Doyle 1996)—and, as noted above, may never have existed, even if the angiosperm–bennettitalean relationship is real. Nonetheless, even in the absence of bona fide morphological mosaics and in the face of distinct differences in key characters, most phylogenetic analyses of morphology-based data sets place angiosperms and the Bennettitales in an anthophyte clade on the basis of the similarities noted above and on their restricted distribution.

With respect to anthophytes in the broader sense, there are, of course, both differences and similarities between the angiosperms and the other taxa. The differences might support a different hypothesis of phylogeny, such as those generated by nucleic acid sequence–based analyses, while the similarities support the anthophytes, such as in all existing phylogenetic analyses based on morphological characteristics. However, taking into account that the differences between the Gnetales and angiosperms/Bennettitales may be significant, given the results of most nucleic acid sequence–based analyses, and that the differences between the Cycadales and Bennettitales have been exaggerated in existing matrices, it is interesting to experiment with morphology-based data bases to see how analyses of such matrices might be affected by some changes in character codings that seem warranted by close inspection of character variation in living taxa and by inclusion of additional bennettitalean terminals to capture variation in key characteristics within the order.

Table 10.6. Characters of the Bennettitales and Angiosperms

Character	Bennettitales	Angiosperms
Bisexual reproductive structures	Present	Present
Carpels	Absent	Present
Ovules borne on	Modified sporophyll(s)	Modified sporophylls?
Nucellus	Free from integument	Fused to integument
Tubular micropyle	Present	Absent
Ovule integument	Single	Single and double
Interseminal scales	Present	Absent
Pollen morphology	Monosulcate	Monosulcate and other types
Pollen wall	Granular	Granular and other types
Endexine	Lamellated	(Usually) Nonlamellated
Unequivocal axillary branching	?	Present
Stomata	Paracytic (syndetocheilic?)	Syndetocheilic and other types
Pitting of tracheary elements	Primarily scalariform	Scalariform and other types
Vessels in wood	Absent	Present
Leaf arrangement	Helical	Helical and other types
Leaf form	Pinnate	Pinnate and other types
Dichotomous leaf venation	Absent	Occasionally present
Leaf venation orders hierarchical	Absent	Present
Sheathing leaf bases	Absent	Occasionally present
Origin of leaf trace bundles	Variable	Variable

Revised Character Coding and Analysis

In this chapter, revised changes in character codings reflect additional information from the literature, as well as direct examination and reexamination of specimens, including those of petrographic thin sections or peels of the genera *Cycadeoidea* and *Williamsonia* (as reported by Rothwell et al. 2009), transmission electron micrographs of *Williamsoniella* pollen (Crepet, unpublished work), and direct examination of specimens of *Williamsoniella* from the University of Connecticut Paleobotanical Collection. New codings are conservative within the context of observed structural and morphological variation for the relevant taxa (Appendices 10.1–10.3). For example, the consistent occurrence of girdling leaf traces in the Cycadales (Table 10.1) has generally been accepted as a character distinguishing the Cycadales from the Bennettitales, even though *Stangeria eriopus* has traces that run directly (if obliquely) from the vascular column to the leaf (Wordsdell 1898; Norstog and Nicholls 1997), and cycadalean leaf homologs—cataphylls and sporophylls—lack girdling traces. In fact, Norstog and Nicholls (1997), citing this evidence, caution against assuming fossils are cycadeoids simply because they appear to lack girdling leaf traces. But such new character information has not been integrated into the matrix in the form of revised character codings, and newly identified characters have also not been added to the matrix at this time, even if they appear to be uniquely shared by the Bennettitales and

Cycadales (e.g., petiole vasculature and dorsal guard cell thickenings) in the interests of being as conservative as possible in the absence of really adequate comparative anatomical studies (e.g., variation in stomatal development in the Cycadales is not well enough understood, and many cycad characters listed in Table 10.4 have yet to be investigated in the Bennettitales). The state of the field of paleobotany also places certain limitations on establishing morphology-based databases as a result of missing values for some relevant fossil taxa and sometimes inadequate studies of relevant fossils. Thus, in this exercise, the number of modified character codings is relatively low. Although certain characters and character codings have been adjusted on the basis of direct investigations and reexaminations of fossils, as well as the literature review mentioned above for the analysis of Rothwell et al. (2009), two other characters or groups of characters will be pursued here in greater depth: leaf cuticular features and the ovulate receptacle.

Leaf Cuticular Features

Rothwell et al. (2009) do not discuss cuticular features that are often used to separate foliage of the Bennettitales from Cycadales, yet these differences may typically have been overstated. Within the Zamiaceae, there are subsidiary cell configurations (two subsidiary cells) that might be characterized as paracytic, which has been known since Thomas and Bancroft first discussed these cuticles in their oft-cited paper of 1913 and later observed by Greguss (1968). Moreover, distinctive dorsal guard cell thickenings so characteristic of bennettitalean guard cells (Sincock and Watson 1988) are also found in the Cycadales and are rarely reported in other taxa (Sincock and Watson 1988). Finally, sinuous anticlinal walls in leaf cuticles are often cited as characters that help distinguish the leaves of fossil Cycadales from those of the Bennettitales (e.g., Sincock and Watson 1988), yet such anticlinal walls also occur in certain cycads (e.g., *Stangeria*, Greguss 1968; *Macrozamia* and *Ceratozamia*, Pant and Nautiyal 1963) as do nonrandomly oriented stomata (Greguss 1968).

The Ovulate Receptacle

One character in particular, the ovulate receptacle, is perhaps foremost among those that distinguish bennettitaleans and is essentially unique to them, depending one's perspective on *Pentoxylon* relationships and its ovulate cone structure. Ovulate receptacles are known from petrified fossils of the Cycadeoidaceae and Williamsonia, as well as from well-preserved compressions of certain other taxa (e.g., *Williamsoniella*), although, as a result of the preservational mode, the latter fossils do not include some anatomical details. Ovulate cone structure is apparently uniform throughout the Bennettitales (with the possible exception of *Westersheimia*, discussed below) and consists of apparently radially symmetrical fleshy receptacles bearing numerous ovules and interseminal scales (generally considered to be homologous with ovules; Wieland 1906). Although usually conical, within the genus *Cycadeoidea*, there are two principal forms: those with small dome-shaped receptacles, and those with larger conical receptacles (Delevoryas 1963, 1968a; Crepet 1974). Ovulate cones have been interpreted as axes bearing reduced megasporophylls, with each ovule and pedicel representing a megasporophyll, so that thousands would be represented by the typical conical ovulate receptacle. Despite their essentially unique nature and significance in defining the Bennettitales, not enough attention has been focused on details of receptacle anatomy and its possible significance in establishing cone homology.

Ovulate receptacles in *Cycadeoidea* spp.

In fact, ovulate receptacle vascularization has not been sufficiently investigated and has sometimes been misrepresented in illustrations, even in classical studies (e.g., Wieland 1906). The vasculature is often represented as a eustele from which ascending traces enter the ovules and interseminal scales directly (e.g., Wieland 1906, fig. 65). On closer inspection, the vasculature of *Cycadeoidea* spp. with conical ovulate receptacles is often dramatically different. There is a central ovulate receptacle circular arrangement of numerous vascular strands (approximately 10–15) surrounding a "pith"; however, the vascular bundles do not give rise to ovule or interseminal scale traces directly. The vascular bundles proceed acropetally and then sharply reflex; in longitudinal sections, they can be seen to run almost parallel to the surface of the ovulate receptacle until they terminate near its base (Figs. 10.3, 10.4A). In cross section, this creates an additional and concentric whorl of vascular traces outside of the more centrally located traces that is also observable in cross sections of well-preserved ovulate receptacles (cf. Rothwell et al. 2009, fig. 9, which illustrates a similar situation in *Williamsonia* sp.). As these vascular strands descend, they give rise to traces that dichotomize (sometimes, and which in some taxa then anastomose) and enter the pedicels of ovules and bases of interseminal scales (Fig. 10.4B). This is unusual because it implies that vascular traces to ovules emanate from the interior of what have been considered eustelar vascular bundles (which are inverted distally because they are reflexed), contrary to the normal site of origination of the traces of various lateral appendages.

The vasculature of the ovulate receptacle is thus, at least in some taxa, a ring of reflexed vascular traces that produces numerous dichotomously branching strands pectinately or pinnately. Although a number of hypotheses of homology might be offered in explanation, the origination of the ovule vasculature from the interior side of the recurved traces is perhaps easier to rationalize in terms of appendicular structures, such as the vascularization of pinnae from a frond rachis, than in terms of axial anatomy. How can this observation be rationalized in the context of the often presumed homology between the ovulate receptacle and a lateral branch? There are a number of possible explanations. First,

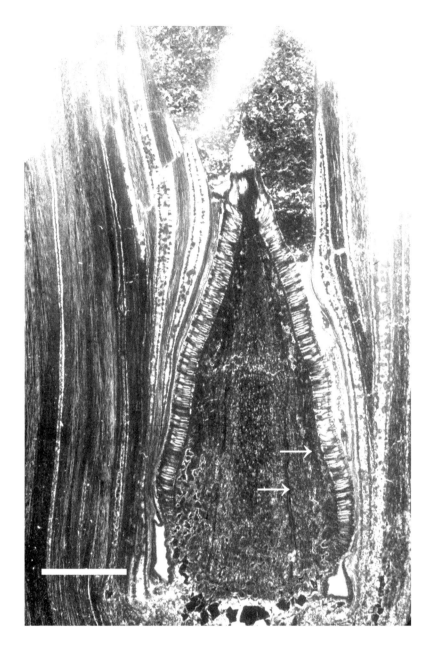

Fig. 10.3. Ovulate receptacle of *Cycadeoidea* with disintegrating microsporophylls preserved distally. The conical ovulate receptacle bears numerous ovules among interseminal scales and illustrates both interior and exterior vascular strands (arrows). Crepet (Yale) slide NS 155. Scale bar = 6.6 mm.

such vascularization could simply be anomalous, with no connotation of homologies and no phylogenetic implications. Such anomalous vascularization would exist at two levels: at the level of cone-bearing axis vascularization, where vasculature originates from leaf traces within the cortex, as discussed above and by Andrews (1943) and Delevoryas (1968b), and at the level of vascularization of the ovules and interseminal scales. Alternatively, if the reflexed vascular bundles in the ovulate receptacle and the pattern of vascularization of ovules and interseminal scales have evolutionary significance, it raises the possibility that ovulate receptacles might be homologous with a modified and ontogenetically fused whorl

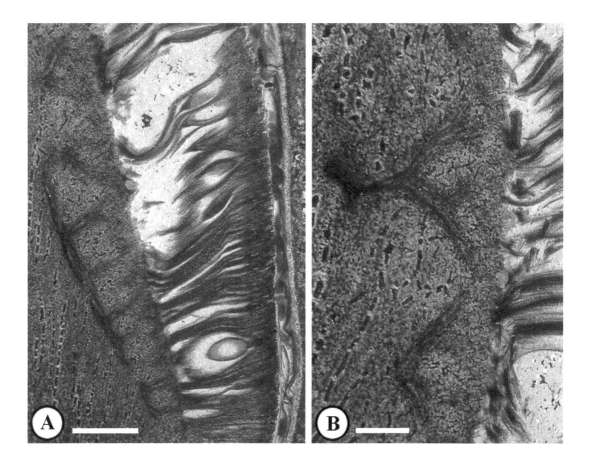

Fig. 10.4. (A) Longitudinal section of *Cycadeoidea* sp. showing traces near the periphery of the ovulate receptacle from which vascular bundles are diverging pectinately to ovule pedicels and interseminal scale base. Crepet (Yale) slide NS 74.2. Scale bar = 1 mm. (B) Ovulate receptacle of *Cycadeoidea* sp. showing dichotomizing traces extending toward bases of ovule pedicels and interseminal scales. Crepet (Yale) slide NS 74.2. Scale bar = 0.5 mm.

or tight spiral of megasporophylls rather than with a terminal branch axis bearing hundreds or even thousands of lateral appendages reduced to pedicels and ovules or interseminal scales (this "reduced megasporophyll" hypothesis was discussed by Delevoryas 1968b). This possibility is consistent with vascularization of the cones axes from leaf traces in the trunk cortex, as pointed out by Delevoryas (1968b), and given the reflexed nature of the receptacle vascular bundles, such a model would imply that the neotonously modified (putatively homologous) megasporophylls composing the ovulate receptacle would themselves have been reflexed rather than incurved. Is this plausible when pinnate leaves, presumed general homologs, often fold inwardly (Stevenson 1981)?

Interestingly, leaves in certain taxa of the Cycadales (e.g., *Dioon*) do exhibit some distal recurvature during ontogeny (Stevenson 1981; Norstog and Nicholls 1977), adding plausibility to the proposed homology of the ovulate receptacle in *Cycadeoidea*, which also has folded pinnate leaf ontogeny (Wieland 1906). There are other characteristics of preserved bennettitalean ovulate receptacles that may support homology with a whorl of megasporophylls. Ovulate receptacle apices in *Williamsoniella* (and even on occasions in *Cycadeoidea*, Yale slide 54.6), for example, are segmented, raising the possibility that this configuration might be a vestige of the individual megasporophylls that have been neotonously transformed into the conical receptacle. Moreover, sometimes fossil ovulate receptacles are longitudinally divided and even bear ovules and interseminal scales on both the inner and outer sides of the segments, possibly reflecting anomalous incomplete fusion of ovulate receptacle "megasporophyllous segments" during development, but it is also possible that such a configuration is a preservational anomaly (Crepet, unpublished data, e.g., Yale Paleobotanical Collection, trunk 54, slide 54.6). It is also interesting to note that neotony is observed within the bennettitalean lineage in evolution of the microsporangiate apparatus, which, in *Cycadeoidea*, has become segmented with permanently (in this case, incurved) microsporophylls that are partially fused and have become modified through proliferation of abaxial rachial tissue (Crepet 1974).

In another scenario raised by Delevoryas (1968b), cones of *Cycadeoidea* might be regarded as homologous with even a single leaf. Cone axis vascularization could have been plausibly derived from typical cycadeoid petioles, which have numerous bundles arranged in omega-like to more or less circular patterns (Wieland 1906, figs. 32, 33), and the ovulate receptacle could be viewed as having evolved through the neotonous fusion of ovule-bearing pinnae in a three-dimensional sporophyll rather than from the fusion of individual megasporophylls on an axis (see Delevoryas 1968b for additional discussion about this possibility). It is important to point out that a foliar origin of the cone of *Cycadeoidea* is compatible with evidence on cone vascularization. Cones of *Cycadeoidea* are vascularized in the cortex from leaf traces (Andrews 1943; Delevoryas 1968b), and there is little in the anatomy of *Cycadeoidea* to suggest that cones terminate axillary branches.

Of course, the suggestion that the cycadeoid ovulate receptacle is somehow homologous to a whorl of pinnate megasporophylls raises a question: Are there transitional fossils that would support this hypothesis? There might be. The compression fossil *Westersheimia*, although insufficiently studied, is most interesting and germane in this context. *Westersheimia* occurs in the Triassic of Lunz, Austria, and is known from only a few fossils (Kräusel 1949). The most complete fossil (Fig. 10.5) is a compressed axis bearing three categories of abscission scars on its surface. The axis, which appears to have at least one branch scar (Kräusel 1949), is associated with the bennettitalean leaf genus (*Pterophyllum longifolium*), as are each of the other *Westersheimia* fossils. Although once thought to be attached to the axis of the specimen in Kräusel's paper (Fig. 10.5), the leaves are actually only associated with the axis (Kräusel 1949). Nonetheless, the diameters of the associated leaf bases and spirally arranged dehiscence scars on the axis are consistent with the possibility that they represent the same taxon implied by the regular association of these fossils. There are also smaller, spirally arranged abscission scars of more or less the same shape on the axis, and Kräusel suggests that these may represent bracts (cataphylls?).

Another class of abscission scars is more round than rhomboidal, and these are arranged apparently intermittently in flat spirals or whorls.

Fig. 10.5. *Westersheimia. Redrawn from Kräusel (1949).*

Most interesting is a pinnate structure attached to one of these whorls. There are apparently at least nine upright pinnae, four on each side, with one terminal pinna (evidence of a missing proximal attachment can be seen in Fig. 10.5). The pinnae are interpreted as having been more or less conical/cylindrical structures in three dimensions. They bear typically bennettitalean ovules plus interseminal scales. Are these pinnate structures megasporophylls? The whorl/flattened spiral of abscission scars in the same whorl and the still-attached pinnate structure suggest that they abscised, a possibility that is also supported by the discovery of isolated pinnate ovulate organs of *Westersheimia*. Although Crane (1988) is appropriately circumspect about the homology of these pinnate fertile structures (i.e., branch versus megasporophyll), and reexamination of these interesting fossils seems warranted, available evidence is consistent with interpreting them as pinnate megasporophylls. Moreover, because they occur in apparent intermittent whorls, they are evocative of the indeterminate megasporangiate "cones" of *Cycas* (Norstog and Nicholls 1997). Finally, it is interesting to note that the reflexed nature of the vasculature discussed above is also known in some cycads such as *Stangeria* (Matte 1904; Stevenson 1990).

In order to assess the implications of changes in our understanding of character variation and character states discussed, changes in appropriate characters and character codings were made to the morphology/structure-based matrix that, upon analyses with NONA, consistently resolved an anthophyte clade (analysis 3 matrix of Rothwell et al. 2009). The Rothwell et al. (2009) matrix was originally assembled by combining the Hilton and Bateman (2006) matrix with the Rothwell and Serbet (1994) matrix and edited to remove redundant characters. The Hilton and Bateman (2006) matrix is largely based on the Doyle (1996) matrix, which in turn incorporates some generally accepted additions or changes suggested by Nixon et al. (1994), as well as other updates. Thus, the analysis 3 matrix of Rothwell et al. (2009) is representative of a number of different perspectives, but has undergone recent refinement (cf. Rothwell et al. 2009). Results of the various analyses of the Rothwell et al. (2009) matrix, while slightly different with respect to the placement of *Ginkgo*, reflect generally accepted seed plant relationships based on morphological characters (Rothwell et al. 2009, fig. 30), except that the position of *Caytonia* has changed as a result of coding changes aimed at removing implicit hypotheses discussed by Nixon et al. (1994) and Rothwell et al. (2009). However, in the present context, this matrix is intended to serve as a template for experimenting with character states and taxa that are relevant to the Bennettitales rather than to refute or corroborate any particular hypothesis of seed plant relationships. In order to do so, we have modified certain character states and added additional taxa to make the matrix more up to date while representing the taxon of interest (Bennettitales) in greater detail.

Specifically, the analysis 3 matrix of Rothwell et al. (2009) matrix has been manipulated in the following fashion. First, characters that appear to have been erroneously coded in a way that exaggerates the difference between the Cycadales and Bennettitales, which have not already addressed in the analysis 3 matrix of Rothwell et al. (2009) (i.e., those noted in Table 10.3), have been recoded accordingly but conservatively. Second, the bennettitalean taxa *Cycadeoidea*, *Williamsonia*, *Williamsoniella*, and *Westersheimia* have replaced "Bennettitales" in the matrix. Character coding for these taxa mirrors codings in other analyses and is noncontroversial (see matrix). Separate analyses with character 52 (megasporangium/ovule-bearing structure) coded as unknown following Hilton and Bateman (2006) or with this character coded as state 1 ("simple paddlelike megasporangia not in two rows") were conducted with similar results (Fig. 10.6). These two analyses (herein grouped for reference purposes under analysis 1) also included the character coding of 0 for character 52 (pinnate) for *Westersheimia* reflecting the discussion above. Finally, an additional analysis (analysis 2) of the matrix included changes noted in the first two points above, but with a single difference: character 52 (megasporangium/ovule-bearing structure) was changed from pinnate (0) to unknown for the genus *Westersheimia*.

While, as noted above, additional characters shared by the Cycadales and Bennettitales have been identified during the course of this study,

Fig. 10.6. Strict consensus tree resulting from analysis 1 testing relationships among the Bennettitales and putatively related clades. Note the Bennettitales have been removed to a clade that includes the Cycadales and other seed ferns. See text for details.

for example, the aforementioned unusual petiole trace configuration that seems to be common to only the Cycadales and Bennettitales and the dorsal cuticular thickenings on guard cells, there are many others worth exploring (Table 10.4). These characters, however, have not been added to the matrices for analysis. To add these characters, many of which may well be shared by the Bennettitales and Cycadales, without a careful review of the comparative morphology of key characters in both living and fossil taxa would be contrary to the goals of this preliminary exercise. However, they demonstrate the need and provide the basis for such a thorough study in light of recent works on the Bennettitales and fossil Cycadales and their implications for seed plant relationships.

Results

Results of Analysis 1

This matrix, using NONA, as in the analyses of the matrices by Rothwell et al. (2009), produced only four most parsimonious trees of 327 steps (CI = 50, RI = 78), and only three branches collapsed in the strict consensus tree (Fig. 10.6). Overall, the phylogeny conforms more or less to the broad topology of the original analysis 3 of Rothwell et al. (2009), but the monophyletic group that includes *Caytonia* also includes the Cycadales and Bennettitales. This new group (if durable in the face of future and more rigorous analyses) perhaps could be characterized as a nouveau Cycadofilicales and includes three subclades: Cycadales (Cycadaceae + Zamiaceae), (*Glossopteris* + (Corystosperms + *Caytonia*)), and the Bennettitales plus *Pentoxylon* (Fig. 10.6).

This topology is rather stable, and changes to various character codings have surprisingly little effect on the outcome of the analysis. Changing character 52 from 1 to unknown for *Cycadeoidea*, *Williamsonia*, and *Williamsoniella*, for example, and changes in epidermal characters of Zamiaceae and Cycadaceae, which might be expected to affect the placement of Bennettitales versus Cycadales, also have little effect. A striking exception is change in character 52 for *Westersheimia*.

Results of Analysis 2

If character 52 is changed to unknown from pinnate (0) for *Westersheimia*, 128 equally parsimonious trees also of 327 steps result and 17 nodes collapse in the strict consensus tree with a corresponding loss of resolution that includes dissolution of the aforementioned (Cycadaceae + Zamiaceae) + (*Glossopteris* + (Corystosperms + *Caytonia*)) + (Pentoxylon + (Bennettitales)) clade (cf. Figs. 10.6, 10.7).

This exercise suggests that there is still a need for attention to details of comparative morphology in many seed plants (fossil and modern) if we are to resolve the apparent discrepancy between nucleic acid sequence–based and most morphology-based trees and to move on to understanding seed plant phylogenetic relationships in greater detail. Characters and taxa incorporated in analysis 1 resulting in an unusual latter-day "cycadofilicalean" clade were rather conservatively coded (depending, of course, on one's point of view), given the apparent actual distribution of characters. The analysis is preliminary in the sense that more data are needed on variation in critical characters in living and fossil taxa (but note, too, that the analyses do not include recently identified additional characters shared by the Cycadales and Bennettitales, adding yet another measure of conservatism to the results). Moreover, confirmation is necessary for the nature of the vascularization of the cycadeoidalean ovulate receptacle, which appears to vary within the Bennettitales. Although such vascularization is also present in cones of *Williamsonia* (Rothwell et al. 2009), it is not clear whether this pattern of vascularization is uniformly distributed within the genus *Cycadeoidea* itself. It is also obvious that further corroborative studies are required for definitively coding the characters of *Westersheimia*. Finally, the Erdtmanithecales and the "charcoalified

Concluding Discussion

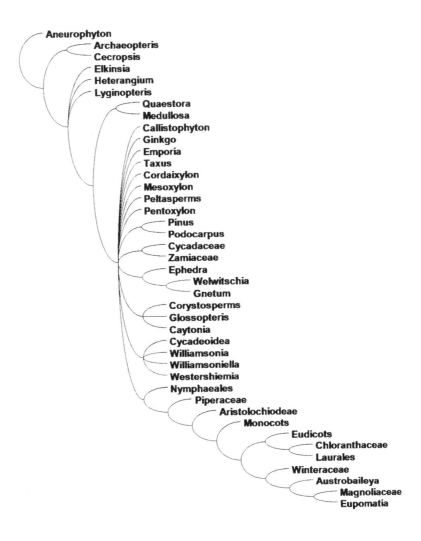

seeds" of Friis et al. (2007) have not yet been included in the above analyses, pending the availability of further information on these taxa for reasons discussed by Rothwell et al. (2009). Nonetheless, this preliminary investigation draws attention to the critical phylogenetic importance of ovulate cone structure for assessing bennettitalean homologies and phylogenetic significance. The concordance of these results with the results of nucleic acid–based phylogenetic analyses (e.g., Chaw et al. 2000) that also remove the Gnetales from Bennettitales is interesting.

This is, in fact, only a rough starting point for a more detailed investigation that may well lead to a different outcome, and the above congruity regarding the phylogenetic separation of the Gnetales and Bennettitales may yet turn out to be coincidental. Meanwhile, there is a perfect symmetry in these results because in 1968, Ted Delevoryas suggested that the Bennettitales (like the Cycadales) were likely most closely related to pteridosperms (and not to the Gnetales). Although the issue is far from settled, the results of the present exercise raise interesting questions and signal the beginning of a new chapter in the quest to achieve a consensus in this enduring dialogue. The incorporation of new comparative observations on extant and fossil cycads and Bennettitales in addition to pteridosperms may shed new light on what has been enigmatic.

Acknowledgments

We thank Gar Rothwell and Kevin Nixon for discussions relevant to this chapter, and Michael Rothman for certain reconstructions.

References Cited

Andrews, H. N. 1943. On the vascular anatomy of the cycadeoid cone axis. *Annals of the Missouri Botanical Garden* 30: 421–427.

Arber, E. A. N., and J. Parkin. 1907. On the origin of angiosperms. *Botanical Journal of the Linnean Society* 38: 29–80.

Arnold, C. A. 1947. *An Introduction to Paleobotany.* New York: McGraw-Hill.

Artabe, A. E., and D. W. Stevenson. 1999. Fossil Cycadales of Argentina. *Botanical Review* 65: 219–239.

Artabe, A. E., A. B. Zamuner, and D. W. Stevenson. 2004. Two new petrified cycad stems, *Brunoa* gen. nov. and *Wordsellia* gen. nov., from the Cretaceous of Patagonia (Bajo de Santa Rosa, Province of Río Negro), Argentina. *Botanical Review* 70: 121–133.

———. 2005. A new genus of Late Cretaceous cycad stem from Argentina, with reappraisal of known forms. *Alcheringa* 29: 87–100.

Bierhorst, D. W. 1971. *Morphology of Vascular Plants.* New York: Macmillan.

Chamberlain, C. J. 1920. The living cycads and the phylogeny of seed plants. *American Journal of Botany* 7: 146–153.

Chaw, S. M., C. L. Parkinson, Y. Cheng, T. M. Vincent, and J. D. Palmer. 2000. Seed plant phylogeny inferred from all three plant genomes: monophyly of extant gymnosperms and origin of Gnetales from conifers. *Proceedings of the National Academy of Sciences U.S.A.* 97: 4086–4091.

Crane, P. R. 1985. Phylogenetic analysis of seed plants and the origin of angiosperms. *Annals of the Missouri Botanical Garden* 72: 716–793.

———. 1988. Major clades and relationships in "higher" gymnosperms. In C. B. Beck (ed.), *Origin and Evolution of Gymnosperms,* pp. 218–272. New York: Columbia University Press.

Crepet, W. L. 1974. Investigations of North American cycadeoids: the reproductive biology of *Cycadeoidea. Palaeontographica,* Abt. B, 148: 144–169.

Cúneo, R. N., I. Escapa, L. Villar de Seoane, A. Artabe, and S. Gnaedinger. 2010. Review of the cycads and bennettitaleans from the Mesozoic of Argentina. In C. T. Gee (ed.), *Plants in Mesozoic Time: Morphological Innovations, Phylogeny, Ecosystems,* pp. 187–212. Bloomington: Indiana University Press.

Delevoryas, T. 1959. Investigations of North American cycadeoids: *Monanthesia. American Journal of Botany* 46: 657–666.

———. 1960. Investigations of North American cycadeoids: trunks from Wyoming. *American Journal of Botany* 47: 778–786.

———. 1963. Investigations of North American cycadeoids: cones of *Cycadeoidea. American Journal of Botany* 50: 45–52.

———. 1965. Investigations of North American cycadeoids: microsporangiate structures and phylogenetic implications. *The Palaeobotanist* 14: 89–93.

———. 1968a. Investigations of North American cycadeoids: structure, ontogeny and phylogenetic considerations of cones of *Cycadeoidea. Palaeontographica,* Abt. B, 121: 122–133.

————. 1968b. Some aspects of cycadeoid evolution. *Botanical Journal of the Linnean Society* 61: 137–146.

————. 1971. Cycadeoidales. In *McGraw-Hill Encyclopedia of Science and Technology*, 3rd ed., 689. New York: McGraw-Hill Book Company.

Delevoryas, T., and R. C. Hope. 1971. A new Triassic cycad and its phyletic implications. *Postilla* 150: 1–21.

Dower, B. L., R. M. Bateman, and D. W. Stevenson. 2004. Systematics, ontogeny, and phylogenetic implications of exceptional anatomically preserved cycadophyte leaves from the Middle Jurassic of Bearreraig Bay, Skye, Northwest Scotland. *Botanical Review* 70: 105–120.

Doyle, J. A. 1996. Seed plant phylogeny and the relationships of Gnetales. *International Journal of Plant Sciences* 157: S3–S39.

Doyle, J. A., and M. J. Donoghue. 1986. Seed plant phylogeny and the origin of angiosperms: an experimental cladistic approach. *Botanical Review* 52: 321–431.

————. 1987. The origin of angiosperms: a cladistic approach. In: E. M. Friis, W. G. Chaloner, and P. R. Crane (eds.), *The Origins of Angiosperms and Their Biological Consequences*, pp. 17–50. Cambridge: Cambridge University Press.

————. 1992. Fossils and seed plant phylogeny reanalyzed. *Brittonia* 44: 89–106.

Friis, E. M., and K. R. Pedersen. 1996. *Eucommiitheca hirsuta*, a new pollen organ with *Eucommiidites* pollen from the Early Cretaceous of Portugal. *Grana* 35: 104–112.

Friis, E. M., P. R. Crane, K. R. Pedersen, S. Bengston, P. C. J. Donoghue, G. W. Grimm, and M. Stampanoni. 2007. Phase contrast X-ray microtomography links Cretaceous seeds with Gnetales and Bennettitales. *Nature* 450: 549–552.

Gifford, E. M., and A. S. Foster. 1988. *Morphology and Evolution of Vascular Plants*, 3rd ed. New York: W. H. Freeman.

Gould, R. E. 1971. *Lyssoxylon grigsbyi*, a cycad trunk from the Upper Triassic of Arizona and New Mexico. *American Journal of Botany* 58: 239–248.

Greguss, P. 1968. *Xylotomy of the Living Cycads with a Description of Their Leaves and Epidermis*. Budapest: Akademiai Kiadó.

Hermsen, E. J., T. N. Taylor, E. L. Taylor, and D. W. Stevenson. 2006. Cataphylls of the Triassic cycad *Antarcticycas schopfii* and new insights into cycad evolution. *American Journal of Botany* 93: 724–738.

————. 2007. Cycads from the Triassic of Antarctica: permineralized cycad leaves. *International Journal of Plant Science* 168: 1099–1112.

Hilton, J., and R. M. Bateman. 2006. Pteridosperms are the backbone of seed-plant phylogeny. *Journal of the Torrey Botanical Society* 133: 119–168.

Kräusel, R. 1949. Koniferen und andere Gymnospermen aus der Trias von Lunz, Nieder-Österreich. Untersuchungen zur mesozoischen Florengeschichte des alpinen und süddeutschen Raumes III. *Palaeontographica*, Abt. B, 84: 35–82.

Kvaček, J., and B. Pacltová. 2001. *Bayeritheca hughesii* gen. et sp. nov., a new *Eucommiidites*-bearing pollen organ from the Cenomanian of Bohemia. *Cretaceous Research* 22: 695–704.

Mathews, S. 2009. Phylogenetic relationships among seed plants: persistent questions and the limits of molecular data. *American Journal of Botany* 98: 228–236.

Matte, H. 1904. Recherches sur l'appareil libero-ligneux des Cycadacées. *Mémoires de La Société Linnéenne Normandie* 22: 1–233.

Nixon, K. C., W. L. Crepet, D. W. Stevenson, and E. M. Friis. 1994. A reevaluation of seed plant phylogeny. *Annals of the Missouri Botanical Garden* 81: 484–533.

Norstog, K. J., and T. J. Nicholls. 1997. *The Biology of the Cycads.* Ithaca, N.Y.: Cornell University Press.

Osborn, J. M., and T. N. Taylor. 1995. Pollen morphology and ultrastructure of the Bennettitales: in situ pollen of *Cycadeoidea*. *American Journal of Botany* 82: 1074–1081.

Pant, D., and D. Nautiyal. 1963. Cuticle and epidermis of certain Cycadales. Leaves, sporangia and seeds. *Senckenbergiana Biologia* 44: 257–348.

Pedersen, K. R., P. R. Crane, and E. M. Friis. 1989. Pollen organs and seeds with *Eucommiidites* pollen. *Grana* 28: 279–294.

Rothwell, G. W., and R. Serbet. 1994. Lignophyte phylogeny and the evolution of the spermatophytes: a numerical cladistic analysis. *Systematic Botany* 19: 443–482.

Rothwell, G. W., and R. A. Stockey. 2002. Anatomically preserved *Cycadeoidea* (Cycadeoidaceae), with a reevaluation of the systematic characters for the seed cones of Bennettitales. *American Journal of Botany* 89: 1447–1458.

———. 2010. Independent evolution of seed enclosure in the Bennettitales: evidence from the anatomically preserved cone *Foxeoidea connatum* gen. et sp. nov. In C. T. Gee (ed.), *Plants in Mesozoic Time: Morphological Innovations, Phylogeny, Ecosystems*, pp. 51–64. Bloomington: Indiana University Press.

Rothwell G. W., W. L. Crepet, and R. A. Stockey. 2009. Is the anthophyte hypothesis alive and well? New evidence from the reproductive structures of Bennettitales. *American Journal of Botany* 96: 296–322.

Scott, A. C. 1909. *Studies in Fossil Botany*, vol. 2, 2nd ed. London: Adam and Charles Black.

Sincock, C. A., and J. Watson. 1988. Terminology used in the description of bennettitalean cuticle characters. *Botanical Journal of the Linnean Society* 97: 179–188.

Stewart, W. N., and G. W. Rothwell. 1993. *Paleobotany and the Evolution of Plants.* New York: Cambridge University Press.

Stevenson, D. W. 1980. Radial growth in Cycadales. *American Journal of Botany* 67: 465–475.

———. 1981. Observations on ptyxis, phenology and trichomes in Cycadales and their systematic implications. *American Journal of Botany* 68: 1104–1114.

———. 1988. Strobilar ontogeny in the Cycadales. In S. Leins, S. C. Tucker, and P. K. Endress (eds.), *Aspects of Floral Development*, pp. 205–224. Berlin: Cramer.

———. 1990. Morphology and systematics of the Cycadales. *Memoirs of the New York Botanical Garden* 57: 8–55.

Stockey, R. A., and G. W. Rothwell. 2003. Anatomically preserved *Williamsonia* (Williamsoniaceae): evidence for bennettitalean reproduction in the Late Cretaceous of western North America. *International Journal of Plant Sciences* 164: 251–262.

Taylor, T. N., E. L. Taylor, and M. Krings. 2009. *Paleobotany: The Biology and Evolution of Fossil Plants*, 2nd ed. San Diego: Academic Press.

Thomas, H. H., and N. Bancroft. 1913. On the cuticles of some Recent and fossil cycadean fronds. *Transactions of the Linnean Society London* 8: 155–204, pls. 61–77.

Thompson, W. P. 1912. Anatomy and relationships of the Gnetales. I. The genus *Ephedra. Annals of Botany* 26: 1077–1104.

Wieland, G. R. 1906. *American Fossil Cycads.* Publication No. 34. Washington, D.C.: Carnegie Institution of Washington.

———. 1916. *American Fossil Cycads*, vol. 2, *Taxonomy.* Publication No. 34. Washington, D.C.: Carnegie Institution of Washington.

———. 1934. Fossil cycads, with special reference to *Raumeria reichenbachiana* Göppert sp. of the Zwinger of Dresden. *Palaeontographica*, Abt. B, 79: 83–130.

Wordsdell, W. C. 1898. The comparative anatomy of certain genera of the Cycadaceae. *Botanical Journal of the Linnean Society* 33: 437–457.

```
                   0    5   10   15   20   25   30   35   40   45   50   55   60
                   |    |    |    |    |    |    |    |    |    |    |    |    |

Aneurophyton       0-000-00000002-00----00-00000-000--00000-0-0-00-00-0000-0------
Archaeopteris      1-010-000010020000-0-10-10000-000--100000000-00-20-001200------
Cecropsis          1-110-00-01-0-000000-10-10000-00---100000000-00-20-001200------
Elkinsia           1-121-00-01-200000---01-00000-000--000001000-00-01-0020010-0000
Heterangium        1-121-00-01-20000000001-10000-0000-000001100-00-01-002-02000000
Lyginopteris       1-121-00101120000000002-10000-0000-000001100-00-01-002002000100
Callistophyton     1-121-001011300000--003-10000-0000-000001100-00-11-002103101001
Quaestora          1-121-00102---200----03-10000-0000-0--00-----------0-2--2000001
Medullosa          1-121-00101-22200000003-11001-0000-000001100--0-01-002--2000001
Ginkgo             1012101010110211000000001010110-0010040011150A--1-?00022003-01001
Emporia            1-121-00101012010000-11-10100-001--611211500--1-20-022013102000
Pinus              101210A010113401000001 3010110-00101611111501--1-10002221320-100
Podocarpus         1A121000101134010000013010110-00100611121501--1-1000-221320-100
Taxus              10121000101134010000003010110-001006011?1500-01-1000-2003-01100
Cordaixylon        1-1210-010113201000001 2010100-0010-702211500-01-00-0-2003101001
Mesoxylon          1-12100020113200000001-10100-0010-702211500--1-00-0-2003101001
Corystosperms      1-12101010112 0?0-000012-10100-001--400001100--0-1000B2203-11001
Peltasperms        1-121------130?--0-0--------0------000001100--0-1000-2103--10--
Glossopteris       1-121--0101131000200-10-10-000001--000001100-00-B0-012202-01100
Caytonia           1-121-1-1011?0??-210----10--0------000001100-00-110-02202-0100-
Cycadaceae         1012100001113010100011201110-000000200001400-00-100002002001101
Zamiaceae          1012100001113010100A11201010-000000300001500-00-110002005000101
Cycadeoidea        1-121-00011030010001-0201010-00010-500000-10-10-?10012004000020
Williamsonia       1-121-000A1A3A---001---------------500001--0-10---0012004000020
Williamsoniella    1-121-00011031----01---------------500000--0--0---00120040000-0
Westersheimia      --121-00-11-30---0------------------100001-----------02004-000-0
Pentoxylon         1-121-101-1031100000010-11---00010-100001210-00-000022002-01001
Ephedra            111210002-2032110-0001001011-001000701210510-11-010022004-1-110
Welwitschia        111210002-2032?1021111301011-011000701210-10-11-010022004-11110
Gnetum             111210-02-203111021111001011-0A1000701210----1--0-0022004-10110
Magnoliaceae       111210001-1031?10111A1411010-001011-00000-000021111102212-1-000
Eupomatia          1-1210001-1031?1011101411010-001011-00000-000021311102212-1-000
Austrobaileya      111210001-2031?1011101411010-001011-00000-000021211102212-1-000
Chloranthaceae     1112100A1-2031?1011101411010-001011-00000--0--B13110-2202-1-000
Laurales           111210001-2031?1011101411010-001011-00000-1000203110-2212-1-000
Winteraceae        111210001-0031?1011111411010-000011-00000-100021311002212-1-000
Eudicots           1112100A1-1033?1011A01011010-001011-00000-100020311002212-1-000
Aristolochioideae  111210011-1033?1011001411010-011011-00000-10102-311002212-1-000
Piperaceae         111210011-1033?1011-01411010-001011-00000-101020311002212-1-000
Nymphaeales        1-1211011-1033?10110-1011010--00-11-00000-00A0212110022-2-1-000
Monocots           111211011-103C?10-1101D-1010--01-11-00000-101020311002212-1-000
```

Polymorphism Key

A = 0 1

B = 0 2

C = 2 3

D = 2 4

Appendix 10.2.
Characters 63–110

```
                           63   68   73   78   83   88   93   98   103  108
                            |    |    |    |    |    |    |    |    |    |

Aneurophyton      -0----00---0006-------000000------0-0-----------
Archaeopteris     -0----00---0006-------0000000-----00------------
Cecropsis         -0----00---0006-------0000000-----00------------
Elkinsia          01000000---1006-------0000000-----0000--------0-
Heterangium       01000000---1006-------0000000-----00000-------0-
Lyginopteris      01000000---1006-------0000000-----00000-------0-
Callistophyton    12100110---0006-------1111000000---0-00-------0-
Quaestora         02100110---0006------------------0000--------0-
Medullosa         02A00110-----06-------0100000-0--0000000------0-
Ginkgo            12100101---0006-------11010000--101000000-000010
Emporia           02100-00---0006-------011100000----0----------1-
Pinus             12111101---0006------01111000000021000010-010011
Podocarpus        1211110----2006------01111000000021000010-010011
Taxus             1211110----3006------12202000-02-21100010-010011
Cordaixylon       12100110---0006-------0111000-00---00-00------0-
Mesoxylon         12100110---0006-------0111000-00---00-00------0-
Corystosperms     12110-10---2106-------111100000----1------------
Peltasperms       -2---------0006-------110-0000-----0------------
Glossopteris      121111-1---2006-------111101000----0-0----------
Caytonia          10111--1---2106-------111100000----1------------
Cycadaceae        12101110---0006------011010000001110000000000010
Zamiaceae         12101110---0006------011010000001110000000000001-
Cycadeoidea       0011111A---0006-------11020000----11000-------1-
Williamsonia      0011111----0006-------110-0000------------------
Williamsoniella   001111-----0006-------11020000------------------
Westersheimia     00---------0006---------------------------------
Pentoxylon        101-1111----006-------110200000----1------------
Ephedra           12111110---0106------02102010-00021100011-1100-1
Welwitschia       10111110---0106------01102010001-2-11111--1101-1
Gnetum            12111100---0106------022020-1-01-2-111111-1101-1
Magnoliaceae      1-1111010014014000001A1102000013-2110201111010-1
Eupomatia         1-11110100140141000-00-002000013-2110201111010-1
Austrobaileya     1-111101001401400000001103100123-2110201111010-1
Chloranthaceae    1-111101-004010-1-12-01103100123-2110201111010--
Laurales          1-111101-00401011010002202-01--3-2110201111010--
Winteraceae       1-111101110401-00000001003100023-2110201111010-1
Eudicots          1-1111011A0401301A0AA0-003---123-2110201111010-1
Aristolochioideae 1-111101004015101011011031000-3-2110201111010--
Piperaceae        1-111101100401-02102-0110300A123-211A201101010-1
Nymphaeales       1-11110111040140010AA0110200-023-2110201101010-0
Monocots          1-111101--04012001011-1103-0-013-2110201111010-1


Polymorphism Key

A = 0 1
B = 0 2
C = 2 3
D = 2 4
```

	Character	State
0	Life cycle	0, homosporous; 1, heterosporous
1	Maule reaction in lignin	0, absent; 1, present
2	Functional megaspores	0, more than one; 1, one
3	Meio-/megasporangial dehiscence	0, along one side; 1, over apex and along part or entire length of both sides; 2, indehiscent [nonadditive]
4	Indehiscent megasporangium	0, absent; 1, present
5	Radicle	0, persistent; 1, replaced by adventitious roots
6	Vegetative short shoots	0, absent; 1, present
7	Habit	0, woody; 1, (semi) herbaceous
8	Branching	0, apical or adventitious; 1, single bud axillary; 2, multiple buds [nonadditive]
9	Persistent leaf bases	0, absent; 1, present
10	Leaves	0, leaves absent; 1, alternate helical; 2, opposite whorled [nonadditive]
11	Cataphylls	0, absent; 1, present
12	Dichotomies in vegetative leaves	0, in all; 1, in some; 2, at base but not distally; 3, in none [nonadditive]
13	Morphology of vegetative leaves	0, pinnately compound; 1, simply pinnate veined or dissected into parallel veined segments; 2, linear to dichotomous with two or more veins; 3, palmately veined; 4, linear with one vein, rarely two, may fork apically [nonadditive]
14	Leaf trace divergence	0, from one cauline protoxylem (px) or sympodium at one level; 1, two or more cauline px strands or cauline bundles at one level; 2, two or more cauline px strands over length of stem [nonadditive]
15	Leaf traces	0, mesarch; 1, endarch
16	Girdling leaf traces	0, absent; 1, present
17	Laminar venation	0, one; 1, reticulate hierarchical; 2, reticulate anastomosing [nonadditive]
18	Guard cell poles	0, raised; 1, level with aperture
19	Stomata	0, anomocytic; 1, paracytic
20	Foliar astrosclereids	0, absent; 1, present
21	*Sparganum/Dictyoxylon* cortex	0, absent; 1, present
22	Secretory structures	0, isolated cells or groups of cells; 1, cavities; 2, canals; 3, absent; 4, oil cells [nonadditive]
23	Apical meristem	0, without tunica; 1, with tunica
24	Protoxylem architecture	0, single centrarch cauline strand; 1, two or more sympodial strands near primary margin
25	Polystele	0, absent; 1, present
26	Primary xylem	0, mesarch; 1, endarch
27	Metaxylem	0, with scalariform pitting; 1, without scalariform pitting
28	Secondary xylem	0, external only; 1, external internal
29	Tertiary spiral thickenings in tracheids	0, absent; 1, present
30	End wall pit or vessel perforations	0, multiple; 1, simple
31	Vessels	0, absent; 1, present
32	Rays	0, at least some multiseriate; 1, all uniseriate or biseriate
33	Companion cells in phloem	0, absent; 1, present
34	Sieve tube plastid inclusions	0, starch; 1, PI type

	Character	State
35	Sporangium-bearing structures	0, not aggregated; 1, aggregated but not determinate strobili; 2, ovulate not aggregated, micro forming simple cones; 3, ovulate aggregated, microsporangia as simple; 4, ovulate and microsporangiate forming simple cones; 5, aggregated ovulate cones, simple pollen cones; 6, compound ovulate cones, simple microsporangiate cones; 7, compound ovulate and microsporangiate cones [nonadditive]
36	Symmetry of megasporangium bearing or ovuliferous fertile shoot	0, radial; 1, bilateral dorsiventral
37	Megasporangium-bearing or ovuliferous bract shoot complexes	0, absent; 1, helical; 2, vertical rows [nonadditive]
38	Megasporangium-bearing or ovuliferous fertile shoot complexes	0, absent; 1, with distinct appendages; 2, without distinct appendages [nonadditive]
39	Bract and axillary female shoot complexes	0, absent; 1, with bract free from shoot; 2, with bract partly or completely fused to shoot [nonadditive]
40	Megasporangia and microsporangia	0, intermixed or aggregated as bisporangiate structure; 1, separated into monosporangiate structures
41	Megasporophylls and microsporophylls	0, dichotomous only; 1, dichotomous and or pinnate; 2, megasp. dichotomous or simple micropinnate or laterally branched; 3, megasp. dichotomous or simple, microsp. simple; 4, megasp. pinnate, microsp. simple; 5, megasp. and microsporophylls simple [nonadditive]
42	Microsporophylls	0, helically arranged; 1, whorled
43	Meio-/microsporangia per sporophyll	0, more than two; 1, two
44	Stamen numbers	0, various; 1, multiples of three
45	Microsporophylls	0, free; 1, basally fused
46	Microsporophylls	0, pinnate or paddlelike; 1, simple, one veined, scalelike; 2, simple, one-(rarely)-three veined, with two pairs of lateral sporangia [nonadditive]
47	Stamen form	0, filamentous; 1, laminar
48	Position of meio-/microsporangia	0, terminal; 1, abaxial; 2, adaxial; 3, unnamed state [nonadditive]
49	Meio-/microsporangia	0, free; 1, fused at least basally
50	Meio-/microsporangia dehiscence	0, ecto/endokinetic; 1, endothecial
51	Inner staminodes	0, absent; 1, present
52	Megasporangium/ovule-bearing structure	0, pinnate; 1, simple, paddlelike megasporangia not in two rows; 2, simple, stalklike, with one megasporangium or megasp. sessile [nonadditive]
53	Meio- or megasporangia	0, dehiscent along one side; 1, dehiscent over apex and along part or entire length of sides; 2, indehiscent [nonadditive]
54	Position of attachment of megasporangium-ovule	0, terminal or lateral; 1, abaxial; 2, adaxial [nonadditive]
55	Ovule–megasporangium orientation	0, erect; 1, inverted
56	Integument	0, absent; 1, lobate preintegument; 2, with simple apex; 3, with bifid apex; 4, with tubular micropyle; 5, unnamed state [nonadditive]
57	Integumentary vascularization	0, numerous bundles one in each lobe; 1, two bundles dividing in major plane; 2, unvascularized [nonadditive]
58	Integument	0, with sclerotesta and sarcotesta; 1, simple

	Character	State
59	Anatomical symmetry of ovule	0, radial; 1, 180 degree rotational; 2, bilateral [nonadditive]
60	Fusion of integument to nucellus	0, free; 1, fused more than fifty percent of length
61	Tubular micropyle	0, missing; 1, composed of two integumentary layers; 2, composed of all integumentary layers [nonadditive]
62	Sarcotesta	0, absent or uniseriate; 1, multiseriate
63	Integument apex sealing after pollination	0, absent; 1, present
64	Pollen chamber	0, absent (cellular apex solid); 1, hydrasperman; 2, nonhydrasperman (no membranous floor with central column) [nonadditive]
65	Central column	0, present; 1, absent
66	Salpinx (=nucellar beak)	0, present; 1, absent
67	Lagenostome–salpinx–nucellar beak	0, present; 1, absent
68	Pollen chamber sealing after pollination	0, not sealed; 1, sealed
69	Megasporangium/nucellus vascularization	0, not vascularized; 1, vascularized
70	Nucellar cuticle	0, thin; 1, thick
71	Testa	0, multiplicative; 1, nonmultiplicative
72	Exotesta	0, normal; 1, palisade
73	Ruminations in the seed coat	0, absent; 1, present = 1
74	Ovule or homolog	0, not enclosed; 1, enclosed in one vascularized structure composed of telomes; 2, enclosed in one vascularized structure composed of laminae; 3, enclosed in one unvascularized structure layer of tissue; 4, enclosed in two layers of tissue [nonadditive]
75	Meio- or megasporangium or ovule	0, not surrounded by subtending leaf homologs; 1, surrounded by one or two pairs of subtend leaf homologs
76	Closed carpel with stigmatic pollen germination	0, absent; 1, present
77	Carpel number	0, one; 1, two; 2, three; 3, four/five; 4, numerous; 5, four/six; 6, none [nonadditive]
78	Hypanthium	0, absent; 1, present
79	Placentation	0, marginal/laminar; 1, apical; 2, basal [nonadditive]
80	Carpels	0, spiral or irregular; 1, whorled
81	Ovules per carpel	0, several; 1, one, apical
82	Perianth	0, more than two whorls, spiral irregular; 1, two whorls; 2, none [nonadditive]
83	Perianth symmetry	0, various; 1, trimerous
84	Microspore/pollen cytokinesis	0, simultaneous; 1, successive
85	Microspore/pollen with	0, proximal tetrad scar; 1, distal sulcus of round germination area; 2, no aperture [nonadditive] [deactivated]
86	Microspore/pollen symmetry	0, radial; 1, bilateral; 2, global [nonadditive]
87	Microspores/pollen	0, nonsaccate or subsaccate; 1, saccate
88	Infratectal structure	0, massive or spongy alveolar; 1, honeycomb alveolar; 2, granular; 3, columellar [nonadditive]
89	Tectum	0, continuous or finely perforate; 1, foveolate-reticulate
90	Exine striations	0, absent; 1, present
91	Supratectal spinules	0, absent; 1, present

	Character	State
92	Aperture membrane	0, smooth or weakly sculptured; 1, conspicuously sculptured
93	Endexine	0, uniformly thick lamellated; 1, absent; 2, thin nonlamellated [nonadditive]
94	Microgametophytes with	0, five or more nuclei; 1, four nuclei, tube nucleus produced by the second division; 2, four nuclei tube nuclei produced by first division, no prothallus; 3, three nuclei [nonadditive]
95	Sterile cell	0, colinear with other microgametophyte cells; 1, ring shaped
96	Sperm	0, small, flagellate and zooidogamous; 1, large, flagellate, and zooidogamous; 2, nonflagellate [nonadditive]
97	Megaspore tetrad	0, tetrahedral; 1, linear
98	Megaspore wall	0, thick; 1, thin
99	Megagametophyte	0, monosporic; 1, tetrasporic
100	Megagametophyte	0, large, cellular, normal archegonia; 1, large, apical part and egg free nuclear; 2, eight- nucleate, central part free-nuclear, egg cellular without neck cells [nonadditive]
101	Megagametophyte cellularization	0, enclosing single nuclei, resulting in uninucleate cells; 1, enclosing several nucleli, resulting in multinucleate polyploid cells
102	Sperm transfer	0, zooidogamous; 1, siphonogamous
103	Fusion of	0, only one sperm with female gametophyte nucleus; 1, regular fusion of both sperm
104	Nutritive tissue in seed	0, endosperm plus perisperm; 1, endosperm only
105	Embryo	0, derived from several free nuclei; 1, from a single uninucleate cell by cellular divisions
106	Proembryo	0, not tiered; 1, tiered
107	Secondary suspensor (embryonal tubes)	0, present; 1, absent
108	Feeder in embryo	0, absent; 1, present
109	Postzygotic quiescence and/or seed dormancy	0, absent; 1, present
110	Seed germination	0, hypogeal; 1, epigeal

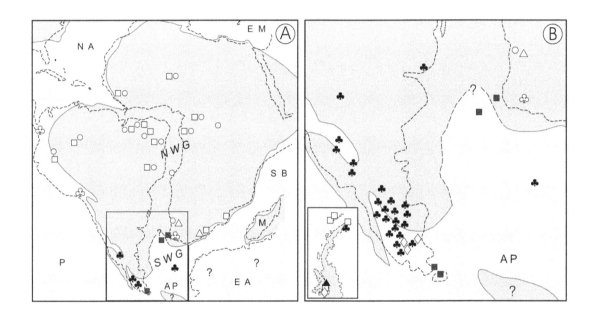

Fig. 11.1. (A) Generalized map of Western Gondwana (adapted and simplified from Owen and Mutterlose 2006) show-ing the major localities from which fossil conifers have been reported. (B) Detail of box in (A) and insert of present-day Antarctic Peninsula. Dark gray indicates the distribution of landmasses during the Aptian at the onset of the separation between Africa and South America. Dashed lines show the outlines of present-day landmasses. NA = North Atlantic; EM = East Mediterranean; NWG = Northwestern Gondwana; SWG = Southwestern Gondwana; SB = Sudan Basin; M = Mad-agascar; EA = East Antarctica; AP = Antarctic Peninsula; P = Paleopacific Ocean. Symbols: white squares = Araucariaceae; black squares = endemic Araucariaceae; circles = Cheirolepidiaceae; white triangles = Podocarpaceae; black triangles = endemic Podocarpaceae; black trefoil = endemic Araucariaceae, Cheirolepidiaceae, and Podocarpaceae; white trefoil = Araucariaceae, Cheirolepidiaceae, and Podocarpaceae; white diamonds = Cupressaceae.

ENDEMISM OF EARLY CRETACEOUS CONIFERS IN WESTERN GONDWANA

11

Sergio Archangelsky and Georgina M. Del Fueyo

Early Cretaceous conifers occur in many parts of Western Gondwana. The families represented are the Araucariaceae, Cheirolepidiaceae, Podocarpaceae, and the taxodiaceous Cupressaceae. On the genus level, many taxa have a global distribution; however, some of them are restricted to the southern regions. In regard to species level, a great number of endemics are located in southwestern Gondwana. This paleophytogeographical area extends from Patagonia to southern Africa in the east, to mid–South America in the north, and probably as far as the Antarctic Peninsula in the south. These endemic taxa are characterized and discussed in this chapter. Endemism is especially strong among the podocarps and araucarians.

Introduction

We present here a survey of Early Cretaceous conifers from a vast area of Western Gondwana that includes most of South America, southern Africa, and the Antarctic Peninsula (Fig. 11.1). The Early Cretaceous was a tectonically unstable epoch as a result of the breakup of South America and Africa, which gave rise to the Atlantic Ocean (Pankhurst et al. 2006; Owen and Mutterlose 2006; Upchurch 2007). Paleoclimates were strongly influenced by high atmospheric carbon dioxide, and organisms had to adapt to greenhouse conditions (Passalia 2008). The vegetation flourished in subpolar regions (Antarctic Peninsula), at warm temperate latitudes (Patagonia and southern South Africa), and in a broad, hot, and partially xeric equatorial zone (northern Argentina, Uruguay, and Brazil). The most diverse paleofloras were located in mid to high paleolatitudes where the greatest number of species and abundance of fossils occur (Fig. 11.1). Because Patagonia and the Antarctic Peninsula have the richest store of fossil plants, this chapter is based mainly on megafossils from these areas. However, we have also included relevant palynological data here when necessary.

Patagonia is now a well-defined geographic region spread across the southern tip of South America (Fig. 11.1). Several endemic conifers grow in the Patagonian Andes, including trees of various sizes, one of which is among the largest living forms on earth: *Pilgerodendron uviferum* (D. Don) Florin.

During the Late Paleozoic, Patagonia was most likely an isolated terrane that collided with Gondwana during the Permian (Pankhurst et al. 2006), a time when similar, mixed paleofloras occurred in southern South Africa and southern Patagonia (Archangelsky and Arrondo 1975).

This paleofloristic similarity continued until the Cretaceous, when Africa and South America began to separate.

A distinct Early Cretaceous floristic area with many endemic taxa developed in the southwestern corner of Gondwana. This led to the recognition of the Patagonian Floristic Province (Vakhrameev 1988; Archangelsky 1996) and the palynological *Cyclusphaera* Province (Volkheimer 1980).

In this chapter, we will review the major paleofloristic elements, whether or not endemic to this province, that pertain to the families Araucariaceae, Cheirolepidiaceae, Podocarpaceae, and the taxodiaceous Cupressaceae. We will also discuss fossil conifer taxa occurring to the north of Western Gondwana, an area with few endemics.

Materials

For this review, we consulted the pertinent literature and examined many of the original specimens cited in the text. The illustrated specimens, microscope slides, and scanning electron microscopy samples that we looked at pertain to original material deposited in the following Argentinian paleobotanical collections: Lillo Institute, Tucumán University (LIL PB); Argentine Natural Sciences Museum "Bernardino Rivadavia" (BA Pb, BA Pb MEB); Paleontological Museum Egidio Feruglio (MPEF Pb), and La Plata Natural Sciences Museum (LP Pb).

Araucariaceae

During the Mesozoic, the Araucariaceae were widespread in both hemispheres (Gee and Tidwell 2010; Hotton and Baghai-Riding 2010), although they are for the most part restricted to the Southern Hemisphere today (Florin 1940; Dettmann and Clifford 2005; Kunzmann 2007). Of the three living genera of the family—*Agathis* Salisbury, *Araucaria* Jussieu, and *Wollemia* Jones, Hill et Allen—only *Araucaria* is present in the fossil record during the Early Cretaceous in Western Gondwana.

Abundant megafossil remains of the family are found in this area of Gondwana during the Berriasian but mostly from the Hauterivian to Albian. The fossils consist of mainly compressions and a few impressions of leafy twigs, some of which occur in organic connection to pollen and seed cones. Isolated ovuliferous scales, as well as wood, are also frequently found.

Two key genera were described from the Baqueró Group (Aptian) in Santa Cruz Province. Both are based on pollen cones connected to leafy twigs. *Nothopehuen brevis* Del Fueyo (Fig. 11.2A; Plate 10A) is based on pollen cones 2 mm long and 1.8 mm wide, which are attached to branches with small, scalelike leaves of the *Brachyphyllum* type. These leaves are amphistomatic with abundant distal papillae and stomata forming irregular rows. On the ultrastructural level, the cuticle consists of three layers; the external and internal layers are compact and homogeneous, while the middle layer is spongy. The microsporangia contain pollen of the *Araucariacites* type (Del Fueyo 1991). N. *brevis* shares some features with section *Eutacta* of *Araucaria*. However, its pollen cones differ too much in shape, form, and number of pollen sacs to be included with certainty in this section.

The second key genus, *Alkastrobus peltatus* Del Fueyo and Archangelsky (2005), consists of pollen cones 10 mm long and 5 mm wide (Fig. 11.2B; Plate 10B) and is associated with twigs with *Brachyphyllum mirandai* leaves and cuticle (Archangelsky 1963). The microsporophylls of the pollen cones are spirally arranged, peltate with a distal rhomboidal head, and bear more than six abaxial elongate pollen sacs with the pollen of *Cyclusphaera psilata* Volkheimer et Sepúlveda (1976). These pollen grains (Fig. 11.2C; Plate 10C) were found embedded in the peritapetal membrane of the microsporangium (Fig. 11.2D; Plate 10D).

A third fertile araucarian plant from the same locality consists of microsporangiate cones 7 mm long and 2 mm wide, which were organically attached to twigs with leaves and cuticle of *Brachyphyllum irregulare* Archangelsky. These cones contain pollen of *Balmeiopsis limbatus* (Balme) Archangelsky (Archangelsky 1963, 1979; Archangelsky and Gamerro 1967). The pollen of *Balmeiopsis* Archangelsky is also found at other Gondwanan localities and is not endemic to its southwestern region. Other Patagonian pollen cones bear pollen of *Araucariacites* Cookson; they are found in Berriasian–Valanginian strata associated with *Pagiophyllum feistmantelii* (Halle) Townrow (Baldoni 1979; Gee 1989).

Ovuliferous scales assigned to *Araucarites* Presl are found in the Baqueró Group of Patagonia: *A. baqueroensis* Archangelsky and *A. minimus* Archangelsky. These two species were referred to the *Eutacta* section of *Araucaria* (Archangelsky 1966). *A. chilensis* Baldoni is another araucarian cone scale found associated with *Pagiophyllum feistmantelii* in the Berriasian–Valanginian strata of Patagonia (Baldoni 1979).

Araucaria grandifolia (Feruglio) Del Fueyo et Archangelsky (Fig. 11.2E; Plate 10E), which is based on branches with leaves, was collected in the Punta del Barco Formation (Baqueró Group) of Aptian age (Del Fueyo and Archangelsky 2002) and placed in the section *Araucaria*. The branches show secondary xylem with araucarioid cross-field pitting. The leaves are 11 cm long and 1–2 cm wide, densely packed, imbricate, and spirally arranged; they are triangular–lanceolate with multiple parallel (14–22) veins and an acute apex. Stomata are present on both sides of the leaves and form regular rows. The stomatal apparatus is elliptical, monocyclic with four subsidiary cells, two lateral and two small polar (Fig. 11.2F; Plate 10F). The ultrastructure of the cuticle shows a unique and compact layer, which is mostly composed of parallel fibrils alternating with amorphous and less electron-dense areas.

Araucarian remains have also been recorded in the Santana Formation of the Araripe Basin in the Aptian of northeast Brazil. *Araucaria cartellei* Duarte is based on a single, oval–lanceolate, multiveined leaf impression (Duarte 1993), which may pertain to the section *Araucaria*. Poorly preserved impressions of cuneiform ovulate scales referred to *Araucarites vulcanoi* Duarte were recognized in the same unit (Duarte 1993).

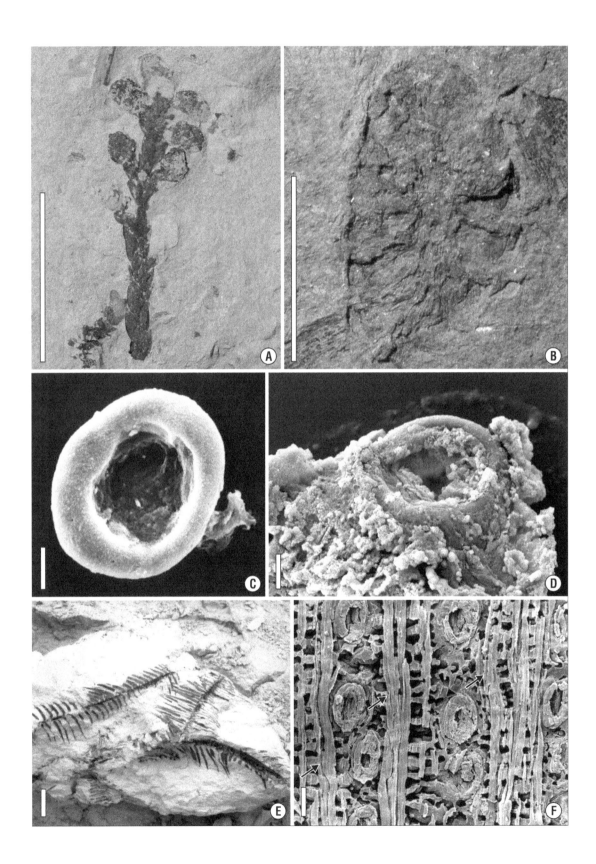

Reproductive and vegetative remains rarely occurring in amber from the Crato Formation of probable Aptian age in the same basin were attributed to cf. *Agathis* and cf. *Wollemia* by Martill et al. (2005).

In the Algoa Basin of South Africa, leaves, pollen, seed cones, and isolated ovuliferous scales were assigned to *Araucaria rogersii* (Seward) Anderson et Anderson (1985) from the Kirkwood Formation, which is Berriasian to early Valanginian in age. The leaves of *A. rogersii* are imbricate, of the *Brachyphyllum* type, and measure up to 5 cm long. Its pollen cones are single or in clusters at the ends of its branches; they are small, 8 mm long and 4 mm wide. Its ovoid seed cones with woody ovuliferous scales are commonly found detached (Anderson and Anderson 1985). According to Brown (1977), large pieces of coniferous wood from the same formation may represent the trunks of *A. rogersii*. *Araucaria mkuziensis* Anderson et Anderson from the Makatini Formation of Zululand, which is Barremian to late Aptian in age, is based on impressions of leaved branches, pollen cones, and ovuliferous scales (Anderson and Anderson 1985). These two South African taxa have not been assigned to any of the extant sections of *Araucaria* because of the poor preservation of their morphological and anatomical features.

In the Early Cretaceous of the Antarctic Peninsula, compressions and impressions of leafy twigs were described by Cantrill and Falcon-Lang (2001) from the Neptune Glacier Formation of Alexander Island. *Araucaria alexandrensis* Cantrill et Falcon-Long consists of branches with helically imbricate, lanceolate–ovate leaves that are 18–40 mm long and 8–12 mm wide, and bear up to 10 parallel veins. *Araucaria chambersii* Cantrill et Falcon-Long is a leafy plagiotrophic shoot with helically linear lanceolate leaves, 45–105 mm long and 8–15 mm wide. These leaves each have multiple parallel veins, a twisted, decurrent base, and an acute apex. Also described from the same site (Cantrill and Falcon-Long 2001) was a cone axis 70 mm long, with helically arranged, cuneate scales, 25 mm long and 8 mm wide, and still attached at the base, as well as two isolated ovuliferous scales, *Araucarites wollemiaformis* Cantrill et Falcon-Long (in close association with *Araucaria alexandrensis*) and *Araucarites citadelbastionensis* Cantrill et Falcon-Long.

Impressions of *Araucarites* cf. *baqueroensis* (Hernández) and *Araucarites* spp. (Césari et al. 1999; Cantrill 2000) were recorded from the Cerro Negro Formation in Livingston Island (Byers Peninsula) and in Snow Island (President Head). Another scale referred to *Araucaria antarctica* Gee was described from the Hope Bay Flora of putative Late Jurassic–Early Cretaceous age (Gee 1989).

Several pollen types have been referred to the Araucariaceae. *Araucariacites australis* Cookson had a broad distribution in Gondwana and other areas (Brenner 1968; Herngreen and Chlonova 1981; Pons 1988; Prámparo and Batty 1994; Campos et al. 1998; Dino et al. 1999). From the Aptian–Albian of northwestern Brazil, Lima (1979) described *A. guianensis* van der Hammen et Burger, while Quattrocchio et al. (2006) mentioned *A. fissus* Reiser et Williams from the Springhill Formation

Fig. 11.2. (A) *Nothopehuen brevis* Del Fueyo, pollen cones in organic connection to ultimate-order branches, BA Pb 11.418 (holotype). Scale bar =1 cm. (B) *Alkastrobus peltatus* Del Fueyo et Archangelsky, pollen cone in longitudinal view showing the thick central axis and peltate microsporophylls; LIL Pb 2562 (holotype). Scale bar = 2.5 mm. (C) *Cyclusphaera psilata* Volkheimer et Sepúlveda, grain in polar view, BA Pb MEB 202. Scale bar = 10 μm. (D) *Cyclusphaera psilata* Volkheimer et Sepúlveda, grain embedded in the internal tissues of a microsporangium, BA Pb MEB 202. Scale bar = 10 μm. (E) *Araucaria grandifolia* Feruglio, general view showing branches and imbricate, spiral arrangement of leaves, MPEF Pb 129. Scale bar = 4.3 cm. (F) *Araucaria grandifolia* Feruglio, cuticle in internal view showing stomata in longitudinal rows and hypodermic fibers (arrows), BA Pb MEB 151. Scale bar = 100 μm.

of Patagonia. *Balmeiopsis limbatus* (Balme) Archangelsky is a common pollen species found in Early Cretaceous sediments of Gondwana and is particularly abundant in the southwestern sector of the continent, that is, in southern Patagonia, the Falkland Plateau, the Antarctic Peninsula, South Africa, and rarely in Brazil (Scott 1976; Lima 1979; Kotova 1983; Riding et al. 1998; Del Fueyo et al. 2007). This pollen type was found in cones organically attached to *Brachyphyllum irregulare* twigs in the Aptian of Patagonia (Archangelsky and Gamerro 1967). *Cyclusphaera* pollen is abundant in Early Cretaceous deposits in Patagonia, where it was found in cones closely associated to *Brachyphyllum mirandai* twigs in the Aptian Anfiteatro de Ticó Formation and referred to the Araucariaceae (Del Fueyo and Archangelsky 2005). *Cyclusphaera psilata* Volkheimer et Sepúlveda is a widely distributed species in Western Gondwana too, but as far we know, it has not been reported from other paleofloristic provinces. It occurs in the Early Cretaceous of Patagonia (Del Fueyo et al. 2007), in the Albian of Antarctica (Baldoni and Medina 1989), offshore on the Falkland Plateau (Kotova 1983), and in the Valanginian–Hauterivian of South Africa (Zavada 1987). *C. radiata*, *C. patagonica*, and *C. crassa* are other endemic species that are found mainly in the Albian strata of Patagonia (Archangelsky et al. 1984).

Early Cretaceous fossil conifer woods were studied from several localities in Western Gondwana. The Antarctic Peninsula has abundant permineralized trunks referable to the genus *Agathoxylon* Hartig, which has araucarian tracheid pitting, araucarioid cross-field pits, and axial parenchyma. However, in addition to the Araucariaceae, this type of wood may pertain to the Cheirolepidiaceae or even to pteridosperms (Philippe et al. 2004). At least eight *Agathoxylon* species left in open nomenclature were described from different localities, mostly from South Shetland Islands (e.g., the mid-Aptian Cerro Negro Formation and the Valanginian–Hauterivian President Head locality) but also from Alexander Island (e.g., the late Albian Neptune Glacier Formation) and James Ross Island (e.g., the Albian Gustav Group) (Herbst et al. 2007). *Agathoxylon arayii* Torres, Valenzuela et González, and *A. floresii* Torres et Lemoigne were identified from the Lower Cretaceous of Livingston Island (Philippe et al. 2004).

Wood attributed to *Agathoxylon* was also found in the Valanginian–Hauterivian Kirkwood Formation and Sunday River Formation of the Algoa Basin in South Africa (Philippe et al. 2004).

Cheirolepidiaceae

This exclusively Mesozoic family has long known to be geographically widespread and highly diverse in its number of taxa as well as its ecological constraints (Alvin 1982; Watson 1988). In Western Gondwana, the Cheirolepidiaceae were well established in South America, western Antarctica, and South Africa by the Early Cretaceous, as confirmed by palynological data based on the distinctive pollen of the family, *Classopollis* (Pflug) Pocock et Jansonius. In Patagonia, *Classopollis* has a continuous record from the Berriasian to the Albian and includes species such as *C. classoides* (Pflug) Pocock et Jansonius, *C. simplex* Reiser et Williams, *C. torosus* Couper, and *C. chateaunovii* Reyre (Vallati 1993; Prámparo 1994; Del Fueyo et al. 2007). In the Aptian–Albian of northeastern Brazil, *Classopollis* accounts for up to nearly 93% of the palynoassemblages (Lima 1976). In the late Albian of Antarctica, Baldoni and Medina (1989) found *C. intrareticulatus* Volkheimer and *C. simplex*. A sequence from the CK1/68 borehole in South Africa also showed the dominance of *Classopollis* pollen in groups I to IV (Scott 1976). This taxon also dominates the palynological assemblages (approximately 80%) at the Deep Ocean Drilling Project Site 511 of the Falkland Plateau (Kotova 1983).

Megafossil remains referred to the Cheirolepidiaceae have been found in the Mesozoic of southern South America, especially in Patagonia (Del Fueyo et al. 2007), but have not been reported from Antarctica and occur with some doubt in South Africa (Bamford and Corbett 1994). In Patagonia, Early Cretaceous cheirolepidiaceans are mostly known from the Baqueró Group (Aptian) in Santa Cruz Province. This material consists of twig compressions, seeds and pollen cones, well-preserved leaf cuticles, bracts, and ovuliferous dwarf shoots.

Tomaxellia biforme was found with both seed and pollen cones attached to leafy twigs in the Aptian of the Anfiteatro de Ticó Formation (Archangelsky 1968). It has dimorphic leaves borne on the same branch: the short *Brachyphyllum* type, which is scalelike and ovate, and which has a decurrent base, and the long *Elatocladus* type, which is linear–falcate with its free distal end spreading at almost 90 degrees to the axis. Because of this dimorphism, *T. biforme* was not included in Alvin's (1982) and Watson's (1988) cheirolepidiaceous groups. Vegetative remains of *T. biforme* were also found in the Brazilian Aptian Crato Formation by Kunzmann et al. (2006). The ultrastructure of the leaf cuticle shows four distinct layers (Villar de Seoane 1998). The seed cones are cylindrical and elongate, and they are borne at the tips of its branches (Archangelsky 1968). The cone scales are spirally and compactly inserted on the cone axis. They consist of an ovate bract that subtends an ovate ovuliferous dwarf shoot with six scales that are fused to one another except at the distal ends. Ovuliferous dwarf shoots of *T. biforme* are fused at the base of their bracts and have scattered stomata at the distal end of the abaxial and adaxial surfaces. Each ovuliferous dwarf shoot bears two lateral ovules. The pollen cones are very small, only 2–3 mm long and 1 mm wide, and they are borne terminally on lateral branches (Archangelsky and Gamerro 1967). Information on the sporophylls and microsporangia is not available because of the heavy coalification of the cones; however, it is clear that they contain abundant pollen that is usually preserved in tetrads and can be identified as *Classopollis classoides*. According to Archangelsky (1968), isolated ovuliferous dwarf shoots of *T. biforme* are common in the fossiliferous beds, suggesting that they were shed from the cone axis at a mature stage while the bracts remained persistent. This same reproductive strategy characterizes the northern members of the Cheirolepidiaceae such as *Hirmeriella*

Fig. 11.3. (A) *Kachaikestrobus acuminatus* Del Fueyo, Archangelsky, Llorens et Cúneo, seed cones in organic connection to a branched twig, MPEF Pb 859A (holotype). Scale bar = 1 cm. (B) *Tarphyderma glabrum* Archangelsky et Taylor, abaxial epidermis in inner view showing stomatal apparatus with tubelike suprastomatal chamber, and elongate subsidiary cell. Note guard cells at the bottom of stomata, BA Pb MEB 148. Scale bar = 0.1 mm. (C) *Trisacocladus tigrensis* Archangelsky, short pedunculate pollen cone borne in the axil of lanceolate leaves, LP Pb 5826 (holotype). Scale bar =1 cm. (D) *Callialasporites* Sukh Dev, grain showing colpus and incipient air sacs, PB Pal. (still under study). Scale bar = 25 µm. (E) *Squamastrobus tigrensis* Archangelsky et Del Fueyo, pollen cone (arrow) connected to branch with *Brachyphyllum* type of leaves (arrowheads). Note conspicuous microsporangia, BA Pb 11311. Scale bar = 1 cm. (F) *Squamastrobus tigrensis* Archangelsky et Del Fueyo, seed cone, BA Pb 11341 (paratype). Scale bar = 1 cm. (G) *Morenostrobus fertilis* Del Fueyo, Archangelsky et Taylor, lateral pollen cones (arrows) borne in the axil of *Elatocladus* type of leaves (arrowheads), BA Pb 11341 (holotype). Scale bar = 1 cm. (H) *Athrotaxis ungeri* (Halle) Florin, seed cone in longitudinal section. Note thick central axis and spatulate distal end of the cone scales, MPEF Pb 859 B. Scale bar = 1 cm.

muensteri (Schenk) Jung (Clement-Westerhof and van Konijnenburg-van Cittert 1991), *Pseudohirmeriella delawarensis* (Arndt) Axsmith, Andrews et Franser (Axsmith et al. 2004), *Alvinia bohemica* (Velenovský) Kvaček (Kvaček 2000), and *Frenelopsis ramosissima* (Fontaine) Watson (Axsmith and Jacobs 2005).

Tomaxellia degiustoi from the Anfiteatro de Ticó Formation (Archangelsky 1963, 1966) is another cheirolepidiacean with branches that bear helically arranged, long and narrow, falcate leaves with acute apices. These leaves are amphistomatic with monocyclic or imperfectly dicyclic stomata that form irregular rows or are randomly distributed.

A cheirolepidiacean of the "brachyphyll" group is *Kachaikestrobus acuminatus* (Del Fueyo et al. 2008a); it was found in the early Albian Kachaike Formation of Patagonia and is based on branched, ultimate-order twigs, each bearing a single seed cone (Fig. 11.3A; Plate 11A). The twigs have helically arranged leaves of the *Brachyphyllum* type with a distinct acuminate apex. These leaves are amphistomatic with randomly placed stomata and abundant papillae on both leaf surfaces. The seed cones of *K. acuminatus* have an overall morphology similar to those of *T. biforme*, but are somewhat smaller in general size. Approximately 40 cone scales are densely and helically arranged on each cone axis. Each cone scale has an ovate acuminate bract that subtends a broadly elliptical ovuliferous dwarf shoot. Bracts and ovuliferous dwarf shoots are fused only at their bases, and both are amphistomatic. The ovuliferous dwarf shoots consist of eight scales, two smaller and six larger. Two of the scales are fertile, while the remaining six are sterile. Distal end of scales are markedly crenulate with abundant papillae on both the adaxial and abaxial epidermis. Pollen adhering to the scales are of the *Classopollis* type. The epimatium emerges as an outgrowth of two adaxial scales; it has two layers composed of rectangular–elongate cells that cover the ovules. These seed cones are thought to be incompletely mature. In regard to the evolutionary trends in Mesozoic conifer ovulate structures (Miller 1999), *K. acuminatus* has some of the most ancestral characters among the cheirolepidiaceans, such as the largest number of scales and bracts longer than the ovuliferous dwarf shoots (Del Fueyo et al. 2008a). Ultrastructural study of the cuticles of the leaves and cone scales revealed that periclinal walls consist of three layers: the external layer with fibrillar elements in a reticulate arrangement, the middle layer with fibrils that are parallel, perpendicular, and compactly arranged, and the innermost layer with dense and parallel fibrillar elements.

Tarphyderma glabrum Archangel et Taylor, a putative cheirolepidiacean, is an unusual conifer from the Anfiteatro de Ticó Formation at the Bajo Grande locality (Archangelsky and Taylor 1986, 1991). It consists of leafy shoots with helically and tightly adpressed leaves of the *Brachyphyllum* type. The leaves are hypostomatic with adaxial cuticle that is thicker (15 µm) than the abaxial cuticle (5–7 µm) as well as closely spaced stomata arranged in irregular rows. The most striking feature of *Tarphyderma* is its stomatal apparatus that has up to 17

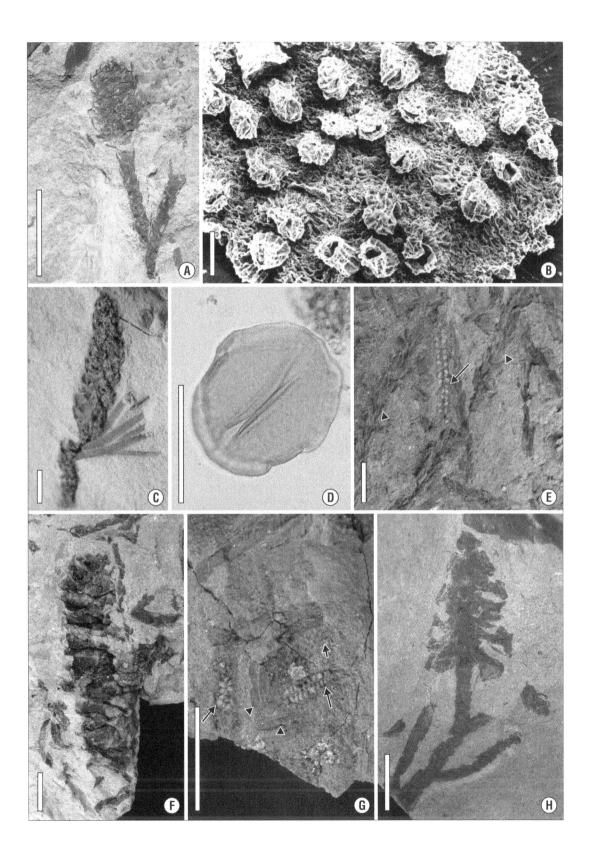

polycyclic subsidiary cells and deeply sunken guard cells at the bottom of an elongate, somewhat tubelike suprastomatal chamber 250 µm deep (Fig. 11.3B; Plate 11B). The ultrastructure of the leaf cuticle consists of an outer lamellate layer, a middle uniformly electron-dense layer, and an inner mostly spongy layer.

The large amount of dispersed *Classopollis* pollen occurring in the same sediments as *Tomaxellia biforme* Archangelsky (2003) and *Kachaikestrobus acuminatus* Del Fueyo, Archangelsky, Llorens et Cúneo (Gamerro 1982) is consistent with an anemophilous pollination syndrome in the Cheirolepidiaceae (Alvin 1982). However, an insect pollination syndrome was also proposed for some cheirolepidiaceans (Labandeira et al. 2007) on the basis of a stigmalike pollen reception area with trichomes on scale distal lobes in the Northern Hemisphere *Alvinia bohemica* (Velenovský) Kvaček (Kvaček 2000). On the other hand, the fact that *Classopollis* pollen was found on the external surfaces of ovuliferous dwarf shoots of the southern *T. biforme* (Archangelsky 1968) and *K. acuminatus*, in addition to the northern *Hirmeriella muensteri* (Schenk) Jung (Clement-Westerhof and van Konijnenburg-van Cittert 1991), suggests that pollen germination occurred on the scale surfaces, from which pollen tubes could have grown toward the ovules. Alvin (1982) and Clement-Westerhof and van Konijnenburg-van Cittert (1991) have pointed out that this advanced pollination mechanism is analogous to those known in the Podocarpaceae (*Saxegothaea*), Pinaceae (*Tsuga*), and Araucariaceae (*Araucaria, Agathis*) (Doyle 1945).

We know little about the growth forms of the Patagonian cheirolepidiaceans, only that *Tomaxellia* was described as a woody plant with up to four orders of branching (Archangelsky 1966). *T. glabrum* and *K. acuminatus* are based on isolated and fragmentary leafy shoots (Archangelsky and Taylor 1986; Del Fueyo et al. 2008a). No trunks have been found in situ with these taxa that would point to a tall, evergreen, arborescent habit similar to that suggested for the Northern Hemisphere trees of *Hirmeriella muensteri* and *Pseudofrenelopsis parceramosa* (Fontaine) Watson, nor has evidence been found for the low, arborescent to shrubby habit of other cheirolepidiaceans (Alvin 1982). Also, *Frenelopsis* and *Pseudofrenelopsis* remains from the Aptian Crato Formation (Kunzmann et al. 2006) and the Early Cretaceous Lima Campos Basin (Mussa et al. 1991) in Brazil, respectively, show no indication whether these frenelopsids were shrubs or large trees.

On the basis of the mid-Cretaceous records of *Classopollis* pollen and megafossil remains, cheirolepidiaceans lived under variable environmental conditions in Western Gondwana during the Early Cretaceous. In northeast Brazil, within a paleolatitude from 15°S to 20°S, it has been proposed that the cheirolepidiaceans inhabited the undulating lands of the nearby shoreline in a hot arid to semiarid climate (Herngreen 1975; Lima 1976; Regali and Santos 1999). Similar coastal vegetation was reconstructed for South Africa, at about a 40°S paleolatitude by Scott (1976). However, Kunzmann et al. (2006) suggested that the cheirolepidiaceans in northeast Brazil probably grew near the brackish environment of a hypothetical lagoon.

In Antarctica, the cheirolepidiaceans grew under a humid temperate climate at roughly a 60°S paleolatitude (Baldoni and Medina 1989). In Patagonia, dispersed *Classopollis* pollen suggests that the Cheirolepidiaceae were widely distributed and grew in arid to humid climates, in either continental or seashore environments, between 40° and 50°S paleolatitudes (Del Fueyo et al. 2007). On the other hand, we know that *Kachaikestrobus acuminatus* lived in a delta floodplain environment, with fluvial dominance and a humid, temperate climate (Del Fueyo et al. 2008a). *Tomaxellia* spp. and *Tarphyderma glabrum* lived under similar environmental conditions, but in a warm to temperate climate (Archangelsky 1968; Archangelsky and Taylor 1986). Following Alvin (1982) and Watson (1988), all Patagonian cheirolepidiaceans could be considered thermophilous. It should be noted that the extremely xerophytic features that leaves of *Tarphyderma glabrum* have are most probably due to the recurrent ash fall that occurred in Patagonia in the Early Cretaceous (Archangelsky and Taylor 1986; cf. Cúneo et al. 2010). On the other hand, it was suggested that the morphological and anatomical features of *Frenelopsis* and *Tomaxellia biforme* from the Crato Formation are adaptations to a warm and seasonally dry climate (Kunzmann et al. 2006).

We still have no clues to explain why and when the Cheirolepidiaceae became extinct in Western Gondwana. One cause may have been the constant changes in the climatic conditions caused by extensive volcanic activity (Archangelsky 2001). The last occurrence of the Cheirolepidiaceae in Patagonia is probably the early Paleogene (Archangelsky and Romero 1974).

Podocarpaceae

The Podocarpaceae are the best-represented conifers in the Southern Hemisphere with 18 genera and about 200 species. *Saxegothaea* Lindl., *Lepidothamnus* Quinn, *Retrophyllum* Page, three species of *Prumnopitys* Phil., and approximately 17 species of *Podocarpus* L'Heritier are native to South America, while *Afrocarpus falcata* (Thunb.) Page and three species of *Podocarpus* grow in South Africa.

The oldest members of the family appeared in the Triassic, but the Podocarpaceae were abundant during the Jurassic and especially the Early Cretaceous. Podocarps are essentially Gondwanan, although there are some reports of the family in northern regions. Well-preserved members of this family (compressions or mummifications) are found in eocretacic strata of Patagonia. Some genera were described on the basis of detailed cuticular studies of vegetative and fertile remains in organic connection. *Trisacocladus* Archangelsky (Fig. 11.3C; Plate 11C) is named for twigs of various orders of branching that bear spirally inserted leaves basally twisted and disposed in one plane; they are long and have a decurrent base, entire margins, and a single vein (Archangelsky 1966). The axillary pollen cones are shortly pedunculate, oval, and elongate, and they have a central axis that bears helically arranged microsporophylls at right angles; the microsporophylls are distally expanded and have at least two oval pollen sacs. The pollen is trisaccate with short and poorly developed sacs that fold distally over the central body. Dispersed grains of this type are referred to as the genus *Trisaccites* Cookson et Pike (also known as *Trichotomosulcites* Couper). Ultrastructural studies of this pollen in situ were made by Baldoni and Taylor (1982). The ovuliferous cones of *Trisacocladus* are laterally inserted in the axils of sterile leaves. They have a central axis that bears erect ovules that are irregularly (or bilaterally) arranged; each ovule has three cuticular membranes corresponding to the inner megaspore, nucellus, and integument. The seeds developed a stony layer. *Trisaccites* pollen has been recorded in Cretaceous and Tertiary strata in other parts of Gondwana, although it is unknown among the living podocarps.

Apterocladus Archangelsky is another Early Cretaceous Patagonian genus that is based on woody plants with branches of penultimate order that bear bifacial and homomorphic leaves that are spirally arranged but flattened in one plane (Archangelsky 1966). The leaves of *Apterocladus* are coriaceous, lanceolate, single veined, basally constricted, and hypostomatic, with stomata forming poorly defined bands and rows. The stomata tend to be oriented longitudinally, and they share or have adjacent subsidiary cells. The pollen cones are shortly pedunculate and laterally inserted on twigs. They are ovate and have a central axis with spirally arranged, compact microsporophylls, each with a rhombic distal head; each microsporophyll bears more than one pollen sac that contains discoidal pollen of circular equatorial outline with usually (but not always) three poorly developed air sacs (Archangelsky 1966). This in situ pollen shows great morphological variation within the same pollen sac (Gamerro 1965); in its dispersed state, it is referred to *Callialasporites* Sukh Dev, a pollen type with a cosmopolitan distribution (cf. Hotton and Baghai-Riding 2010). The Patagonian material was referred to the podocarps on the basis of the three incipient pollen sacs that develop by the separation of the outer exine layer; this is evident in ultrastructural studies that differentiate *Callialasporites* pollen from those of superficially similar morphology such as *Araucariacites*, *Cyclusphaera*, or *Balmeiopsis* (Archangelsky 1994). However, it has also been suggested that this pollen (in the dispersed state) may pertain to the Araucariaceae (van Konijnenburg-van Cittert 1971). Recent studies (S.A., unpublished work) have shown that one of the species referred to *Callialasporites* has a distinct distal colpus with thickened margins (Fig. 11.3D; Plate 11D), an uncommon type of aperture in either living or fossil podocarp or araucarian pollen. Both hypotheses are plausible and warrant further investigations. In one case, it implies that the Araucariaceae had rudimentary vesiculate pollen; in the other, the past phytogeographic distribution of podocarps would have covered vast regions of both hemispheres.

Squamastrobus Archangelsky et Del Fueyo is a third Patagonian podocarp from the Early Cretaceous that is based on sterile and fertile organs found in organic connection (Archangelsky and Del Fueyo 1989). The plants were probably small, with branching up to the fourth order. The radially and irregularly inserted branches bore helically arranged, adpressed, and squamiform leaves (of the *Brachyphyllum* type). The leaves were amphistomatic with monocyclic to imperfectly dicyclic stomata forming rows. Elongate pollen cones (Fig. 11.3E; Plate 11E) were laterally inserted and had a central axis bearing helically arranged microsporophylls with pollen sacs containing the bisaccate grains of the *Podocarpidites* type. The ovuliferous cones were terminal and elongate, and they had a central axis bearing helically scale–bract complexes (Fig. 11.3F; Plate 11F). Bracts were (possibly partially) fused to scales, which have cuticle with haplocheilic stomata.

Morenostrobus Del Fueyo, Archangelsky et Taylor, is also an Early Cretaceous Patagonian podocarp (Del Fueyo et al. 1990, 2009). The branches bore lanceolate leaves that were helically arranged and single veined, and of the *Elatocladus* type (Fig. 11.3G; Plate 11G). The leaves were twisted at their base, suggesting a distichous arrangement; they were amphistomatic with monocyclic and sunken stomata in files. The pollen cones were cylindrical in shape, narrow, and grouped apically; the microsporophylls were inserted helically and bore abaxial pollen sacs with the bisaccate pollen of the *Podocarpidites* type. The ultrastructure of the leaf cuticle and pollen exine in both *Morenostrobus* and *Squamastrobus* has been studied (Archangelsky and Del Fueyo 1989; Del Fueyo et al. 1990).

In South Africa, podocarp remains correspond to fossil wood genera of *Podocarpoxylon* Gothan and *Mesembrioxylon* Seward found in the Early Cretaceous Namaqualand Continental Shelf (Bamford and Corbett 1994), as well as to impressions of leafy twigs of *Podocarpus* sp. from the Kirkwood Formation (Anderson and Anderson 1985).

Antarctica holds several records of the Podocarpaceae. *Bellingshausium willeyi* Cantrill et Falcon-Lang pertains to seed cones in organic connection to branches from the late Albian Neptune Formation. These branches have scale leaves that grade from falcate to awl shaped and are spirally arranged. The seed cones are borne laterally with a fleshy axis made up of fused elements, each bearing an inverted ovule (Cantrill and Falcon-Lang 2001). The remains of arborescent vegetation growing at a paleolatitude of 62°S within the Antarctic circle are also represented mainly by several species of *Podocarpoxylon* from the late Albian Neptune Glacier and the mid-Aptian Cerro Negro Formations (Cantrill and Falcon-Lang 2001; Falcon-Lang and Cantrill 2001).

Bisaccate and trisaccate pollen of the Podocarpaceae is abundant in the Early Cretaceous deposits of Western Gondwana, mostly in the southern paleolatitudes from which a great variety of bisaccates, including several new species, have been described (Archangelsky and Villar de Seoane 2005). One example is *Gamerroites* Archangelsky, a bisaccate pollen grain, which is a common and endemic component of Early Cretaceous assemblages of southwestern Gondwana (Archangelsky 1988; Campos et al. 1998).

Cupressaceae

This cosmopolitan family has the largest number of living genera (30) among the conifers, with approximately 140 species that occur in nearly all habitats (Farjon 2005).

The only living members of the Cupressaceae in southern South America and South Africa represent endemic genera. In the former area, *Pilgerodendron uviferum*, *Austrocedrus chilensis* (D. Don) Florin et Boutelje and *Fitzroya cupressoides* (Molina) I. M. Johnston inhabit Argentina and Chile. At the present day, these three relict genera have a restricted distribution. *Fitzroya* and *Pilgerodendron* occur in humid areas on nutrient-poor soils in the Valdivian, Patagonian, and Magellanic rain forests, mainly at elevations from about 100 to 1200 m, while *Austrocedrus* inhabits open woodlands at dry sites and in cool temperate forests from about 500 to 1800 m (Veblen et al. 1995; Correa 1998). *Pilgerodendron* does not have a megafossil record. *Fitzroya* has been found in Eocene and early Oligocene sediments in Argentina (Berry 1938) and in Tasmania (Hill and Whang 1996), respectively, while *Austrocedrus* has been reported from the Eocene of Río Negro Province, Argentina (Berry 1938), and from the Oligocene of Tasmania (Paull and Hill 2008).

Extant Cupressaceae in South Africa are represented by the endemic African genus *Widdringtonia* Endl. with four species (*W. cedarbergensis* Marsh, *W. cupressoides* Endl., *W. schwarzii* [Marloth] Mast, *W. whytei* Rendle), which have their center of distribution in the Cape Province, associated with the fynbos and sometimes in the ecotone between the fynbos and evergreen forest (Midgley et al. 1995). Another taxon that is not considered a typically southern conifer is *Juniperus excelsa* Bieb., which extends south of the equator in North and East Africa (Midgley et al. 1995).

In Western Gondwana during the Early Cretaceous, the Cupressaceae were represented solely by megafossil remains, while palynomorphs with an affinity to this family have not been found so far. The plant megafossils correspond to *Athrotaxis* D. Don, a genus which currently has three species (*A. cupressoides* D. Don, *A. selaginoides* D. Don, and *A. laxifolia* Hook) that are confined to Tasmania (Hill et al. 1993).

Only one fossil species has been reported from Western Gondwana as *Athrotaxis ungeri* (Halle) Florin (Fig. 11.3H; plate 11H). In Patagonia and Antarctica, this species had an important paleobiogeographic distribution. It has been found in the Aptian Anfiteatro de Ticó Formation (Archangelsky 1963), in the early Albian Piedra Clavada Formation (Piatnitzky 1938), in the Kachaike Formation (Halle 1913; Del Fueyo et al. 2008a), and in the late Albian of Alexander Island (Cantrill and Falcon-Lang 2001). The branches of *Athrotaxis ungeri* have leaves that are spirally arranged and squamiform, and that bear a cuticle. This cuticle is characterized by having few stomata located at the base of the leaves, similar to the extant species of *Athrotaxis*. Fertile branches have terminal ovuliferous cones and may bifurcate. These cones are cylindrical and oval, and they measure up to 18 mm long. Their bracts and ovuliferous scales are fused and helically inserted on a central axis. Each cone scale is peltate;

it has a cuneate base broadly attached to the axis and a spatulate distal portion that ends in a short, acute appendage (Del Fueyo et al. 2008a).

Other remains of the family include the fossil wood of *Taxodioxylon* sp., recorded from the late Albian Neptune Glacier Formation in Antarctica (Falcon-Lang and Cantrill 2000).

Several conifer taxa on the genus and species levels were endemic to the Western Gondwana region during the Early Cretaceous. This includes megafossils and palynomorphs that have not been recorded from other parts of the world. Patagonia is the region where most of these endemic fossil conifers are concentrated, and it may have been a center of endemism for other plant groups as well (Del Fueyo et al. 2007). This fact was underscored by the previous recognition of a paleofloristic province, the Patagonian Province, in the Western Gondwana Realm, which was based on plant megaflora fossils (Vakhrameev 1988; Archangelsky 1996) and plant microfossils (Volkheimer 1980). Many of the taxa mentioned in this chapter were found as compressions with cuticle of vegetative organs connected to fertile ovuliferous or pollen-bearing cones that made possible the study of their fine structures with scanning and transmission electron microscopy. This highly detailed level of characterization makes these taxa nearly unique and often difficult to compare to similar forms found in other paleofloristic provinces.

Among the Araucariaceae, *Nothopehuen brevis* has *Brachyphyllum*-like leaves and organically attached male cones with *Araucariacites* pollen. *Alkastrobus peltatus*, however, also has branches bearing *Brachyphyllum* type leaves and pollen cones with *Cyclusphaera* grains. These two taxa are both endemic araucarian genera in the Patagonian Province. All *Brachyphyllum* species described from Western Gondwana are known only from this province, such as *B. irregulare* which was found organically attached to a cone with pollen of *Balmeiopsis limbatus* type. Ovuliferous scales referred to *Araucarites* have been found associated with *Nothopehuen*, *Alkastrobus*, and *Brachyphyllum irregulare*, which underscores their araucarian affinity. Vegetative remains (leafy twigs) from Brazil, Patagonia, South Africa, and the Antarctic Peninsula were referred to new species of *Araucaria*; they illustrate the variable leaf morphology found in the Western Gondwanan araucarians during the Early Cretaceous.

The Cheirolepidiaceae were widely distributed in Western Gondwana as evidenced by the abundance of *Classopollis*, a pollen type exclusive to this family. However, megafossil remains show that some of the southern cheirolepidiaceans differ from northern taxa referred to the genera *Frenelopsis* or *Hirmeriella*. The genus *Tomaxellia*, known from the Aptian of the Patagonian Province with a single record in northern Brazil, has twigs with either homomorphic or dimorphic leaves that are organically attached to seed cones and pollen cones (with *Classopollis* grains), and that are associated with ovuliferous scales. Its reproductive strategy of shedding its ovuliferous dwarf shoots at cone maturity is

Discussion and Conclusions

similar to that observed in the Northern Hemisphere cheirolepidiaceans. Another endemic member of the family is the Albian *Kachaikestrobus acuminatus* from southern Patagonia. This cheirolepidiacean belongs to the "brachyphyll" group (i.e., those with squamiform leaves) and consists of twigs bearing seed cones with primitive characters—a large number of scales and bracts that are longer than the ovuliferous dwarf shoot. Thick-cuticle leaves with a unique type of deeply sunken stomata were referred to the new genus *Tarphyderma*, which is also known from fossil floras in the Northern Hemisphere.

The Podocarpaceae are the prevailing conifers of the Southern Hemisphere today, underscoring the important role they have played in Cretaceous plant assemblages of Gondwana. *Trisacocladus tigrensis* has twigs with elongate leaves, axillary pollen cones with *Trisaccites* type grains, and lateral ovuliferous cones inserted in the axils of sterile leaves. The pollen type of *Triasaccites* was endemic to the Gondwana Realm during the Cretaceous. *Apterocladus lanceolatus* has branches with bifacial leaves flattened in one plane (*Elatocladus* type) with shortly pedunculate pollen cones laterally inserted on twigs. Its in situ pollen is of the *Callialasporites* type and is referred to this family. This pollen type and its peculiar morphology is considered by some authors to be related to the araucarians rather than the podocarps. However, this remains an open question as far as the Patagonian material is concerned. *Apterocladus* is also endemic in Western Gondwana if all of the characters of the vegetative and fertile organs found in organic connection are considered. *Squamastrobus tigrensis* is another endemic podocarp with branches bearing *Brachyphyllum*-like leaves, laterally inserted pollen cones with in situ bisaccate grains of the *Podocarpidites* type, and terminal ovule-bearing cones with bracts that may be partially fused to the scales. Also endemic for this region is *Morenostrobus fertilis*, which has cones with *Podocarpidites* type pollen but with branches bearing twisted lanceolate leaves resembling the *Elatocladus* type.

Finally, the taxodiaceous Cupressaceae are represented in the Patagonian Province by *Athrotaxis ungeri*, an endemic species of a genus now exclusive to Tasmania. The branches of *Athrotaxis* bear squamiform leaves with a cuticle with a characteristic distribution of stomata. Its fertile branches bear terminal ovuliferous cones with bracts fused to scales. Fossil woods referred to the family (*Taxodioxylon*) have also been described from Antarctica.

Cyclusphaera is a common dispersed pollen type in Early Cretaceous assemblages, which made its first appearance at the J/K boundary. It is mostly found in southern South America and Southern Africa, and becomes rare to the north, in Peru, and to the south, on the Antarctic Peninsula. It became extinct in the Late Cretaceous. *Araucariacites* shows some resemblance to the pollen of extant *Araucaria* and *Agathis* (Dettmann and Jarzen 2000; Del Fueyo et al. 2008b) and has a widespread global distribution. It survives today in South America (Argentina, Brazil, and Chile) and is found in pollen cones of the extant *Araucaria angustifolia*

(Bert.) O. Kuntze and *Araucaria araucana* (Mol.) K. Koch. (Archangelsky 1994; Del Fueyo and Archangelsky 2005). Both of these living taxa pertain to the section *Araucaria* of the genus *Araucaria*. During the Early Cretaceous in Western Gondwana (South America, South Africa, and Antarctica), three of the six sections of *Araucaria* (except sections *Intermedia*, *Perpendiculares*, and *Yezonia*) were represented: *Araucaria*, *Bunya*, and *Eutacta*. At the end of the Mesozoic, the *Bunya* and *Eutacta* sections became extinct in Western Gondwana, while section *Araucaria* has survived to the present day.

Finally, the limited knowledge we have on Early Cretaceous floras in many areas of Western Gondwana should be emphasized, especially in South Africa and northern South America, where thick sedimentary sequences of this age do occur. At present, the main flow of data is coming from the Antarctic Peninsula and the Patagonian Province. Further studies in all parts of Western Gondwana are needed to bring together more information that will help us understand the role that conifers played in Cretaceous plant assemblages and to elucidate their relationship to conifers in floras elsewhere on the globe.

Acknowledgments

We thank C. Gee for her useful suggestions. We also thank A. González for drafting Figure 11.1. This work was supported by CONICET grant PIP 5093 and ANPCYT grant 32320.

References Cited

Alvin, K. L. 1982. Cheirolepidiaceae: biology, structure and paleoecology. *Review of Palaeobotany and Palynology* 37: 71–98.

Anderson, J. M., and H. M. Anderson. 1985. *Paleoflora of Southern Africa. Prodromus of South Africa Megafloras Devonian to Lower Cretaceous.* Rotterdam: A. A. Balkema.

Archangelsky, S. 1963. A new Mesozoic flora from Ticó, Santa Cruz Province, Argentina. *Bulletin of the British Museum (Natural History), Geology* 8: 4–92.

———. 1966. New gymnosperms from the Ticó Flora, Santa Cruz Province, Argentina. *Bulletin of the British Museum (Natural History), Geology* 13: 261–295.

———. 1968. On the genus *Tomaxellia* (Coniferae) from the Lower Cretaceous of Patagonia (Argentina) and its male and female cones. *Botanical Journal of the Linnean Society* 61: 153–165.

———. 1979. *Balmeiopsis*, nuevo nombre genérico para el palinomorfo *Inaperturopollenites limbatus* Balme. *Ameghiniana* 14: 122–126.

———. 1988. *Gamerroites*, nuevo género de polen bisacado del Cretácico de Patagonia, Argentina. *Boletín de la Asociación Latinoamericana de Paleobotánica y Palinología* 11: 1–6.

———. 1994. Comparative ultrastrucutre of three Early Cretaceous gymnosperm pollen grains: *Araucariacites*, *Balmeiopsis* and *Callialasporites*. *Review of Palaeobotany and Palynology* 83: 185–198.

———. 1996. The Jurassic and Cretaceous vegetation of the Patagonian Province. In M. A. Akhmetiev and M. P. Doludenko (eds.), *Memorial Conference Dedicated to V. A. Vakhrameev*, pp. 8–10. Moscow: Russian Academy of Sciences.

————. 2001. The Ticó flora (Patagonia) and the Aptian extinction event. *Acta Paleobotanica Polonica* 42: 115–122.

————. 2003. La flora cretácica del Grupo Baqueró, Santa Cruz, Argentina. *Monografías del Museo Argentino de Ciencias Naturales* 4: 1–14 + CD-ROM.

Archangelsky, S., and O. G. Arrondo. 1975. Paleogeografía y plantas fósiles en el Pérmico Inferior Austrosudamericano. *Actas I Congreso Argentino de Paleontología y Bioestratigrafía* 1: 479–496.

Archangelsky, S., and G. M. Del Fueyo. 1989. *Squamastrobus* gen. nov., a fertile podocarp from the early Cretaceous of Patagonia, Argentina. *Review of Palaeobotany and Palynology* 59: 109–126.

Archangelsky, S., and J. C. Gamerro. 1967. Pollen grains found in coniferous cones from the Lower Cretaceous of Patagonia (Argentina). *Review of Palaeobotany and Palynology* 5: 179–182.

Archangelsky, S., and E. J. Romero. 1974. Polen de gimnospermas coníferas (del Cretácico Superior y Paleoceno de Patagonia). *Ameghiniana* 11: 217–236.

Archangelsky, S., and T. N. Taylor. 1986. Ultrastructural studies of fossil plant cuticles. II. *Tarphyderma* gen. n., a Cretaceous conifer from Argentina. *American Journal of Botany* 73: 1577–1587.

————. 1991. *Tarphyderma punctactum* (Michael) Archangelsky & Taylor comb. nov., an early Cretaceous conifer. *Taxon* 40: 319–320.

Archangelsky, S., and L. Villar de Seoane. 2005. Estudios palinológicos del Grupo Baqueró (Cretácico Inferior), provincia de Santa Cruz, Argentina. IX. Polen bisacado de Podocarpáceas. *Revista Española de Paleontología* 20: 37–56.

Archangelsky, S., A. Baldoni, J. C. Gamerro, and J. Seiler. 1984. Palinología estratigráfica del Cretácico de Argentina Austral. 3. Distribución de las especies y conclusiones. *Ameghiniana* 21: 15–33.

Axsmith, B. J., and B. F Jacobs. 2005. The conifer *Frenelopsis ramosissima* (Cheirolepidiaceae) in the Lower Cretaceous of Texas: systematic, biogeographical, and palaeocological implications. *International Journal of Plant Sciences* 166: 327–337.

Axsmith, B. J., F. M. Andrews, and N. C. Fraser. 2004. The structure and phylogenetic significance of the conifer *Pseudohirmerella delawarensis* nov. comb. from the Upper Triassic of North America. *Review of Palaeobotany and Palynology* 129: 251–263.

Baldoni, A. M. 1979. Nuevos elementos paleoflorísiticos de la tafoflora de la Formación Springhill, límite Jurásico–Cretácico, subsuelo de Argentina y Chile Austral. *Ameghinina* 16: 103–119.

Baldoni, A. M., and F. Medina. 1989. Fauna y microflora del Cretácico, en Bahía Brandy, isla James Ross, Antártida. *Serie Científica del Instituto Nacional Antártico Chileno* 39: 43–58.

Baldoni, A. M., and T. N. Taylor. 1982. The ultrastructure of *Trisaccites* pollen from the Cretaceous of Southern Argentina. *Review of Palaeobotany and Palynology* 38: 23–33.

Bamford, M. K., and I. B. Corbett. 1994. Fossil wood of Cretaceous age from the Namaqualand Continental Shelf, South Africa. *Palaeontographica Africana* 31: 83–95.

Berry, E. W. 1938. Tertiary flora from the Rio Pichileufu, Argentina. *Geological Society of America. Special Papers* 12: 1–149.

Brenner, G. J. 1968. Middle Cretaceous spores and pollen from Northeastern Peru. *Pollen et Spores* 10: 341–383.

Brown, J. T. 1977. On *Araucarites rogersii* Seward from the Lower Cretaceous Kirkwood Formation of the Algoa Basin, Cape Province, South Africa. *Palaeontographica Africana* 20: 47–51.

Campos, C., M. J. García, R. Dino, G. Veroslavsky, A. R. Saad, and V. J. Fulfaro. 1998. Registro palinológico dos poços SL9-C1 e SL12-SB, Formaçao Castellanos, no porção norte da bacía de Santa Lucía. Cretáceo do Uruguai. *Segundo Congreso Uruguayo de Geología (Punta del Este, Uruguay), Actas*, pp. 173–176.

Cantrill, D. J. 2000. A Cretaceous (Aptian) flora from President Head, Snow Island, Antarctica. *Palaeontographica*, Abt. B, 253: 153–191.

Cantrill, D. J., and H. J. Falcon-Lang. 2001. Cretaceous (Late Albian) Coniferales of Alexander Island, Antarctica. 2. Leaves, reproductive structures and roots. *Review of Palaeobotany and Palynology* 115: 119–145.

Césari, S., C. Parica, M. Remesal, and F. Salani. 1999. Paleoflora del Cretácico Inferior de península Byers, islas Shetland del Sur, Antártida. *Ameghiniana* 36: 3–22.

Clement-Westerhof, J. A., and J. H. A. van Konijnenburg-van Cittert. 1991. *Hirmeriella muensteri*: new data on the fertile organs leading to a revised concept of the Cheirolepidiaceae. *Review of Palaeobotany and Palynology* 68: 147–179.

Correa, M. N. 1998. *Flora Patagónica*. Buenos Aires: Colección Científica del INTA.

Cúneo, R. N., I. Escapa, L. Villar de Seoane, A. Artabe, and S. Gnaedinger. 2010. Review of the cycads and bennettitaleans from the Mesozoic of Argentina. In C. T. Gee (ed.), *Plants in Mesozoic Time: Morphological Innovations, Phylogeny, Ecosystems*, pp. 187–212. Bloomington: Indiana University Press.

Del Fueyo, G. M. 1991. Una nueva Araucariaceae Cretácica de Patagonia, Argentina. *Ameghiniana* 28: 149–161.

Del Fueyo, G. M., and A. Archangelsky. 2002. *Araucaria grandifolia* Feruglio from the Lower Cretaceous of Patagonia, Argentina. *Cretaceous Research* 23: 265–277.

Del Fueyo, G. M., and S. Archangelsky. 2005. A new araucarian pollen cone with in situ *Cyclusphaera* Elsik from the Aptian of Patagonia, Argentina. *Cretaceous Research* 26: 757–768.

Del Fueyo, G. M., S. Archangelsky, and T. N. Taylor. 1990. Una nueva podocarpácea fértil (Coniferal) del Cretácico inferior de Patagonia. *Ameghiniana* 27: 63–73.

———. 2009. *Morenostrobus*, a new substitute name for *Morenoa* Del Fueyo et al. 1990, non La Llave 1824. *Ameghiniana* 46: 215.

Del Fueyo, G. M., L. Villar de Seoane, A. Archangelsky, V. Guler, M. Llorens, S. Archangelsky, J. C. Gamerro, E. Musacchio, M. Passalia, and V. Barreda. 2007. Biodiversidad de las paleofloras de Patagonia Austral durante el Cretácico Inferior. *Ameghiniana* 50° *Aniversario, Publicación Especial* 11: 101–122.

Del Fueyo, G. M., S. Archangelsky, M. Llorens, and R. Cúneo. 2008a. Coniferous ovulate cones from the Lower Cretaceous of Santa Cruz Province, Argentina. *International Journal of Plant Sciences* 169: 799–813.

Del Fueyo, G. M., M. A. Caccavari, and E. Dome. 2008b. Morphology of the pollen cone and pollen grain of the *Araucaria* species from Argentina. *BioCell* 32: 49–60.

Dettmann, M. E., and H. T. Clifford. 2005. Biogeography of Araucariaceae. In J. Dargavel (ed.), *Australia and New Zealand Forest Histories. Araucaria Forests*, pp. 1–9. Occasional Publication 2. Kingston: Australian Forest History Society.

Dettmann, M. E., and D. M. Jarzen. 2000. Pollen of extant *Wollemia* (Wollemi pine) and comparisons with pollen of other extant and fossil Araucariaceae. In M. M. Harley, C. M. Morton, and S. Blackmore (eds.), *Pollen and*

Spores: Morphology and Biology, pp. 187–203. London: Royal Botanic Gardens, Kew.

Dino, R., O. Braga da Silva, and D. Abrahão. 1999. Caracterização palinológica e estratigráfica de estratos Cretáceos da Formação Alter do Chão, Bacia do Amazonas. *Boletim do 5° Simposio sobre o Cretáceo do Brasil. UNESP—Campus de Rio Claro/SP,* 557–565.

Doyle, J. 1945. Developmental lines in pollination mechanisms in the Coniferales. *Scientific Proceeding of the Royal Society of Dublin* 24: 43–63.

Duarte, L. 1993. Restos de Araucariáceas da Formação Santana-Membro Crato (Aptiano), NE do Brasil. *Anais da Academia Brasileira de Ciências* 65: 357–362.

Falcon-Lang, H. J., and D. J. Cantrill. 2000. Cretaceous (late Albian) Coniferales of Alexander Island, Antarctica. 1. Wood taxonomy: a quantitative approach. *Review of Palaeobotany and Palynology* 111: 1–17.

———. 2001. Gymnosperm woods from the Cretaceous (mid-Aptian) Cerro Negro Formation, Byres Peninsula, Livingston Island, Antarctica: the arborescent vegetation of a volcanic arc. *Cretaceous Research* 22: 277–293.

Farjon, A. 2005. *A Monograph of Cupressaceae and Sciadopitys.* London: Royal Botanic Gardens, Kew.

Florin, R. 1940. The Tertiary fossil conifers of south Chile and their phytogeographical significance. *Kungliga Svenska Vetenskapsakademiens Handlingar* 19: 1–106.

Gamerro, J. C. 1965. Morfología del polen de *Apterocladus lanceolatus* Archang. (Coniferae) de la Formación Baqueró, Provincia de Santa Cruz. *Ameghinina* 4: 133–136.

———. 1982. *Informe palinológico del perfil arroyo Caballo Muerto, Santa Cruz.* Buenos Aires: Yacimientos Petrolíferos Fiscales.

Gee, C. T. 1989. Revision of the Late Jurassic/Early Cretaceous flora from Hope Bay, Antarctica. *Palaeontographica,* Abt. B, 213: 149–214.

Gee, C. T., and W. D. Tidwell. 2010. A mosaic of characters in a new whole-plant *Araucaria, A. delevoryasii* Gee sp. nov., from the Late Jurassic Morrison Formation of Wyoming, U.S.A. In C. T. Gee (ed.), *Plants in Mesozoic Time: Morphological Innovations, Phylogeny, Ecosystems,* pp. 67–94. Bloomington: Indiana University Press.

Halle, T. 1913. The Mesozoic flora of Graham Land. *Wissenschaftliche Ergebnisse Schwedischen Südpolar-Expedition, 1901–1903,* 3: 1–123.

Herbst, R., M. Brea, A. Crisafulli, S. Gnaedinger, A. Lutz, and L. Martínez. 2007. La paleoxilología en la Argentina. Historia y desarrollo. *Ameghiniana* 50° *Aniversario, Publicación Especial* 11: 57–71.

Hernández, P., and V. Azcárate. 1971. Estudio paleobotánico preliminar sobre restos de una tafoflora de la península Byres (Cerro Negro), Isla Livingston, Islas Shetlands del Sur, Antárctica. *Serie Científica del Instituto Antártico Chileno* 2: 15–50.

Herngreen, G. F. W. 1975. Palynology of Middle and Upper Cretaceous strata in Brazil. *Mededelingen Rijks Geologische Dienst, Nieuwe Serie* 26: 39–91.

Herngreen, G. F. W., and A. F. Chlonova. 1981. Cretaceous microfloral provinces. *Pollen et Spores* 23: 441–555.

Hill, R. S., and S. S. Whang. 1996. A new species of *Fitzroya* (Cupressaceae) from Oligocene sediments in north-western Tasmania. *Australian Systematic Botany* 9: 867–875.

Hill, R. S., G. J. Jordan, and R. J. Carpenter. 1993. Taxodiaceous macrofossils from the Tertiary and Quaternary sediments in Tasmania. *Australian Systematic Botany* 6: 237–249.

Hotton, C. L., and N. L. Baghai-Riding. 2010. Palynological evidence for conifer dominance within a heterogeneous landscape in the Late Jurassic

Morrison Formation, U.S.A. In C. T. Gee (ed.), *Plants in Mesozoic Time: Morphological Innovations, Phylogeny, Ecosystems*, pp. 295–328. Bloomington: Indiana University Press.

Kotova, I. Z. 1983. Palynological study of Upper Jurassic and Lower Cretaceous sediments, Site 511, Deep Sea Drilling Project Leg 71 (Falkland Plateau). *Initial Reports of the Deep Sea Drilling Project* 71: 879–906.

Kunzmann, L. 2007. Araucariaceae (Pinopsida): aspects in palaeobiogeography and palaeobiodiversity in the Mesozoic. *Zoologischer Anzeiger* 246: 257–277.

Kunzmann, L., B. A. R. Mohr, M. E. C. Bernardes-de Olivera, and V. Wilde. 2006. Gymnosperms from the Early Cretaceous Crato Formation (Brazil). II. Cheirolepidiaceae. *Fossil Record* 9: 213–225.

Kvaček, J. 2000. *Frenelopsis alata* and its microsporangiate and ovuliferous reproductive structures from the Cenomanian of Bohemia (Czech Republic, Central Europe). *Review of Palaeobotany and Palynology* 112: 51–78.

Labandeira, C. C., J. Kvaček, and M. B. Mostovski. 2007. Pollination drops, pollen, and insect pollination of Mesozoic gymnosperms. *Taxon* 56: 663–695.

Lima, M. R. de. 1976. O gênero *Classopollis* en as bacias mesozoicas do nordeste de Brasil. *Ameghiniana* 13: 226–234.

———. 1979. Palinologia da Formação Santana (Cretáceo do Nordeste do Brasil). II. Descrição sistemática dos esporos da subturma Zonotriletes e turma Monolotes, e dos polens da turmas Saccites e Aletes. *Ameghiniana* 16: 27–63.

Martill, D. M., R. F. Loveridge, J. A. Ferreira Gomes de Andrade, and A. Herzog Cardoso. 2005. An unusual occurrence of amber in laminated limestones: the Crato Formation lagerstätte (Early Cretaceous) of Brazil. *Palaeontology* 48: 1399–1408.

Midgley, J. J., W. J. Bond, and C. J. Geldenhuys. 1995. The ecology of Southern African conifers, pp. 64–80. In N. J. Enright and R. S. Hill (eds), *Ecology of Southern Conifers*. Melbourne: Melbourne University Press.

Miller, C. N. 1999. Implication of fossil conifers for the phylogenetic relationships of living families. *Botanical Review* 65: 239–277.

Mussa, D., M. E. C. Bernardes-de-Oliveira, R. Dino, and M. Arai. 1991. A presenta do gênero *Pseudofrenelopsis* Nathorst da bacia mesozóico de Lima Campos, Estado do Ceará, Brasil. *12 Congreso Brasileiro de Paleontología Resumos São Paulo*, 115.

Owen, H. G., and J. Mutterlose. 2006. Late Albian ammonites from offshore Suriname: implications for biostratigraphy and palaeobiogeography. *Cretaceous Research* 27: 717–727.

Pankhurst, R. J., C. W. Rapela, C. M. Fanning, and M. Márquez. 2006. Gondwanide continental collision and the origin of Patagonia. *Earth-Science Reviews* 76: 235–257.

Passalia, M. G. 2008. Estudio de cutículas fósiles de Patagonia: su potencial como estimador paleoatmosférico de CO_2. Ph.D. diss., Buenos Aires University, Buenos Aires.

Paull, R., and R. S. Hill. 2008. Oligocene *Austrocedrus* from Tasmania (Australia): comparisons with *Austrocedrus chilensis*. *International Journal of Plant Sciences* 169: 315–330.

Philippe, M., M. Bamford, S. McLoughlin, L. S. R. Alves, H. J. Falcon-Lang, S. Gnaedinger, E. G. Ottone, M. Pole, A. Rajanikanthi, R. E. Shoemaker, T. Torres, and A. Zamuner. 2004. Biogeographic analysis of Jurassic–Early Cretaceous wood assemblages from Gondwana. *Review of Palaeobotany and Palynology* 129: 141–173.

Piatnitzky, A. 1938. Observaciones geológicas en el oeste de Santa Cruz (Patagonia). *Boletín de Informaciones Petroleras* 165: 45–95.

Pons, D. 1988. *Le Mésozoïque de Colombie. Macroflores et Microflores.* Cahiers de Paléontologie. Paris: Éditions du Centre National de la Recherche Scientifique.

Prámparo, M. B. 1994. Lower Cretaceous palynoflora of the La Cantera Formation, San Luis Basin: correlation with other Cretaceous palynofloras of Argentina. *Cretaceous Research* 15: 193–203.

Prámparo, M. B., and M. Batty. 1994. Primeros datos palinológicos del Cretácico Inferior de la Cuenca de Arequipa, Sur de Perú. *Zentralblatt für Geologie und Paläontologie* 1: 413–425.

Quattrocchio, M. E., M. A. Martinez, A. Carpinelli Pavisich, and W. Volkheimer. 2006. Early Cretaceous palynostratigraphy, palynofacies and palaeoenvironments of well sections in northeastern Tierra del Fuego, Argentina. *Cretaceous Research* 27: 584–602.

Regali, M. S. P., and P. R. S. Santos. 1999. Palinoestratigrafia e Geocronologia dos sedimentos Albo–Aptianos das Bacias de Sergipe e de Alagoas–Brasil. *Boletim do 5° Simposio sobre o Cretáceo do Brasil. UNESP—Campus de Rio Claro/SP*, pp. 414–419.

Riding, J. B., J. A. Crame, M. E. Dettmann, and D. J. Cantrill. 1998. The age of the base of the Gustav Group in the James Ross Basin, Antarctica. *Cretaceous Research* 19: 87–105.

Scott, L. 1976. Palynology of Lower Cretaceous deposits from the Alagoa Basin (Republic of South Africa). *Pollen et Spores* 18: 563–609.

Upchurch, P. 2007. Gondwanan break-up: legacies of a lost world? *Trends in Ecology and Evolution* 23: 229–236.

Vakhrameev, V. A. 1988. [Jurassic and Cretaceous floras and climates of the earth]. *Transactions of the Russian Academy of Sciences* 230: 1–214. [In Russian.]

Vallati, P. 1993. Palynology of the Albornoz Formation (Lower Cretaceous) in the San Jorge Gulf Basin (Patagonia, Argentina). *Neues Jahrbuch für Geologie und Paläontologie* 187: 345–373.

van Konijnenburg-van Cittert, J. H. A. 1971. In situ gymnosperm pollen from the Middle Jurassic of Yorkshire. *Acta Botanica Neerlandica* 20: 1–97.

Veblen, T. T., B. R. Burns, T. Kitzberger, A. Lara, and R. Villalba. 1995. The ecology of the conifers of southern South America. In N. J. Enright and R. S. Hill (eds.), *Ecology of Southern Conifers*, pp. 120–155. Melbourne: Melbourne University Press.

Villar de Seoane, L. 1998. Comparative study of extant and fossil conifer leaves from the Baqueró Formation (Lower Cretaceous), Santa Cruz Province, Argentina. *Review of Palaeobotany and Palynology* 99: 247–263.

Volkheimer, W. 1980. Microfloras del Jurásico Superior y Cretácico Inferior de América Latina. *II Congreso Argentino de Paleontología y Bioestratigrafía y I Congreso Latinoamericano de Paleontología, Actas V*, 121–136.

Volkheimer, W., and E. Sepúlveda. 1976. Biostratigraphische Bedeutung und microfloristische Assoziation von *Cyclusphaera psilata* n. sp., einer Leitform aus der Unterkreide des Neuquen-Beckens (Argentinien). *Neues Jahrbuch für Geologie und Paläontologie, Monatshefte* 2: 97–108.

Watson, J. 1988. The Cheirolepidiaceae. In C. B. Beck (ed.), *Origin and Evolution of Gymnosperms*, pp. 382–447. New York: Columbia University Press.

Zavada, M. S. 1987. The occurrence of *Cyclusphaera* sp. in Southern Africa. *Actas Séptimo Simposio Argentino de Paleobotánica y Palinología (Buenos Aires 1987)*, 101–105.

Fig. 12.1. Simplified map of the Four Corners area of the southwestern United States showing localities where specimens of *Tempskya* occur that contain stems of dicotyledonous lianas, as well as generalized stratigraphic columns illustrating the rock sequences at each locality. For clarity, the formations containing liana-bearing *Tempskya* specimens are shaded in each column. (Locality 3 is the same as locality 6, and locality 4 is the same as locality 5 in Tidwell and Hebbert 1992.) *Compiled from O'Sullivan et al. (1972), Ash and Read (1976), and Tidwell and Hebbert (1992).*

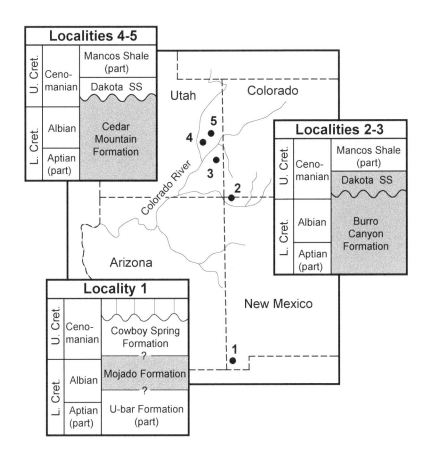

OLDEST KNOWN DICOTYLEDONOUS LIANAS FROM THE EARLY LATE CRETACEOUS OF UTAH AND NEW MEXICO, U.S.A.

12

William D. Tidwell, Sidney R. Ash, and Brooks B. Britt

The remains of more than 15 specimens of the oldest known fossil lianous dicotyledonous stems have been discovered embedded in the false trunks of four species of the tree fern *Tempskya* Corda in mainly early Late Cretaceous (Cenomanian) strata in Utah and New Mexico. These dicotyledonous stems are assigned to *Munzingoxylon delevoryasii* gen. et sp. nov. and *Rodoxylon scheetzii* gen. et sp. nov. Most of the stems are embedded in the false trunks of *T. jonesii* Tidwell et Hebbert from the Cedar Mountain and Burro Canyon Formations of central and southeastern Utah and *T. reesidei* Ash et Read from the Mojado Formation of southwestern New Mexico. The specimen of *T. reesidei* is significant because it contains the remains of lianous stems that are alternately exposed on the surface of the trunk and embedded in it, clearly illustrating the scrambling growth habit of these particular stems. The dicotyledonous stems are compared to several modern families having similar scandent members. Their occurrence in early Late Cretaceous (Cenomanian) strata indicates that the angiospermous flora at the time was more diverse in taxonomy and habit than previously thought.

Introduction

Tempskya Corda (1845) is an unusual genus of tree ferns with a false trunk composed of a dichotomously branching stem system embedded in a thick mantle of small adventitious roots. This fossil was known only from widespread localities in the Northern Hemisphere until recently, when it was reported from the Southern Hemisphere at localities in Argentina (Tidwell and Wright 2003) and Australia (Clifford and Dettman 2005). *Tempskya* is especially common in western North America, and these specimens were long thought to be Early Cretaceous (Albian) in age. However, radiometric dates (Garrison et al. 2007) for the uppermost Cedar Mountain Formation (Mussentuchit Member), a primary source of the taxon, indicate a Cenomanian age of 97.2 ± 0.6 million years. Recently, *Tempskya* has been reported from the Late Cenomanian of the Czech Republic (Mikuláš and Devořák 2002) and the Late Cretaceous (Santonian) of Japan (Nishida 1986, 2001).

Although the permineralized remains of the false trunks of *Tempskya* have been studied for more than 150 years, it was only recently that other types of fern roots and stems of dicotyledonous lianas adhering to and/or embedded in them were discovered and reported (Ash and Read 1976; Tidwell et al. 1977; Schneider and Kenrick 2001). Most of these dicotyledonous stems are embedded in fragments of the false trunks of five species of *Tempskya*. These specimens occur in mainly Early Cretaceous (Albian) strata in Utah and New Mexico (Fig. 12.1). The dicot stems are more fully described here as *Munzingoxylon delevoryasii* gen. et sp. nov. and *Rodoxylon scheetzii* gen. et sp. nov., and their implications are discussed.

The habit of growth of lianas in a 1-m-long obconical trunk of *T. reesideii* Ash et Read (1976) from New Mexico is particularly instructive because the lianas are not only embedded within the trunk itself, but are also exposed for short distances on one side of the trunk. This indicates that at least in this specimen, the lianas may have had a scrambling habit of growth. Furthermore, it is evident that sections of the stems were surrounded by the adventitious roots of the *Tempskya* plant while it, and presumably the stems, were alive.

Stratigraphy of the *Tempskya* Localities

The *Tempskya* specimens bearing the dicotyledonous lianas were collected from two localities in New Mexico and three localities in Utah (Fig. 12.1).

Locality 1 is in the Mojado Formation east of the Animas Mountains, about 140 km southeast of Deming in extreme southwestern New Mexico (Ash and Read 1976). The Mojado Formation consists of a sparsely fossiliferous nonmarine lower part and a marine upper part that is highly fossiliferous in certain layers. The specimen of *T. reesidei* that bears the dicotyledonous stems occurred near the top of the formation in a relatively thin (approximately 10 m thick) bed of shale and claystone separating two somewhat thinner beds of limestone that contain a middle to late Albian marine fauna identified by W. A. Cobban (in Ash and Read 1976). This fauna includes the foraminifera *Cribratina texana*, the ammonite *Engonoceras serpentinum*, and several species of pelecypods, scaphopods, and gastropods.

Locality 2 is in the Dakota Formation about 44 km northwest of Shiprock in northwestern New Mexico, where the formation has been divided into three informal members. The specimens of *Tempskya* sp. were found in what is "believed unequivocally to be the basal sandstone of the Dakota" (O'Sullivan et al. 1972: E13). At this locality, the basal member consists principally of sandstone and conglomerate (Strobell in Ash and Read 1976). The basal member averages about 6 m in thickness in this region and is often overlain by the medial member that consists of a 5–9-m-thick bed of brown to black carbonaceous mudstone, siltstone, coal, and interbedded sandstone stringers. At many places, the medial member is overlain by an upper member composed of pale orange, cross-bedded sandstone that contains an early Late Cretaceous marine fauna. The lower part of the Dakota Formation in the Colorado Plateau is considered to be as old as Aptian by some authors (Cobban and Reeside 1952), whereas others believe that all but the basal beds of the Dakota are of the Late Cretaceous age (O'Sullivan et al. 1972). The presence of *Tempskya* in these beds suggested to O'Sullivan et al. (1972) and Ash and Read (1976) that at least these basal beds of the Dakota are of Albian age, but the *Tempskya* species reported from the Dakota of Utah are considered to be Cenomanian in age (Rushforth 1971; Tidwell and Hebbert 1992). This discrepancy in the age of the fragmentary *Tempskya* specimens from New Mexico might be the result of the reworking of the specimens into the Dakota Formation from still older (Albian?) strata. The *Tempskya* species from the Dakota Formation in eastern Utah and northwestern New Mexico (Fig. 12.1) includes *Tempskya sp.*, *T. jonesii* Tidwell et Hebbert, *T. minor* Read et Brown, and possibly *T. knowltonii* Seward (for details of these occurrences, see Read and Brown 1937; Brown 1950; Ash and Read 1976; Tidwell and Hebbert 1992; Tidwell et al. 2007).

Locality 3 is in the Burro Canyon Formation about 21 km southeast of Moab, Utah (Fig. 12.1). At this locality, the formation consists of mudstone, sandstone, and conglomeritic sandstone. The formation unconformably underlies the Dakota Formation and overlies the Morrison Formation of Late Jurassic age. It contains freshwater invertebrates, which suggests that it is entirely of Early Cretaceous age (Simmons 1957), with the base being Aptian (O'Sullivan et al. 1972). The occurrence of *Tempskya* in the upper part of the unit indicates that those beds are probably Albian because it is associated with marine faunas of that age at several other localities (Ash and Read 1976).

Locality 4 is in the Cedar Mountain Formation about 24 km northwest of Moab, Utah (Fig. 12.1). Here the formation unconformably underlies the Dakota Formation and overlies the Morrison Formation. Radiometric dating indicates that the Cedar Mountain Formation ranges in age between 124 million years for the base of the formation in the Moab area (Burton et al. 2006; Greenhalgh et al. 2006; Greenhalgh and Britt 2007) and 97 million years for the top of the formation on the east flank of the San Rafael Swell south of Interstate 70 (Cifelli et al. 1997; Garrison et al. 2007). Thus, the formation spans about 27 million years, from the earliest Aptian to early Cenomanian, following the timescale of Gradstein et al. (2004). Fossil plants in the formation include false trunks of *Tempskya*, cycadeoids, *Monanthesia*, conifers such as *Frenelopsis* and *Brachyphyllum*, compressions of a platinoid flower, and several coniferous and angiospermous woods. The coniferous woods include *Mesembrioxylon stokesii* Thayn et Tidwell (1984) and *Palaeopiceoxylon thinosus* Tidwell et Thayn (1985). *Paraphyllanthoxylon utahense* Thayn, Tidwell et Stokes (1983) and *Icacinoxylon pittiense* Thayn, Tidwell et Stokes (1985) comprise the angiospermous woods presently described from this formation (Tidwell 1998; Tidwell et al. 2007). Within the formation, three distinct dinosaur faunas are recognized (Kirkland et al.

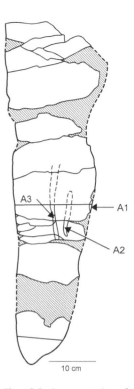

Fig. 12.2. Reconstruction of the basal portion of the false trunk of *Tempskya reesidei*. Missing parts are shaded. The dicotyledonous stems found in this specimen are indicated by the longitudinal solid and dashed lines (A1–A3) in the middle of the fossil. Solid lines denote the remains of the stems preserved both on the surface and within the trunk. Dashed lines represent impressions of the stem on the surface of the trunk. *USNM 167547; modified from Ash and Read (1976).*

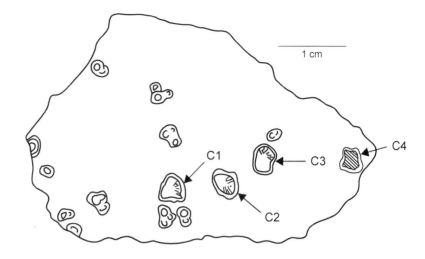

Fig. 12.3. Specimen C containing four dicotyledonous stems in *Tempskya* sp. from northwestern New Mexico. Stems C1–C3 are poorly preserved, although they may represent *Munzingoxylon*. The C4 stem was invaded and destroyed by *Tempskya* roots (USNM 538920).

1997), with the uppermost fauna consisting primarily of teeth and other microfossils (Cifelli et al. 1997).

Tempskya is represented in the Cedar Mountain Formation by nine species of the permineralized false trunks (Tidwell and Hebbert 1992). These are *T. minor, T. grandis* Read et Brown, *T. zelleri* Ash et Read, *T. superba* Arnold, *T. wesselii* Arnold, *T. wyomingensis* Arnold, *T. jonesii, T. stichkae* Tidwell et Hebbert, and *T. readii* Tidwell et Hebbert.

Most specimens of *Tempskya* found in the Cedar Mountain Formation lie parallel to the strata, but east of Castle Dale, Utah, they occur in growth position in pyritic carbonaceous shale, which suggests that they were growing in a swamp (Tidwell and Hebbert 1992). The top of one specimen was surrounded by splay sands, indicating that it grew near a river (Tidwell and Hebbert 1992).

Locality 5 is in the Cedar Mountain Formation about 19 km northeast of Moab in the Yellowcat mining district. *T. zelleri* is the only *Tempskya* fossil currently known from this locality.

Materials and Methods

The remains of 15 woody dicotyledonous stems are preserved in three specimens of *Tempskya* from New Mexico and five specimens of *Tempskya* from Utah.

One of the most significant specimens is the unidentified liana in the paratype of *T. reesidei* from the Early Cretaceous (Albian) Mojado Formation in extreme southwestern New Mexico (Fig. 12.1). This *Tempskya* is an incomplete base of a small, slightly compressed obconical false trunk that resembles the well-known holotype of *T. knowltonii* (Fig. 12.2). The fossil was preserved parallel to the bedding plane. Elsewhere in southwestern New Mexico, *T. reesidei* is closely associated with a late Albian fauna (Ash and Read 1976). This stem has been slightly compressed tangentially and in cross section now measures about 12 mm by 7 mm in diameter. Like the other dicotyledonous stems in this trunk, it is deeply infiltrated by roots, especially the pith, and most anatomical details have been destroyed. Also, the inner edge of the xylem ring has been eroded along most of its original extent, and in only a few places is it possible to determine the original thickness of the ring. The xylem ring is about 1.2 mm thick and consists of diffuse porous xylem strands separated by uni- to multiseriate medullary rays. Clearly the stems in this fossil represent *Rodoxylon scheetzii* gen. et sp. nov.

This fossil is important because it illustrates the growth habit of three dicotyledonous stems directly associated with the false trunk of *Tempskya* in this specimen (Fig. 12.2). The stems occur about halfway between the base and top of the preserved trunk. Two of the stems (A1, A2) are represented by slightly discolored, longitudinal impressions about 2 cm wide on the exterior of the specimen. In places, a thin rim of tissue adheres to the exterior of the trunk. The third example (A3) is the most completely preserved of the dicotyledonous stems. In the lower part of the trunk, the stem is represented by a longitudinal discolored impression about 6 cm long and 1 cm wide. At the upper end of the impression, about 28 cm above the base of the specimen, the dicotyledonous stem A3 penetrates the false trunk at a high angle. At the point where it has reached a depth of about 1.6 cm from the exterior, the stem follows a course that parallels the sides of the false trunk for a distance of about 7 cm. It then bends outward at a high angle and returns to the exterior of the false trunk, where it is represented once again by a discolored longitudinal impression about 1–2 cm wide and 9 cm long (Ash and Read 1976).

The other two New Mexican specimens of *Tempskya* were collected from the surface of basal strata of the Dakota Formation in extreme northwestern New Mexico (locality 2 in Fig. 12.1; Ash and Read 1976). One specimen contains at least six dicotyledonous stems. Unfortunately, all of the stems except B1 are poorly preserved. Stem B1 most likely represents a specimen of the new genus, *Munzingoxylon*. One stem (B2) is represented by a hole surrounded by some stem tissue. Another impression (B3) appears on the weathered side of the specimen. Stem (B4) has been partially replaced by amorphous quartz. The interior of stems B5 and B6 were largely destroyed by *Tempskya* roots. The second specimen (Fig. 12.3) contains three poorly preserved stems (C1–3) that may represent specimens of *Munzingoxylon*. The anatomy of the fourth stem (C4) was completely destroyed by *Tempskya* roots.

Of the Utah specimens, only four are considered here for detailed description. Two are from locality 4 in the Cedar Mountain Formation and two from locality 3 in the Burro Canyon Formation. They were thin sectioned and studied by means of both reflective and incident light microscopes. Repository for the New Mexico specimens is the U.S. National Museum of Natural History (USNM, USGS), and for the Utah specimens, the Brigham Young University Museum of Paleontology (BYUP).

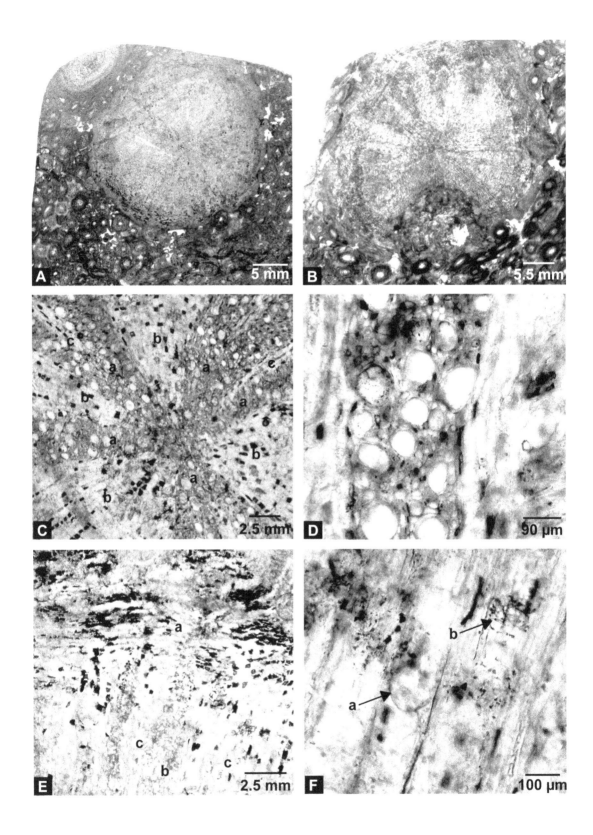

Clade Angiospermae
Clade Eudicots
Family incertae sedis

MUNZINGOXYLON GEN. NOV.

Diagnosis. Stem fluted; lianalike; growth rings absent; vascular strands separated by thin to broad medullary rays radiating from small pith; secondary xylem diffuse porous; vessels numerous, generally solitary; perforation plates mostly simple, rarely scalariform; tyloses lacking; parenchyma abundant, scanty paratrachial, occasionally vasicentric; fibers nonseptate; xylem rays lacking.

Etymology. Named for our good friend, the late George Munzing, acknowledging him for his able assistance and interest in fossil plants over the years.

MUNZINGOXYLON DELEVORYASII SP. NOV.
FIGS. 12.4A–F, 12.5A–F, 12.6A, B

Diagnosis. Stem fluted; growth rings lacking; small pith composed of sclerified parenchyma and fibers; secondary xylem vessels small to large, numerous, 40–65 per square millimeter, diffuse porous, generally solitary, occasionally in tangential pairs, rarely multiples of three, vessel outline irregular, usually round to oval, sometimes polygonal, large vessel radial diameter ranging from 95 to 115 μm, small vessel 50 to 70 μm in diameter, vessel element length 33 to 200 μm; vessel element walls 2 to 3.6 μm wide; perforations simple, very rarely scalariform; tyloses absent; intervascular pits opposite to subopposite, round to oval, occasionally vestured, 5 to 7 μm in diameter, pits widely spaced, not contiguous, often scalariform, alternate; parenchyma abundant, scanty paratrachial, some vasicentric; fibers and fiber tracheids, thick walled, 14 to 16 μm across, lumen 7 μm in diameter, 15 to 19 μm long, nonseptate, pitting generally round, uniseriate to multiseriate, some scalariform; fascicular and interfascicular cambiums several layers thick, cells tangentially elongated; secondary phloem composed of sieve tubes, companion cells, parenchyma, and fibers, cells of secondary phloem often tangentially flattened, mostly crushed; wood rays absent; primary and secondary medullary rays present, medullary rays thin near pith (11 cells wide) to very broad near the interfascicular cambium (33 or more cells wide), composed mostly of square cells, some procumbent; periderm several cells thick.

Repository. BYUP 40 (holotype, Figs. 12.4A, C–F, 12.5 C–F, 12.6 A, B); BYUP 41 (paratype, Fig. 12.4B).

Locality. Figure 12.1, localities 3–4.

Age. Early Late Cretaceous.

Horizon. Upper Cedar Mountain, Burro Canyon, and possibly basal Dakota Formations.

Etymology. The species epithet honors Dr. Theodore Delevoryas for his many important contributions to paleobotany.

Fig. 12.4. *Munzingoxylon delevoryasii* gen. et sp. nov. (A–E) Transverse sections. (A) Overview of holotype (BYUP 40). (B) Overview of paratype (BYUP 41). (C) Close-up of pith area of (A). Holotype (BYUP 40). Note radiating xylem strands (a), medullary rays (b), and secondary rays (c). Holotype (BYUP 40). (D) Close-up of holotype showing portion of a xylem strand with large and small vessels (large cells), and the paratrachial parenchyma (smaller cells). Note the absence of uniseriate rays in the strand. Holotype (BYUP 40). (E) Close-up of holotype showing outer part of the stem illustrating the secondary layers (a), xylem strand (b), and medullary rays (c). Holotype (BYUP 40). (F) Longitudinal section. Vessels with simple perforation plates (a) with some scalariform pitting (b) on their walls. Holotype (BYUP 40).

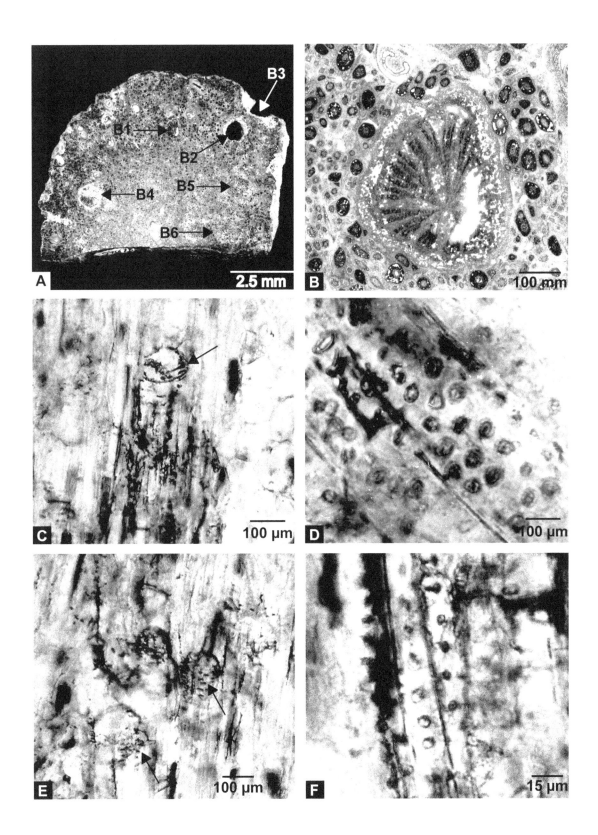

Munzingoxylon delevoryasii stems are round and vary from 10 to 13 mm across (Fig. 12.4A, B). The small, sclerified pith has not been invaded by the surrounding *Tempskya* roots. The xylem strands and the medullary rays begin small near the pith and expand in width, forming a wedge shape toward the outer areas of the stem (Fig. 12.4Cb). The xylem strands are divided by the primary medullary rays, but also may be divided further by wide secondary rays (Fig. 12.4Cc). Interestingly, no wood or xylem rays occur in the strands (Fig. 12.4D). The vessels in the strands are scattered and mostly round to oval in diameter, although some are angular. The latter may be due to strong compression. The vessels are separated by xylem parenchyma, fibers, and fiber tracheids. Some of these cells contain some type of dark-colored material.

Cortex and secondary tissues are mostly eroded away. Secondary tissues, where they occur, are very thin. Periderm is eroded or weathered away and, where present, at least 12 cells wide. The erosion may be due to *Tempskya* roots. Some of the xylem strands and medullary rays extend through the tissues as branching traces (Fig. 12.6A).

A number of very small fecal pellets, 15–20 μm in size, occur in an irregularly shaped area in the paratype, where the thin walled tissue of a portion of the interfascicular cambium and a medullary ray has been destroyed. This is the only area in specimens of this species where fecal pellets occur.

RODOXYLON GEN. NOV.

Diagnosis. Stem fluted; lianalike; growth rings lacking; secondary xylem diffuse porous, vessels sparse, in radial rows, generally solitary, separated by scanty paratracheal or rarely vasicentric parenchyma and fiber tracheids; perforation plates simple; tyloses lacking; fibers generally square; fiber tracheids often tangentially elongated; xylem rays uniseriate, composed of mostly upright, some only procumbent cells or both; tangentially oriented cells constitute most of the center of the broader medullary rays.

Etymology. Named for Dr. Rodney Scheetz, curator of the Brigham Young University Museum of Paleontology in Provo, Utah, for his continued assistance with and interest in fossil plants.

RODOXYLON SCHEETZII SP. NOV.
FIGS. 12.6C–F, 12.7A–E

Diagnosis. Stem fluted; lianalike; growth rings absent; secondary xylem vessels sparse, small to large, occur in radial rows, solitary, rarely in multiples of two or three, separated by paratracheal parenchyma and fiber tracheids, vessel outline irregular, generally round to oval, occasionally angular, large vessel radial diameter 70 to 112 μm, small vessel 40 to 75 μm in diameter, vessel element walls 5 to 7 μm thick; perforations simple, tyloses absent, paratracheal parenchyma scanty to vasicentric; fibers usually square in cross section, 14 μm wide, lumen 7 μm across,

Description

Fig. 12.5. *Munzingoxylon delevoryasii* gen. et sp. nov. (A, B) Transverse sections. (A) Specimen B containing six dicotyledonous stems in *Tempskya* sp. from northwestern New Mexico. B1, stem similar to *Munzingoxylon;* B2, B3, openings indicate where stems were destroyed; B4, poorly preserved stem; B5, B6, remains of stems invaded and destroyed by *Tempskya* roots (USNM 538921). (B) Close-up of stem B1 in (A). (C–F) Longitudinal sections. (C) Vessel with poorly preserved scalariform perforation plate (arrow). Holotype (BYUP 40). (D) Vessels with opposite to subopposite round to oval pitting. (E) Scalariform pitting on vessels (arrows). Holotype (BYUP 40). (F) Round pitting on fiber tracheid walls. Note the tapering cells. Holotype (BYUP 40).

often occur in radial rows separating the vessels; fiber tracheids often tangentially elongated, 7 to 20 μm wide radially by 42 μm wide tangentially, two to three fiber tracheids often separate vessels radially; fascicular and interfascicular cambiums several cell layers thick, tangentially elongated; secondary phloem generally crushed, sieve tubes 70 to 105 μm in diameter; phloem fibers large 90 by 120 μm across, small 30 by 60 μm; phloem parenchyma thin walled and generally crushed; xylem rays uniseriate, 35 to 140 μm wide, mostly upright cells; some procumbent only, others mixed; medullary rays up to 30 cells wide, heterogeneous, one to four uniseriate rays bordering medullary ray, procumbent cells form much of the medullary rays, upright cells mostly square, 36 to 78 μm wide, large, thin to occasionally thick walled, tangentially oriented cells, 42 by 175 μm across, generally constitutes middle of most medullary rays; cortical cells elongated, tangentially flattened and aligned, 14 by 70 μm across, some square shaped; periderm several cells thick, cells tangentially elongated, radially aligned 12 by 60 μm across.

Repository. BYUP 42 (holotype, Figs. 12.6C–F, 12.7A, B, E); BYUP 43 (figured specimen, Fig. 12.7C, D).

Locality. Figure 12.1, locality 4.

Age. Early Late Cretaceous.

Horizon. Upper Cedar Mountain Formation.

Etymology. Same as for the genus.

Description

The specimens of *Rodoxylon scheetzii* are 6 to 9 mm in diameter and consist of secondary xylem in strands separated by uniseriate and broad multiseriate medullary rays (Fig. 12.6C). The pith of the holotype is incomplete as a result of being destroyed by invading *Tempskya* roots (Fig. 12.6D). The vessels are diffuse porous, arranged in radial rows, and separated by one to three uniseriate rays and/or radially arranged strands or fibers (Fig. 12.6F). The vessels are mostly solitary with simple, transverse perforation plates. Their distribution is 38 per square millimeter. Some of the uniseriate rays are homogeneous, whereas others are not. The cells in the center of most medullary multiseriate rays are tangentially elongated (Fig. 12.7A). These are usually thick walled. The medullary rays are added to by the interfascicular cambium and appear prominently in the secondary tissue separating the strands composed of phloem. Most of the phloem cells are crushed, as well as the cortex. According to Carlquist (1991), deformation of the sieve tube elements may be due to torsion affecting their thin walls (Fig. 12.6E). These phloem cells have been squeezed, often resulting in tangentially elongated cells. The cortex appears to be 2 mm thick. The periderm (Fig. 12.7E) is uneven, eroded by the surrounding *Tempskya* roots. The periderm is 17 or more cells thick.

Cavities containing fecal pellets occur along the cambial regions or within the medullary rays (Fig. 12.7A). Small pellets are 30 by 50 μm, and most of the large ones (77 by 150 μm) are nearly twice the width of the vessels. On the basis of the sizes of the fecal pellets in *R. scheetzii*

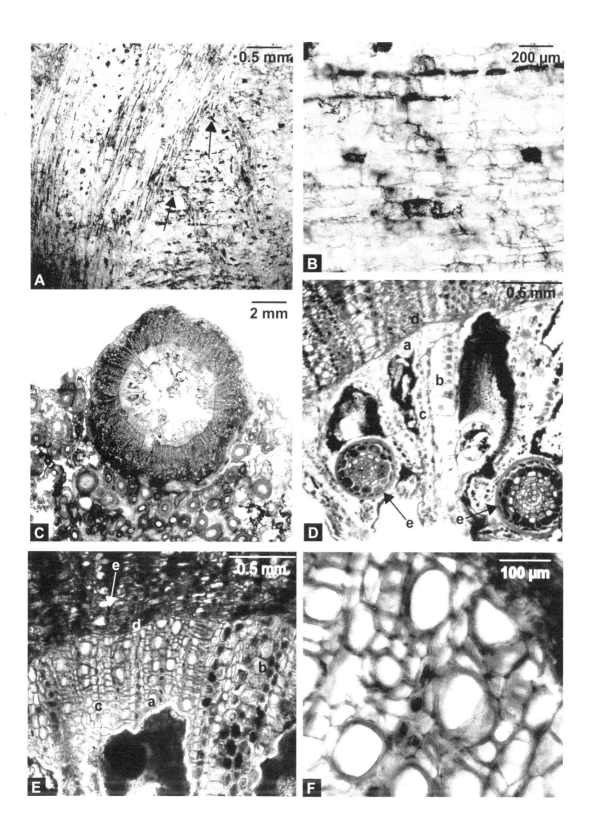

and those reported by Scott and Taylor (1983), the animals may have been mites, collembolan, and/or millipedes. Mites most likely deposited smaller pellets and millipedes the larger (see Ash 2000).

<table>
<tr><td>

Discussion and Comparison

</td><td>

Munzingoxylon and *Rodoxylon* differ from each other in the number and arrangement of vessels. They are numerous and scattered in *Munzingoxylon* (Fig. 12.4D) and uncommon and radially aligned in *Rodoxylon* (Fig. 12.6F). Also, xylem rays are absent and the medullary rays are wider in *Munzingoxylon*. The pith is larger in *Rodoxylon*. Its medullary ray cells are tangentially flattened and for unknown reasons are invaded by *Tempskya* roots and animals (as evidenced by fecal pellets). The latter rarely occurs in *Munzingoxylon*.

</td></tr>
</table>

Fig. 12.6, previous page. *Munzingoxylon delevoryasii* gen. et sp. nov. (A, B) Longitudinal sections. (A) Branching xylem strand (arrows). Note the broad medullary rays between xylem strands. Holotype (BYUP 40). (B) Section through a broad medullary ray. Cells are mostly procumbent. Holotype (BYUP 40). *Rodoxylon scheetzii* gen. et sp. nov. (C–F) Transverse sections. (C) Overview of the holotype (BYUP 42). Note the destroyed pith. (D) Close-up of (C). Note the xylem strand (a), wide (b) and narrow rays (c), cambial layers (d), and *Tempskya* roots in the pith and some of the medullary rays (e). Holotype (BYUP 42). (E) Close-up of (C). Xylem strand (a), medullary ray (b), narrow ray (c), cambial layers (d), and phloem with a row of distorted sieve tubes (e). Holotype (BYUP 42). (F) Close-up of part of a xylem strand illustrating the radial alignment of the vessels separated by fibers, fiber tracheids, or xylem parenchyma. Uniseriate rays occur near the vessels. Some fibers are uniseriate strands as well. Holotype (BYUP 42).

Because of the lianalike anatomy of *Munzingoxylon* and *Rodoxylon*, they are compared to Gnetales and angiospermous families having lianas or vines that are somewhat similar. The gymnosperm order Gnetales is generally unlike *Munzingoxylon* and *Rodoxylon*. The gnetalean fossil record is sparse, consisting mostly of fossil pollen, and extends from the Middle Permian (Wilson 1959) to the present. Most megafossil remains are compressions, such as the slender stems lacking secondary growth with attached leaves of *Drewria* from the Lower Cretaceous Potomac Group (Crane and Upchurch 1987) or the reproductive structures of *Eoanthus* from the Lower Cretaceous of Mongolia (Krassilov 1986). Other possible gnetalean forms have been mentioned from the mid-Cretaceous, but they represent leaf remains resembling *Drewria* or reproductive structures like *Eoanthus* (Crane 1988). The Triassic stem, *Hexagonacaulon minutum* Lacey et Lucas, from the South Shetland Islands has been compared to *Ephedra* and ray details of the permineralized stem of *Schildera* from the Triassic Chinle Formation in Arizona has been compared to *Gnetum* (Daugherty 1941). However, neither of these two genera is similar to *Munzingoxylon* or *Rodoxylon*.

Living *Gnetum* species are mostly lianas. The cambria in these climbing species are successive, thereby giving rise to a series of coaxial, concentric cylinders of xylem and phloem, similar to those in the living cycad *Cycas* (Seward 1919; Sporne 1965), and unlike the stems of *Munzingoxylon* or *Rodoxylon*.

Vascular tissues of many living angiospermous lianas and vines originate as discrete bundles or strands, usually radiating outwardly from their pith, which remain separated by broad, primary medullary rays, and also by secondary rays in some (Stover 1951). The perforation plates of these lianas or vines are mostly simple (Carlquist 1991), similar to those in *Munzingoxylon* and *Rodoxylon*. The rays are 5 to 15 cells wide in the vines of *Menispermum*, *Clematis*, *Psedera*, and *Vitis*, which are not as wide as those in these two new genera. Under certain circumstances, parenchyma adjacent to the primary xylem in living *Menispermum* remains thin walled. With the addition of secondary xylem in this genus, the inner areas of the rays may become compressed, forcing the primary xylem into its pith and crushing it (Stover 1951). This may have also happened to some extent in *M. delevoryasii*.

The families with scandent members that are similar to these *M. delevoryasii and R. scheetzii* include the Lardizabalaceae, Actinidiaceae, Ampelidaceae (Vitaceae), Dilleniaceae, Icacinaceae, Macgraviaceae, Leguminosae, and Aristolochiaceae.

In overall appearance, *Aristolochia*, particularly *A. triangularis* Cham. of the Aristolochiaceae, is similar to *Munzingoxylon* and *Rodoxylon*. The most conspicuous anatomical character of this family is the broad primary medullary rays with broad secondary rays that form in older stems (Metcalfe and Chalk 1950). However, the rays in this family are not as wide as those in either *Munzingoxylon* or *Rodoxylon*. Vessels in some vines, up to 450 μm in *Pararistolochia*, are much larger than those in these two new genera, although the smaller vessels of *Apama siliquosa* Lam. (50–100 μm) are closer in size. The smaller pith and absence of uniseriate xylem rays in *Munzingoxylon* also separate it from members of this family, which have larger piths and uniseriate rays. The distribution of the scanty paratrachial xylem parenchyma occurring in variable, irregular, uniseriate, tangential lines (Dadswell and Record 1936; Metcalfe and Chalk 1950) is different than the xylem parenchyma of *Munzingoxylon* and *Rodoxylon*.

Lardizabaloxylon lardizabaliodes Schönfeld, described as a fossil liana from either Upper Cretaceous or Lower Tertiary strata of Patagonia, was closely compared to *Lardizabala* of the Lardizabalaceae (Schönfeld 1954). The latter is a small family composed mostly of vines (Metcalfe and Chalk 1950). *Lardizabala* differs from *Munzingoxylon* and *Rodoxylon* by having widely spaced vascular strands separated by broad, generally lignified, medullary rays. The rays are smaller than those in these new genera, however. Fibers in members of Lardizabalaceae are sometimes septate. The walls of the member's smaller vessels have spiral thickenings. Further, xylem parenchyma is sparse or lacking in living *Lardizabala biternata* Rhiz. et Pav. (Dadswell and Record 1936). All these features are unlike *Munzingoxylon* or *Rodoxylon*.

Actinidioxylon princeps (Ludwig) Müller-Stroll et Mädel-Angelie (1969) was reported as a liana from the Pliocene ironstone of Dernbach, Germany. The species was described as being close to the modern genus *Actinidia* of the Actinidiaceae. This genus has scattered vessels as in *Munzingoxylon*, but it also has distinct growth rings, small vessels with scalariform perforation plates, uniseriate xylem rays, and fibers with slitlike apertures. These are features not found in *Munzingoxylon*. In *Rodoxylon*, however, the radial arrangement of its vessels, the absence of scalariform plates and growth rings separate it from *Actinidia* and *A. princeps*.

The Ampelidaceae (Vitaceae) are a widely distributed family in temperate to tropical environs containing many typically woody vines. The large and small vessels in *Vitis*, for example, are similar to those in *Munzingoxylon* and *Rodoxylon*. However, the large vessels in the family

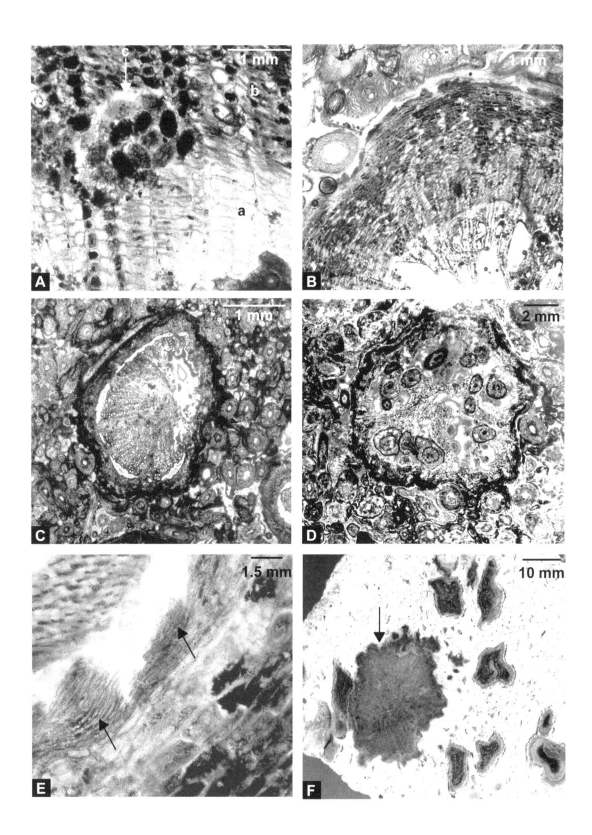

are sometimes zonate. Its numerous vessels are in small multiples and its frequently septate xylem fibers have simple pits. These characters are not common to the two new genera described here.

The lianas of the Dilleniaceae have small to large vessels like *M. delevoryasii* and *R. scheetzii*, but differ from them by having predominantly scalariform perforation plates. The liana of *Doliocarpus* in this family also has included phloem (Metcalfe and Chalk 1950), which does not occur in either *Munzingoxylon* or *Rodoxylon*.

The Icacinaceae are a large tropical family of trees, shrubs, and woody vines. The vessels in this family are small, numerous, and well distributed without any definite pattern. Most of the climbers have simple perforation plates similar to those in *M. delevoryasii* and *R. scheetzii*. The remainder of the family, including *Icacinoxylon pittense*, has scalariform to reticulate perforation plates. Xylem parenchyma is in irregularly metatracheal lines (Metcalfe and Chalk 1950), except in *I. pittense*. These characters are unlike those in *M. delevoryasii* or *R. scheetzii*.

The Macgraviaceae are a small, tropical family of climbing shrubs with numerous small to large scattered, mostly solitary vessels with simple perforation plates that are similar to those of *Munzingoxylon*. Large rays are composed mostly of upright or square cells. Its septate fibers differ from the fibers in both *Munzingoxylon* and *Rodoxylon*.

The Leguminosae consist of three subfamilies, Mimosoideae, and Lotoideae or Papilionoideae. The Mimosoideae consist mostly of trees and shrubs with some lianas. They differ from *Munzingoxylon* and *Rodoxylon* by some species being ring porous or semi–ring porous, paratrachial parenchyma in round or diamond-shaped sheaths, one to nine rays (mostly two to five), cells wide or exclusively uniseriate, and lianas with winged and grooved stems (Metcalfe and Chalk 1950), none of which occurs in either *Munzingoxylon* or *Rodoxylon*.

The Caesalpinioideae are mainly a tropical subfamily of trees and shrubs with a few herbs. In this subfamily, parenchyma is abundant to occasionally scanty paratrachial, in round to diamond-shaped sheaths, often in irregular confluent bands or regular continuous bands, rays are one to seven (usually two to three) cells wide or only uniseriate, fibers may be septate, and the xylem has solitary vessels with some grouped in irregular clusters or small multiples. Some species are ring porous or semi–ring porous, and most species have solid cylindrical xylem. *Bauhinia* is the only genus of the Caesalpinioideae with anomalous secondary growth. Its stems become band shaped during thickening, with wings and ribs arising from the broader sides of the older stems. Its xylem becomes fissured, with the central mass eventually becoming separated. The liana *Entrada* has secondary strands of phloem embedded in the parenchyma of its xylem. Parenchyma forms the largest part of the xylem in *Entradopsis* (Metcalfe and Chalk 1950). Again, these features are unlike those in *Munzingoxylon* or *Rodoxylon*.

The Papilionoideae are widely distributed and consist of trees, shrubs, herbs, and xeromorphs. In the woody genera of this subfamily, the xylem

Fig. 12.7. *Rodoxylon scheetzii* gen. et sp. nov. (A–F) Transverse sections. (A) Close-up of broad medullary ray (a), phloem (b), and cavity containing large fecal pellets. Holotype (BYUP 42). (c). (B) Outer surface of stem illustrating the cortex and periderm next to *Tempskya* roots. Holotype (BYUP 42). (C) Transverse section of *R. scheetzii* embedded in the root mantle of *Tempskya jonesii*. Figured specimen (BYUP 43). (D) Close-up of another stem invaded by *Tempskya* roots from the same transverse section shown in (C) (BYUP 43). (E) Close-up of the periderm. The cells appear to be forming a lenticel. On the upper left is part of a *Tempskya* root. Holotype (BYUP 42). (F) Unidentified dicotyledonous stem (arrow) in *T. zelleri* (BYUP 44).

generally forms a closed cylinder and is commonly ring porous. Stems of its lianas are mostly cylindrical, but band shaped in the taxa *Machaerium* and *Rhynchosia phasoloides* DC. Spiral thickenings are common on vessel walls in various members of the Papilionoideae. Their vessels generally occur in radial multiples or clusters and are particularly large in some lianas species. Parenchyma is predominantly paratrachial or in moderately regular bands. In some genera of this subfamily, round or diamond-shaped sheaths occur, although less commonly than confluent or other banded types. Rays vary from uniseriate to 2 to 3 cells wide to some species with broad rays 20 or more cells wide. The xylem of herbaceous species is generally composed of vascular bundles separated by conspicuous medullary rays, whereas in the woody cylindrical forms, the primary rays are usually narrow (Metcalfe and Chalk 1950). Except for the broad rays in some species in this subfamily, these characters do not occur in *Munzingoxylon* or *Rodoxylon*.

Wheeler and Lehman (2000) noted a wood that as a result of poor preservation, they tentatively classified as the "Javelina Vine" from the Upper Cretaceous Javelina Formation in Big Bend National Park, Texas. The specimen does not show any growth rings. Its vessels are solitary and mostly in radial multiples. The latter feature is not present in either *Munzingoxylon* or *Rodoxylon*.

Conclusions

The presence of lianas embedded in *Tempskya* trunks is certainly unusual (Fig. 12.8), but it is also noteworthy because most of the fossils are early Late Cretaceous (Cenomanian) in age. The growth habit of most of the specimens described here remains unclear. Modern lianas and vines grow in a variety of habitats, but they are often restricted in their occurrence by their climbing structures and mechanisms. These include tendrils, hooks, adhesive pads, twining shoots, leaves, petioles, adventitious roots, a scrambling habit, or any combination of these (Hegarty 1991; Putz and Holbrook 1991). Twining is the most common mode of climbing, particularly for thin stems. At present, there is no conclusive evidence of the type of climbing mechanism or form of attachment in the lianas growing on the trunks of *Tempskya*. However, the liana in the small tree fern of *T. reesidei* probably had a scrambling type of habit because the stems only occur on one side of the trunk, and they do not show any evidence of other forms of attachment of climbing mechanisms, such as twining. Also, there is no obvious evidence of tendrils, hooks, and other structures that could have been used to attach the stems in *T. reesidei* to the outside of the trunk. Of course, once the stems were engulfed by the adventitious roots of the trunks, they would not have required any attachment structures or mechanisms. The other species of lianas within *Tempskya* may have had a different means of climbing. There are traces in a specimen of *Munzingoxylon* that could represent adventitious roots that held the specimen to the *Tempskya*, although they could be leaf traces as well. Because of the limited amount of material, there is no way of knowing for certain whether they are root or leaf traces.

There is no indication that these lianas grew through the false trunk of the *Tempskya* specimens. The root mantle of the false trunks is too dense for another plant to have penetrated it, and there is no distortion of the roots surrounding these scandent stems. The lianas probably climbed around the outside of the *Tempskya* false trunks when the trunks were smaller. They then became incorporated as the trunk expanded, and the adventitious roots of *Tempskya* grew down around the stems, eventually encompassing them.

Lianas and vines in modern woodlands or forests develop strategies to place their foliage in the most productive position. They respond to irregularities in the forest structure by forming aggregations or clumps, usually in disturbed areas in the forest canopy (Hegarty and Caballé 1991). Lianas and vines are often detrimental to the host plant by forming excessive shade, thus lowering the host's ability to photosynthesize. What effect these lianas had on the living *Tempskya* plant is not known, but they had to have had some.

The flowers and leaves of these lianas in *Tempskya* are currently unknown. Among modern lianas, pollination is by medium- to large-sized bees. None are known to be thrip or wind pollinated (Gentry 1991). If the lianas in *Tempskya* occurred in clumps, which they appear to do in the two small specimens from northwestern New Mexico and in *T. reesidei*, they may have been wind pollinated, otherwise being widely separated; wind pollination would be low and vegetative reproduction increased. Of course, primitive bees, whose fossils are yet to be found, may have been their pollinators. Other important flower-visiting insects with Cretaceous fossil records that may have been the pollinators are vespid wasps, flies from various families, and numerous basal Lepidoptera forms (Grimaldi 1999).

According to Young (1960), the Cedar Mountain Formation, including the Burro Canyon Formation, was deposited on flood plains inland from the sea. On the basis of palynology, Carroll (1992) concluded that the Cedar Mountain Formation was deposited in a terrestrial environment associated with a mixed vegetational community of gymnosperms, ferns, and angiosperms. However, the basal Cedar Mountain is considered by some to have been arid during the Early Cretaceous (Smith et al. 2001; Retallack 2005). Toward the top of the formation, the reduction of soil carbonates and the increase in organic material suggest wetter climatic conditions related to the encroaching seaway (Currie 1998; Greenhalgh and Britt 2007). It seems that the environmental conditions under which *Icacioxylon pittiense*, *Paraphyllantoxylon utahense*, *Munzingoxylon delevoryasii*, and *Rodoxylon scheetzii* existed, particularly those of temperature and moisture availability, were relatively uniform because there is little difference in vessel size throughout the individual specimens. Vessels show a tendency toward grouping (multiples) in dry environments and rarely grouped under humid conditions. They

Fig. 12.8. Reconstruction of lianas growing around and in *Tempskya*.

are generally solitary, as in *M. delevoryasii* and *R. scheetzi* (Carlquist 1966; Carlquist and Hoekman 1985; Alves and Angyalossy-Alfonso 2000). Moreover, scalariform perforation plates are more common in temperate and tropical floras than in tropical humid environments (Baas 1976; Alves and Angyalossy-Alfonso 2000), and because *M. delevoryasii* and *R. scheetzii* have predominantly simple perforation plates, they may have very well grown under tropical humid conditions. The lack of growth rings in *I. pittiense*, *P. utahense*, *M. delevoryasii*, and *R. scheetzii* also suggests that they grew under uniform humid tropical or subtropical conditions (Thayn et al. 1985).

Finally, the existence of early Late Cretaceous (Cenomanian) lianas *Munzingoxylon* and *Rodoxylon*, as well as the unidentified lianas from central Utah and northwestern New Mexico, demonstrates that there was more diversity in taxonomy and habit near the time of early angiosperm diversification than is commonly recognized.

Acknowledgments

This chapter is dedicated to Dr. Ted Delevoryas for his interest in and many significant contributions to the study of fossil plants. We thank Harry Cleaveland and the late Frank Lemon of Moab, Utah, the late Jack Strobell, and Steven Gratton of Marlon, New York, for donating some of the specimens used in this study.

References Cited

Alves, E. S., and V. Angyalossy-Alfonso. 2000. Ecological trends in wood anatomy of some Brazilian species. I. Growth rings and vessels. *IAWA Journal* 21: 3–30.

Ash, S. R. 2000. Evidence of oribatid mite herbivory in the stem of a Late Triassic tree fern from Arizona. *Journal of Paleontology* 74: 1065–1071.

Ash, S. R., and C. M. Read. 1976. North American species of *Tempskya* and their stratigraphic significance. *U.S. Geological Survey Professional Paper* 847: 1–12.

Baas, P. 1976. Some functional and adaptive aspects of vessel member morphology. In P. Baas, A. J. Bolton, and D. H. Catling (eds.), *Wood Structure in Biological and Technological Research*, pp. 157–181. Leiden Botanical Series No. 3. The Hague: Leiden University Press.

Brown, R. W. 1950. Cretaceous plants from southwestern Colorado. *U.S. Geological Survey Professional Paper* 221-D: 45–66.

Burton, D., B. W. Greenhalgh, B. B. Britt, B. J. Kowalis, W. S. Elliot Jr., and R. Barrick. 2006. New radiometric ages from the Cedar Mountain Formation, Utah, and the Cloverly Formation, Wyoming—implications for contained dinosaur faunas. *Geological Society of America Abstracts with Programs* 39: 33.

Carlquist, S. 1966. Wood anatomy of the Compositae: a summary, with comments on factors controlling wood evolution. *Aliso* 6: 25–44.

———. 1991. Anatomy of vine and liana stems: a review and synthesis. In F. E. Putz and H. A. Mooney (eds.), *The Biology of Vines*, pp. 53–71. Cambridge: Cambridge University Press.

Carlquist, R. E., and D. A. Hoekman. 1985. Ecological wood anatomy of the woody Southern California flora. *IAWA Bulletin*, n.s., 6: 319–347.

Carroll, R. E. 1992. Biostratigraphy and paleoecology of mid-Cretaceous sedimentary rocks, eastern Utah and western Colorado—a palynological interpretation. Ph.D. diss., Michigan State University.

Cifelli, R. L., J. I. Kirkland, A. Weil, A. L. Deino, and B. J. Kowalis. 1997. High-precision ^{40}Ar/^{39}Ar geochronology and the advent of North America's Late Cretaceous terrestrial fauna. *Proceedings of the National Academy of Sciences U.S.A.* 94: 11163–11167.

Clifford, H. T., and M. E. Dettman. 2005. First record from Australia of the Cretaceous fern genus *Tempskya* and the description of a new species, *T. judithae. Review of Palaeobotany and Palynology* 134: 71–84.

Cobban, W. A., and J. B. Reeside Jr. 1952. Correlations of the Cretaceous formations of the Western Interior of the United States. *Geological Society of America Bulletin* 63: 1011–1044.

Corda, A. J. 1845. *Beiträge der Flora der Vorwelt.* Prague: J. G. Calve'sche.

Crane, P. R. 1988. Major clades and relationships in the "higher" gymnosperms. In C. B. Beck (ed.), *Origin and Evolution of Gymnosperms,* pp. 219–272. New York: Columbia University Press.

Crane, P. R., and G. R. Upchurch. 1987. *Drewria potomacensis* gen. et sp. nov., an Early Cretaceous member of Gnetales from the Potomac Group of Virginia. *American Journal of Botany* 74: 1722–1236.

Currie, B. S. 1998. Upper Jurassic–Lower Cretaceous Morrison and Cedar Mountain Formations, NE Utah–NW Colorado—relationship between nonmarine deposition and early Cordilleran foreland basin development. *Journal of Sedimentary Research* 68: 632–652.

Dadswell, H. E., and S. J. Record. 1936. Identification of woods with conspicuous rays. *Tropical Woods* no. 48: 1–30.

Daugherty, L. H. 1941. *The Upper Triassic Flora of Arizona.* Publication No. 526. Washington, D.C.: Carnegie Institution of Washington.

Garrison, J. R., D. Brinkman, D. J. Nichols, P. Layer, D. Burge, and D. Thayne. 2007. A multidisciplinary study of the Lower Cretaceous Cedar Mountain Formation, Mussentuchit Wash, Utah—a determination of the paleoenvironment and paleoecology of the *Eolambia carolionesa* dinosaur quarry. *Cretaceous Research* 28: 461–464.

Gentry, A. H. 1991. The distribution and evolution of climbing plants. In F. E. Putz and H. A. Mooney (eds.), *The Biology of Vines,* pp. 393–423. Cambridge: Cambridge University Press.

Gradstein, F. M., J. G. Ogg, and A. G. Smith. 2004. *A Geologic Time Scale 2004.* New York: Cambridge University Press.

Greenhalgh, B. W., and B. B. Britt. 2007. Stratigraphy and sedimentology of the Morrison–Cedar Mountain Formation boundary, east-central Utah. In G. C. Willis, M. D. Hylland, D. L. Clark, and T. C. Chidsey Jr. (eds.), *Central Utah, Diverse Geology of A Dynamic Landscape,* pp. 81–100. Publication No. 36. Salt Lake City: Utah Geological Association.

Greenhalgh, B. W., B. B. Britt, and B. J. Kowalis. 2006. New U-Pb age control for the lower Cedar Mountain Formation and an evaluation of the Morrison Formation/Cedar Mountain Formation boundary, Utah. *Geological Society of America Abstracts with Program* 4–8.

Grimaldi, D. 1999. The co-radiations of pollinating insects and angiosperms in the Cretaceous. *Annals of the Missouri Botanical Garden* 86: 373–406.

Hegarty, E. E. 1991. Vine–host interactions. In F. E. Putz and H. A. Mooney (eds.), *The Biology of Vines,* pp. 357–375. Cambridge: Cambridge University Press.

Hegarty, E. E., and G. Caballé. 1991. Distribution and abundance of vines in forest communitites. In F. E. Putz and H. A. Mooney (eds.), *The Biology of Vines,* pp. 337–356. Cambridge: Cambridge University Press.

Kirkland, J. I., B. B. Britt, D. L. Burge, K. Carpenter, R. L. Cifelli, F. L. DeCourten, J. G. Eaton, J. S. Hasiotis, and T. F. Lawton. 1997. Lower to middle Cretaceous dinosaur faunas of the Central Plateau—a key to understanding 35 million years of tectonics, sedimentology, evolution, and biogeography. *Brigham Young University Geology Studies* 42 (2): 69–103.

Krassilov, V. A. 1986. New floral structure from the Lower Cretaceous of Lake Baikal area. *Review of Palaeobotany and Palynology* 47: 9–16.

Metcalfe, C. R., and L. Chalk. 1950. *Anatomy of the Dicotyledons*, 2 vols. Oxford, U.K.: Clarendon Press.

Mikuláš, R., and Z. Devořák. 2002. Vrtby v dřveitých tkáních stromových kapradin rodu *Tempskya* v české křídové pánvi. *Zprávy o geologických výzkumech v roce* 2002: 129–131.

Müller-Stroll, W. R., and E. Mädel-Angelie. 1969. *Actinidioxylon princeps* Ludwig n. comb., ein ianenholz aus dem Pliozän von Dernbach in Westerwald. *Senchenbergiana Lethaea* 50: 105–115.

Nishida, H. 1986. A new *Tempskya* from Japan. *Transactions and Proceedings of the Palaeontological Society of Japan*, n.s., 143: 435–446.

———. 2001. A leptosporangiate fern *Tempskya uemurae*, sp. nov. (Tempskyaceae) from the Upper Cretaceous (Santonian) of Iwate Prefecture, Japan. *Acta Phytotaxon Geobotany* 52: 41–48.

O'Sullivan, R. B., C. A. Repening, E. C. Beaumont, and H. G. Page. 1972. Stratigraphy of the Cretaceous rocks and the Tertiary Ojo Alamo Sandstone, Navajo and Hopi Reservations, Arizona, New Mexico, and Utah. *U.S. Geological Survey Professional Paper* 521E: E1–E65.

Putz, F. E., and N. M. Holbrook. 1991. Biomechanical studies of vines. In F. E. Putz and H. A. Mooney (eds.), *The Biology of Vines*, pp. 73–97. Cambridge: Cambridge University Press.

Read, C. B., and R. W. Brown. 1937. American Cretaceous ferns of the genus *Tempskya*. *U.S. Geological Survey Professional Paper* 186-F: 105–131.

Retallack, G. J. 2005. Pedogenic carbonate proxies for amount and seasonality of precipitation in paleosols. *Geology* 33: 333–336.

Rushforth, S. R. 1971. A flora from the Dakota Sandstone Formation (Cenomanian) near Westwater, Grand County, Utah. *Brigham Young University Science Bulletin* 14: 1–44.

Scott, A. C., and T. N. Taylor. 1983. Plant/animal interactions during the Upper Carboniferous. *Botanical Review* 49: 259–307.

Schneider, H., and P. Kenrick. 2001. An Early Cretaceous root-climbing epiphyte (Lindsaeaceae) and its significance for calibrating the diversification of polypodiaceous ferns. *Review of Palaeobotany and Palynology* 115: 33–41.

Schönfeld, E. 1954. Über eine fossile Liane aus Patagonien mit einigen Bemerkungen über Beobachtungen an breiten Markstrahlen. *Palaeontographica*, Abt. B, 97: 23–25.

Seward, A. C. 1919. *Fossil Plants 4. Ginkgoales, Coniferales, Gnetales.* Cambridge: Cambridge University Press.

Simmons, G. C. 1957. Contact of Burro Canyon Formation with Dakota Sandstone, Slick Rock District, Colorado, and correlation of Burro Canyon Formation. *American Association of Petroleum Geologists Bulletin* 41: 2519–2529.

Smith, E. A., G. A. Ludvigson, R. M. Joeckel, J. I. Kirkland, S. J. Carpenter, L. A. Gonzalez, and S. K. Madsen. 2001. Reconnaissance carbon isotopic chemostratigraphy of pedogenic–palustrine carbonates in the Early Cretaceous Cedar Mountain Formation, San Rafael Swell, eastern Utah. *Geological Society of America Abstracts with Programs* 33: 445.

Sporne, K. R. 1965. *The Morphology of Gymnosperms.* London: Hutchinson University Library.

Stover, E. L. 1951. *An Introduction to the Anatomy of Seed Plants*. Boston: D. C. Heath.

Thayn, G. F., and W. D. Tidwell. 1984. Flora of the Lower Cretaceous Cedar Mountain Formation of Utah and Colorado, part II, *Mesembrioxylon stokesi*. *Great Basin Naturalist* 43: 294–402.

Thayn, G. F., W. D. Tidwell, and W. L. Stokes. 1983. Flora of the Lower Cretaceous Cedar Mountain Formation of Utah and Colorado, part I, *Paraphyllanthoxylon utahense*. *Great Basin Naturalist* 43: 294–402.

———. 1985. Flora of the Lower Cretaceous Cedar Mountain Formation of Utah and Colorado, part III, *Icacinoxylon pittiense* n. sp. *American Journal of Botany* 72: 175–180.

Tidwell, W. D. 1998. *Common Fossil Plants of Western North America*, 2nd ed. Washington, D.C.: Smithsonian Institution Press.

Tidwell, W. D., and N. Hebbert. 1992. Species of the Cretaceous fern *Tempskya* from Utah. *International Journal of Plant Science* 53: 513–528.

Tidwell, W. D., and G. F. Thayn. 1985. Flora of the Lower Cretaceous Cedar Mountain Formation of Utah and Colorado, part IV, *Palaeopiceoxylon thinosus* (Protopinaceae). *Southwestern Naturalist* 30: 525–532.

Tidwell, W. D., and W. W. Wright. 2003. *Tempskya dernbachii* sp. nov. from Neuquén Province, Argentina, the first *Tempskya* species reported from the Southern Hemisphere. *Review of Palaeobotany and Palynology* 134: 71–84.

Tidwell, W. D., N. E. Hebbert, and J. D. Shane. 1977. Petrified angiosperms within *Tempskya* false trunks from the Cedar Mountain Formation, Utah. *Geological Society of America, Abstracts with Programs* 9: 515.

Tidwell, W. D., B. B. Britt, and L. S. Tidwell. 2007. A review of the Cretaceous floras of east-central Utah and western Colorado. In G. C. Willis, M. D. Hylland, D. L. Clark, and T. C. Chidsey Jr. (eds.), *Central Utah, Diverse Geology of a Dynamic Landscape*, pp. 467–482. Publication No. 36. Salt Lake City: Utah Geological Association.

Wheeler, E. A., and T. M. Lehman. 2000. Late Cretaceous woody dicots form the Aguja and Javelina Formations, Big Bend National Park, Texas, U.S.A. *IAWA Journal* 21 (1): 83–120.

Wilson, L. R. 1959. Geological history of the Gnetales. *Oklahoma Geology Notes* 19: 35–40.

Young, R. G. 1960. Dakota Group of Colorado Plateau. *American Association of Petroleum Geologists Bulletin* 44: 144–156.

PART 3

Ecosystems and Mesozoic Plants

Fig. 13.1. Map of sample localities (stars) and limits of Morrison Formation (dashed line). (1) MM-NM, Goat Mountain, New Mexico. (2) MNA199a, Carrizo Mountains, Arizona. (3–5) FR1 (Fruita 1), FR84-2 (Fruita 2), FR88-3 (Fruita 3), Fruita Paleontological Area, Colorado. (6) RV88-1, Mygatt-Moore Quarry, Colorado. (7) DNM88-7 (DNM 96) Dinosaur National Monument, Rainbow Park, Utah. (8) WY2002-7, near Red Gulch Dinosaur Track-site, Wyoming. (9) Belt Cliffs, Montana. More detailed information on each locality (excluding site 9) appears in Table 13.1. *Map redrawn after Dunagan and Turner (2004).*

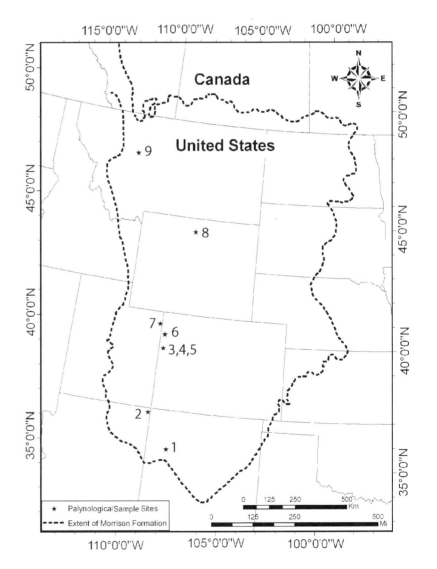

PALYNOLOGICAL EVIDENCE FOR CONIFER DOMINANCE WITHIN A HETEROGENEOUS LANDSCAPE IN THE LATE JURASSIC MORRISON FORMATION, U.S.A.

13

Carol L. Hotton and Nina L. Baghai-Riding

Eight palynological samples from a broad geographical range of the upper part of the Morrison Formation have yielded a rich palynoflora of over 100 morphospecies of bryophytes, lycophytes, sphenophytes, ferns, and seed plants. In striking contrast to Morrison megafloras, conifers are dominant throughout the palynoflora, with ferns diverse but subordinate; other components are uncommon to rare. Palynofloras display increased taxic diversity and increased spore abundance and diversity northward, accompanied by reduction in frequency of the probable xerophytic conifer family Cheirolepidiaceae, suggesting a moisture-driven latitudinal diversity gradient. Facies and taphonomic biases preclude extrapolation to the regional flora and climate from these data, but the palynological data suggest that coniferous trees with an understory of ferns comprised the predominant vegetation overall, alongside more heterogeneous riparian and wetland vegetation, during later Morrison times.

Introduction

The Upper Jurassic Morrison Formation of western North America has been known for over 100 years for its extraordinary dinosaur megafauna and a host of other tetrapods, including fish, amphibia, turtles, crocodiles, and mammals (Foster 2007a). Although the vertebrate fossils have been intensively studied over the years, the types of plant communities that supported this diverse terrestrial fauna remain in dispute, largely due to the poor plant fossil record. Despite the extensive geographical and temporal range of the Morrison Formation, fewer than 30 plant megafossil sites are currently known from it and correlative units in Canada (Ash and Tidwell 1998; Parrish et al. 2004). Two contrasting hypotheses for Morrison plant communities have been proposed.

Drawing primarily from the plant megafossil record, the first hypothesis posits a warm, humid climate supporting a lush forest of conifers and ginkgophytes, with a dense understory of tree ferns, cycads, bennettitaleans, and herbaceous pteridophytes (Tidwell 1990; Tidwell and Medlyn 1992; Ash 1994; Ash and Tidwell 1998). Evidence supporting this view includes the following: the presence of hygrophilous plants such as ferns in the compression and permineralized floras; the presence of *Czekanowskia*, *Ginkgo*, and *Nilssonia*, all considered temperate taxa; the absence of growth rings in some permineralized woods, indicating the absence of periodic drought or seasonality; and certain leaf physiognomic traits, such as thin cuticle and unarmored stomata, in many of the ferns and ginkgophytes (Tidwell et al. 1998). The abundant dinosaur megafauna is also cited as evidence for a lush forest environment on the grounds that xerophytic vegetation would have been insufficient to support a community of such large herbivores (Tidwell 1990; Suess and Schultka 2001).

The second hypothesis draws on abundant sedimentological and paleosol data as well as climate models indicating that tropical subarid and seasonal climates prevailed over most of the temporal and geographical range of the Morrison (Hallam 1994; Parrish 1993; Demko and Parrish 1998; Turner and Peterson 2004). This view holds that the vegetation was sparse and dominated by herbaceous plants, mainly xerophytic ferns, with arborescent plants confined mainly to riparian sites and lake margins (Demko et al. 2004; Turner and Peterson 2004; Parrish et al. 2004). Parrish et al. (2004) adopt a novel approach to circumvent the poor plant megafossil record by defining and comparing plant taphofacies to depositional environments. Their analysis leads them to the conclusion that the general scarcity of plant debris throughout the formation, the paucity of vitrinite in debris layers, and the large disparity in taxonomic diversity between plant megafossil and palynofossil morphospecies all support inference of a subarid, highly seasonal climate supporting open fern-dominated plant communities throughout much of the Morrison.

The plant megafossil record appears to support the first hypothesis of a relatively lush, mesic to hydric vegetation in the Morrison Formation. However, sedimentology and paleosol evidence point to a seasonal and semiarid climate, which would appear unable to support many of the plants commonly found in the Morrison Formation. How do we reconcile these conflicting lines of evidence? We propose that palynology can provide an important additional source of data for inferring both plant communities and climate. Although the data we present here are insufficient to resolve the conflict outlined above, they shed additional light on the question and suggest a resolution of the impasse.

Previous Palynological Analyses

Few palynological analyses of the Morrison Formation have been undertaken, largely because it consists predominantly of highly oxidized and alkaline sediments that are not conducive to palynomorph preservation. However, Litwin et al. (1998) showed that the Morrison could be quite palynologically productive, and to date, their work remains the most comprehensive analysis of Morrison palynofloras. Previous palynological work on the Morrison is ably summarized therein. More recent works that include mention of Morrison palynomorphs include Trujillo (2003), Gee (2006), and Gee and Tidwell (2010). Litwin et al. (1998) focused primarily on the age of the Morrison through the use of stratigraphically significant palynomorphs, but they also presented some paleoecological data in the form of per-sample species richness through the section. Although a few other palynological studies have touched on paleoecology (Gottesfeld in Dodson et al. 1980; Hotton 1986), palynology remains a greatly under-utilized but potentially valuable source of data for reconstructing plant communities in the Morrison.

Geology of the Morrison Formation

The Morrison Formation, along with its equivalents in Canada, extends over 1,000,000 km² (400,000 mi²), from New Mexico to Saskatchewan, Canada, and occurs in 13 states throughout the western United States (Fig. 13.1). Its age is believed to be upper Oxfordian to lower Tithonian (Litwin et al. 1998; Kowallis et al. 1998), although its boundary with overlying formations is still poorly defined, and it has been suggested that the uppermost Morrison may be basal Cretaceous in some areas (Trujillo 2003). The Morrison Formation is almost entirely terrestrial and dominated by low to high sinuosity fluvial, overbank, and lacustrine sediments deposited on a low relief floodplain (Turner and Peterson 2004). Pedosols are common and give the Morrison its characteristic banded appearance. Soil types such as Calcisols and Vertisols (Demko et al. 2004) and evidence of aeolian deposits and evaporites, especially in the lower part of the formation, suggest generally arid to subhumid conditions (Bell 1986; Condon and Peterson 1986; Turner and Fishman 1991). There is evidence of a trend toward decreasing aridity in younger rocks (Turner and Peterson 2004), and similar gradients of higher water tables from south to north and from west to east can be identified from soil profiles and overbank deposits (Demko et al. 2004). Most of the plant megafossils from the Morrison have been recovered from the highest member, the Brushy Basin Member (Ash and Tidwell 1998; Parrish et al. 2004), which we believe is the source of most or all of the samples used in this analysis (see below).

Materials and Methods

The locations of the samples used in this analysis are shown in Figure 13.1. The Arizona sample, from the San Juan Basin, is poorly documented, but we include it here because to our knowledge, it is the only Morrison palynoflorule reported from Arizona, and it geographically anchors the southwestern corner of the formation (Table 13.1). The sample was most likely collected from the upper Morrison because finer-grained and less oxidized units are more common there. The sample from New Mexico, also within the San Juan Basin, was probably collected from the Goat Mountain section reported by Turner-Peterson et al. (1980) and was most likely collected from the Westwater Canyon Member, which is correlative with the lower Brushy Basin Member of the Colorado Plateau (Turner-Peterson 1986). The Westwater Canyon Member includes more fine-grained facies than the overlying Brushy Basin Member at Goat

Mountain, and several gray mudstones from that member have been reported to contain palynomorphs (Turner-Peterson et al. 1980; Tschudy et al. 1981). All other samples are from the Brushy Basin Member of the Morrison Formation. Each of the samples from the Fruita Paleontological Area was collected from individual tetrapod localities that have since been tied into a detailed stratigraphic and sedimentological map (Kirkland 2006) (Table 13.1). The sample from the Mygatt-Moore Quarry was collected about 2.25 m above the base of the bone-bearing unit (see section in Kirkland and Carpenter 1994). The Mygatt-Moore Quarry is believed to be stratigraphically slightly higher than the Fruita Paleontological Area localities (Kirkland and Carpenter 1994; Foster 2007a). The Dinosaur National Monument sample is located approximately 2 m above a microvertebrate locality (DNM 96) reported by Chure and Engelmann (1989), also cited by Engelmann and Callison (1998). This locality occurs in an isolated outcrop, so its precise stratigraphic relationship to other sections is unknown, but it is thought to be approximately equivalent to the Fruita Paleontological Area (G. Engelmann, personal communication to C.L.H., 2008). The Wyoming sample was collected approximately 20 m below the top of the Morrison Formation as marked by a major unconformity, from a section previously reported to have produced Jurassic palynomorphs (see section in Furer et al. 1997). A ninth sample was collected from near Belt, Montana, but yielded only a few poorly preserved bisaccates and several possible *Leptolepidites* species, which were insufficient for quantitative analysis. Further maceration and analysis of this sample may yield more data, but it will not be considered further in this study.

Samples were prepared by standard techniques of hydrochloric and hydrofluoric acid maceration followed by heavy liquid separation, and minimal or no oxidative treatment. Residues were screened through 15-μm screens and mounted in glycerine jelly. Palynomorphs were studied with a Nikon 80i Eclipse compound microscope and photographed through a 60× plan apo VC DIC lens with a Nikon DXM 1200F digital camera, and with an Olympus BH-2 compound microscope mounted with an Olympus C-35 camera with Kodacolor 200 film.

Percentages of important taxa were estimated by approximately 300 grain counts. Slides were then further scanned to an estimated total of 1000 grains (technique for estimation described in Hotton 2003). Hotton (1986) scanned much larger populations of three of these samples, the Arizona, New Mexico, and Fruita 1 localities (estimated 8,000 to 10,000 grains), which yielded somewhat higher per-sample species richness. The taxa from this earlier analysis are marked in Appendix 13.1 with an asterisk, but they were not included in the graphs presented here unless they were encountered in the new analysis. Scanning such a large sample number is certainly not the norm in palynology, and it is unnecessary for many purposes, such as stratigraphy and facies recognition. However, the rationale for surveying such a large population is to create a more accurate and complete picture of overall floral diversity and capture more of the

Table 13.1. Locality Details for Samples Used in This Study

Sample No.	State	County	Locality	Collector	Date	Stratigraphy	Lithology	Munsell No.	Facies
MNA199a	Arizona	Apache	Carrizo Mountain[a]	NA[b]	NA	Morrison Fm.	Fissile claystone	N5, medium gray	NA
MM-NM	New Mexico	McKinley	Goat Mountain[c]	E. Nesbitt	~1979	?Westwater Mbr.	NA	NA	NA
FR1	Colorado	Mesa	Ceratosaurus Pond, Fruita Paleontological Area[d]	G. Callison	~1979	Upper Brushy Basin Mbr.	NA	NA	Pond/marsh[d]
FR84-2	Colorado	Mesa	Tom's Place, Fruita Paleontological Area[d]	G. Callison	1984	Upper Brushy Basin Mbr.	Siltstone with carbonaceous debris	NA	Alkaline marsh[d]
FR88-3	Colorado	Mesa	Tom's Place, Fruita Paleontological Area[d]	C. Hotton	1988	Upper Brushy Basin Mbr.	Siltstone with very fine sandstone stringers, carbonaceous and cuticle debris	5Y 4/1, olive gray	Crevasse splay[d]
RV88-1	Colorado	Mesa	Mygatt-Moore Quarry[e]	C. Hotton	1988	Upper Brushy Basin Mbr.	Blocky mudstone, much carbonaceous debris	N4-N5, medium to dark gray	Water hole or ephemeral overbank[e,f]
DNM88-7	Utah	Uintah	DNM 96 microvertebrate locality[g]	C. Hotton	1988	Upper Brushy Basin Mbr.	Blocky mudstone, sparse fine carbonaceous debris	N4, medium gray	NA
WY2002-7	Wyoming	Big Horn	Near Red Gulch Dinosaur Tracksite[h,i]	C. Hotton	2002	Brushy Basin Mbr.	Fissile mudstone	5Y 7/2m yellowish gray	NA

[a] Obtained by C. L. Hotton in 1981 from the Museum of Northern Arizona; sample labeled "MNA locality 199a, Morrison Formation, exact locality and collector unknown, but probably from the north side of Carrizo Mountains, NE corner of Apache Co., Arizona."

[b] NA = not available or unknown.

[c] Section in Turner-Peterson et al. (1980).

[d] Kirkland (2006).

[e] Kirkland and Carpenter (1994).

[f] Foster (2007b).

[g] Chure and Engelmann (1989), Engelmann and Callison (1998).

[h] Kvale et al. (2001).

[i] Furer et al. (1997).

rare taxa, including those that may occur at greater distances from the depositional source (upland, hinterland, or extrabasinal components).

The only publications illustrating Morrison palynomorphs are the open-file reports of Tschudy et al. (1980, 1981, 1988a, 1988b) and Litwin et al. (1998). For identification of morphospecies, we compared our floras to the Upper Jurassic and Lower Cretaceous treatments of Pocock (1962, 1970), Ricketts and Sweet (1986), Burden and Hills (1989), Fensome (1983, 1987), Singh (1964, 1971), and Brenner (1963) for North America, as well as European treatments such as Couper (1958), Doerhoefer and Norris (1977), and Norris (1969), and Australian treatments such as Dettmann (1963), Filatoff (1975), and Backhouse (1988), among many others. Work on the taxonomy of the palynoflora is ongoing, and the taxon list presented in Appendix 13.1 should be considered preliminary, especially for the notoriously difficult bisaccate taxa. A more complete taxonomic treatment of taxa encountered in this study will be presented elsewhere.

For paleoecological analyses, we classified individual morphospecies according to their likely botanical affinity. Botanical affinity was assessed by a combination of general morphological features, resemblance to living representatives, and most important, in situ sporangia or pollen cones. The bisaccate pollen grains posed particular problems. Pollen forms falling within the circumscription of *Podocarpidites* (sacci larger than central body, central body ornamented), *Phyllocladidites/Pristinuspollenites* and *Microcachrydites* were treated as Podocarpaceae, whereas other bisaccate morphogenera were treated as unknown (but probably primarily associated with the Pinaceae). Given that most of these taxa have not been linked to specific megafossils of recognizable affinities, this classification should be treated as highly approximate. Invaluable treatments of in situ pollen and spore types include Couper (1958), Balme (1995), and van Konijnenburg-van Cittert (1971, 1989, 1991), as well as more recent literature (Batten and Dutta 1997; Archangelsky and Archangelsky 2006; Stockey et al. 2006; Archangelsky and Del Fueyo 2010; Gee and Tidwell 2010).

Results

Taxonomy and Stratigraphy

A list of the morphospecies recognized in this study and their distribution among samples is presented in Appendix 13.1. The composition of the Morrison palynoflora is typical of other Late Jurassic floras reported from Canada, the United Kingdom, Europe, and Australia, and nothing in the flora suggests an Early Cretaceous age (Figs. 13.2–13.4; Plates 12–14; Appendix 13.1). For example, *Cicatricosisporites* and *Pilosisporites* were extremely rare, and spores of *Appendicisporites* were absent (although reported in Litwin et al. 1998); such schizaeaceous spores become more abundant and diverse in Berriasian and younger rocks (Norris 1970; Doerhoefer and Norris 1977; Fensome 1987). Of particular interest is the presence of *Shanbeipollenites quadratus* (Fig. 13.2H; Plate 12H), described

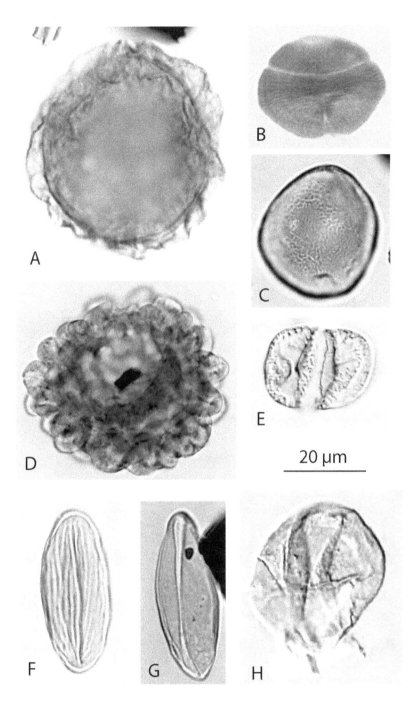

Fig. 13.2. Light micrographs of common palynomorphs. Taxon names and authorities are followed by slide number and England Finder (EF) co-ordinates. (A) *Callialasporites* sp. cf. *C. microvelatus* Schulz: FR88-3-1, EF J48/2. (B) *Classopollis classoides* Pflug: MNA199a-2, EF H40/1. (C) *Exesipollenites tumulus* Balme: FR1-6, EF K39/4. (D) *Cerebropol-lenites mesozoicus* (Couper) Nilsson: MNA199a-2, EF H39/2. (E) *Vitreisporites pallidus* (Reissinger) Nils-son: FR84-2-2, EF C33/4. (F) *Equisetosporites* sp. cf. *E. multicostatus* (Brenner) Norris: FR1-6, EF S52/2. (G) *Cycadopites follicularis* Wilson et Webster: FR1-6, EF X35. (H) *Shanbeipollenites quadratus* (Kumar) Schrank: WY2002-7-4, EF G48.

by Schrank (2004) from the Dinosaur Beds of Tendaguru, Tanzania, which contains a similar dinosaur fauna and which is thought to be approximately coeval and ecologically similar to the Morrison system (Suess and Schultka 2001). Most of the more common species appear throughout the samples, differing primarily in their abundance from sample to sample. There does not appear to be any substantial compositional difference among samples, including the two southernmost samples of more uncertain provenance, suggesting that all of the samples were probably derived from the Brushy Basin Member of the Morrison Formation. The composition of the palynofloras reported by Tschudy et al. (1988a, 1988b) from the lower members of the Morrison generally resemble our known Brushy Basin samples. However, palynology of the lower Morrison is poorly known, so there is not much of a basis for comparison. Although no quantification is presented in the open-file reports, a large variety of bisaccates and other conifers are illustrated, in contrast to a relatively small number of spore taxa figured there (Tschudy et al. 1980, 1981, 1988a, 1988b).

Relative Abundance of Key Groups

Conifers are dominant in every sample, although the dominant family varies from site to site. The New Mexico and Arizona sites display the greatest disparity with respect to taxonomic dominance from the rest of the samples (Fig. 13.5). Species of *Classopollis* (Fig. 13.2B; Plate 12B), representing the extinct conifer family Cheirolepidiaceae, are extremely abundant in both sites. Pollen of the Araucariaceae (*Araucariacites australis*, *Callialasporites* spp., Fig. 13.2A; Plate 12A) is subdominant in the Arizona sample, whereas no other family makes up more than 5% of the total count (Fig. 13.5). In the New Mexico sample, *Exesipollenites* (Fig. 13.2C; Plate 12C), most likely representing the Cupressaceae sensu lato (including the Taxodiaceae), and bisaccates, representing primarily the Podocarpaceae and Pinaceae (Fig. 13.3B, C, E; Plate 13B, C, E) are also relatively abundant (Fig. 13.5).

The abundance of Cheirolepidiaceae drop dramatically in the more northern sites, never comprising more than 4% of the total count (Fig. 13.5). The three Fruita samples are heavily dominated by *Exesipollenites*, with smaller percentages of *Inaperturopollenites* and *Perinopollenites*, all suggesting local dominance by the Cupressaceae. Araucariaceous pollen is subdominant in Fruita 1 and Fruita 3 (Fig. 13.5). *Exesipollenites* and *Callialasporites* often occur in clumps in the Fruita 1 and Fruita 2 samples. The Dinosaur National Monument locality most closely resembles Fruita 1 in dominance of Cupressaceae and subdominance of Araucariaceae, although it more closely resembles Fruita 3 in its high percentage of bisaccates (Fig. 13.5).

The Mygatt-Moore locality differs in a number of respects from the other Colorado Plateau samples. For example, it contains more than 50% bisaccates, especially species of *Podocarpidites* (Fig. 13.3B, C; Plate 13B, C). Other conifer families (e.g., Cheirolepidiaceae) are rare or absent (Fig. 13.5). The Mygatt-Moore is also notable in its high percentage of spores, primarily *Crybelosporites vectensis* (Fig. 13.3A; Plate 13A) and probable microspores of Isoetales (Fig. 13.3D; Plate 13D). (The attribution of *C. vectensis* to *Crybelosporites*, a genus for microspores of the water fern order Marsileales, is probably incorrect [Lupia et al. 2000], so

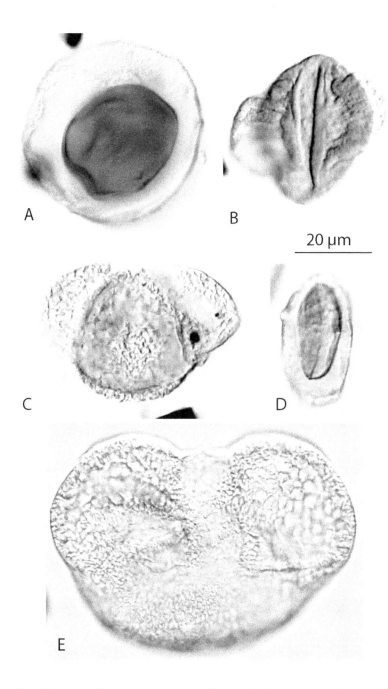

Fig. 13.3. Taxon name and authority followed by slide number and England Finder (EF) coordinates. (A) *Crybelosporites vectensis* Kemp: DNM88-7-1, EF O44. (B) *Pristinuspollenites* sp. M16: WY2002-7-4, EF T42/3. (C) *Podocarpidites* sp. cf. *P. epistratus* Brenner: RV88-1-3, EF U61/. (D) Isoetaceae microspore?: FR3-88-1, EF J57/1. (E) *Cedripites* sp.: DNM88-7-1, EF E61/4.

20 µm

the affinities of this spore type are unknown.) Spiny acritarchs, possibly indicating brackish or alkaline waters, occur at a frequency of about 1% in this sample; elsewhere, they are virtually absent.

The Dinosaur National Monument locality resembles Fruita 1 in possessing relatively high percentages of species of *Exesipollenites* (Cupressaceae) and Araucariaceae, and relatively lower percentages of bisaccates. The most northern of our sites, in Wyoming, on the other hand, displays an abundance of bisaccates, but here, nonpodocarpaceous bisaccates are particularly abundant. The Araucariaceae and

Fig. 13.4. Taxon name and authority followed by slide number and England Finder (EF) coordinates. (A) *Nevesisporites vallatus* de Jersey et Paten: FR84-2-2, EF C32/4. (B) *Leptolepidites proxigranulatus* (Brenner) Doerhoefer: DNM88-7-1, EF X42/3. (C) *Stereisporites* sp. cf. *S. antiquasporites* (Wilson et Webster) Dettmann: DNM88-7-1, EF C37/1. (D) *Cibotiumspora jurienensis* (Balme) Filatoff: FR88-3-1, EF J53/4. (E) *Cyathidites minor* Couper: FR88-3-1, EF M44/1. (F) *Retitriletes austroclavatidites* (Cookson) Doering et al.: WY2002-7-4, N55/2. (G) *Todisporites minor* Couper: RV88-2, EF N57/2. (H) *Concavissimisporites* sp. cf. *C. crassatus* (Delcourt et Sprumont) Delcourt, Dettmann et Hughes: DNM88-7-1 EF Q61/2. (I) *Crassitudisporites problematicus* (Couper) Hiltmann: WY2002-7-3, EF J47/2.

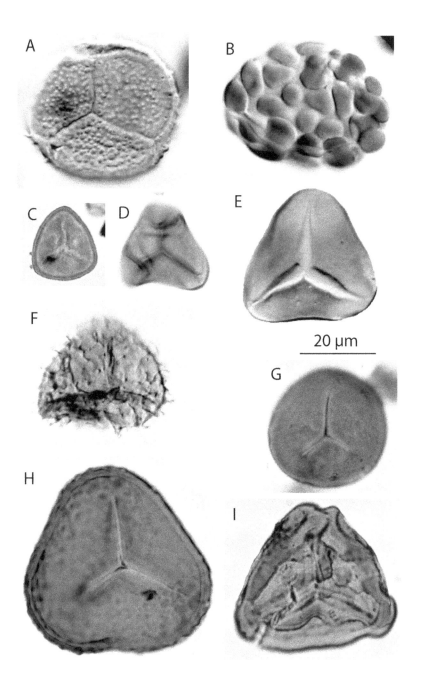

Cupressaceae (*Exesipollenites*) are subdominant in the Wyoming sample (Fig. 13.5).

Other nonconiferous seed plants (Cycadales, Bennettitales, ginkgophytes, Caytoniaceae, gnetophytes) rarely comprise more than 1% of the total in any sample and so are not included in Figure 13.5. Some of these pollen types occur consistently throughout the range of our samples, especially the probable conifer *Cerebropollenites mesozoicus* (Fig. 13.2D; Plate 12D) and the caytoniaceous seed fern *Vitreisporites pallidus* (Fig. 13.2E; Plate 12E). However, gnetophytes, represented by at least two or

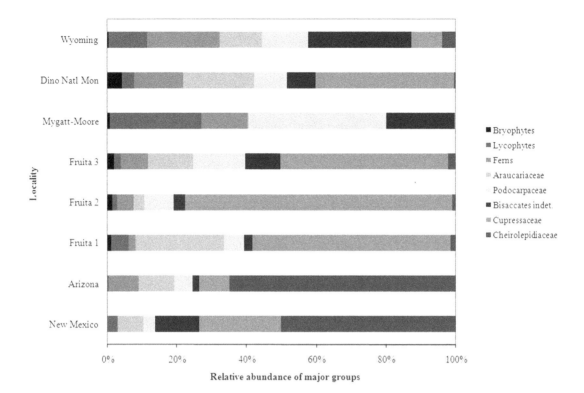

Fig. 13.5. Relative per-sample abundance of key plant groups in an approximately 300 grain count from localities 1–8 (see Fig. 13.1).

three species of *Equisetosporites* (Fig. 13.2F; Plate 12F), appear to be confined to the southern localities and Colorado. It is quite possible, however, that further searching of the more northerly samples will reveal the presence of these distinctive forms. The smooth monosulcate pollen grains produced alike by cycads, Bennettitales, and gingkophytes and given a variety of form generic names (called here *Cycadopites* [Fig. 13.2G; Plate 12G] and *Chasmatosporites*) are notably rare throughout the playnoflora.

Lower plants, represented by bryophytes, lycophytes, sphenophytes, and ferns, are subordinate in every sample, ranging from 2% in the New Mexico sample to about 20% in the Wyoming sample and about 40% in the Mygatt-Moore sample (Fig. 13.5). The sample from New Mexico is especially notable in its low abundance of spores. One of the Fruita samples (the *Ceratosaurus* Pond locality) also displays a low abundance of spores. In both cases, swamping by a few abundant taxa (*Classopollis* or *Exesipollenites*) undoubtedly contributes to the generally low abundance of spores; however, this is probably not the only explanation. The Arizona and Fruita 2 samples also displays high percentages of *Classopollis* and *Exesipollenites*, respectively, but in both cases, spore abundance is roughly twice that of the New Mexico and Fruita 1 samples (Fig. 13.5).

Bryophytes never occur at more than 2% of the total and are limited to the *Nevesisporites* complex (Fig. 13.4A, Plate 14A), most likely derived from the hornwort family Notothylidaceae (Dettmann 1963) and species of *Stereisporites* (Fig. 13.4C; Plate 14C) attributable to the moss family

Sphagnaceae. Fern spores are reasonably abundant and diverse through-out most of the samples and include representatives of the Dipteridaceae (Fig. 13.4D; Plate 14D), the tree fern group consisting of the Cyatheaceae and Dicksoniaceae (Fig. 13.4E; Plate 14E), the ancient leptosporangi-ate family Osmundaceae (Fig. 13.4G; Plate 15G), early members of the Schizaeaceae (Fig. 13.4H; Plate 14H), and one of the earliest members of the Pteridaceae (Fig. 13.4I; Plate 14I). Other trilete spores of unknown but presumably filicalean affinities include species of *Leptolepidites* (Fig. 13.4B; Plate 14B), one of the more common and diverse forms in the flora. Presumed microspores of *Isoetes* occur in almost every sample, sometimes in abundance, but otherwise, lycophytes are rare, and include repre-sentatives of the Lycopodiaceae, such as *Retitriletes austroclavatidites* (Fig. 13.4F; Plate 14F), as well as several forms of *Selaginella* type spores. Sphenophyte spores (*Calamospora*) are extremely rare throughout, but this is most likely due to undercounting (their spores can be difficult to distinguish from debris) or to lower preservation potential because of their thin exine, or both factors.

Relative Species Richness

Overall species richness is relatively high among the eight samples, to-taling roughly 110 morphospecies. Spore morphotaxa comprise approxi-mately half of this total, but individual species tend to be quite rare, a phenomenon also noted by Litwin et al. (1998). This observation raises a common problem in palynology: circumscribing the boundaries of a palynological morphospecies. Many morphospecies display continuous variation that can be observed if a large population is available for com-parison, but if only a few grains are available, they may appear to be dis-tinct. It should be kept in mind, then, that some of the rarer species could represent end points of continuous variation, rather than distinct species.

The pollen samples were not prepared to determine absolute abun-dances (for example, by spiking with *Lycopodium* spores), and certain samples were clearly burdened with overrepresentation of a few types that tend to swamp out other species (for example, the New Mexico and some of the Fruita samples). Despite these potential biases, clear diversity patterns emerge. A distinct south-to-north gradient of increasing species richness can be observed (Fig. 13.6). The New Mexico sample is particularly impoverished in spore taxa, yielding only five distinct types in a 1000-grain search. A higher specimen scan (approximately 8000 grains) yielded only eight additional taxa (Appendix 13.1). In contrast, 14 spore morphotaxa were recognized in the Arizona sample. A higher specimen scan (approximately 8000 grains) yielded only four additional spore taxa (Appendix 13.1), suggesting a greater evenness in taxonomic distribution. Total species richness is higher in the Arizona sample as well; in fact, both spore and overall species richness are comparable to the Colorado samples (Fruita Paleontological Area and Mygatt-Moore) (Fig. 13.6). The

Fruita samples display considerable variation in both relative spore/pollen species richness and overall species richness even though they are from stratigraphically equivalent units and are separated by a few hundred meters, with the Fruita 3 sample displaying the highest spore diversity and highest overall diversity (Fig. 13.6). The Mygatt-Moore Quarry sample appears closest to Fruita 3 in overall species richness but has lower spore diversity. The two most diverse samples are Dinosaur National Monument and the northernmost sample in Wyoming, in terms of both spore diversity and overall diversity. Spore diversity jumps to 23 to 24 morphotaxa, almost twice as many as in any of the other samples (Fig. 13.6).

The palynofloras analyzed here clearly indicate conifer dominance throughout the sampled geographic range of the upper Morrison Formation. Lower plants, comprising ferns, sphenophytes, lycopsids, and bryophytes, are always numerically subordinate but display significant morphotaxonomic diversity. Despite the limited number of sample sites, there is a clear south-to-north gradient of increasing total diversity and increasing spore abundance and diversity. Seed plants other than conifers, such as *Caytonia*, Bennettitales, cycads, and gnetophytes, are notably quite rare in the pollen floras. This appears surprising in light of the fact that Bennettitales and cycads comprise some of the more common elements in the megafloral record (Tidwell 1990; Ash and Tidwell 1998; Chure et al. 2006).

The palynological data support sedimentological evidence of increasing humidity and/or higher water tables northward. The two southern samples are dominated by species of the cheirolepidiaceous conifer *Classopollis*. The Cheirolepidiaceae possess strongly xeromorphic traits, such as sunken stomata, thick cuticle, and reduced leaves (Alvin 1982), and they are associated with sedimentological indicators of aridity (Francis 1984; Watson and Alvin 1996). Together, these traits suggest that *Classopollis* is a reliable indicator of warm and relatively arid conditions, although it likely occupied a wide range of habitats during its zenith in the Jurassic (Vakhrameev 1991). The combination of abundant *Classopollis* and the low abundance and diversity of spores in the southern samples correlates well with the sedimentological evidence of arid conditions in the southern end of the range of the Morrison Formation (Tyler and Ethridge 1983; Condon and Peterson 1986; Peterson 1994). That spores increase in abundance and diversity northward in tandem with decreased abundance of *Classopollis* also underlines the common inference that spores are a reliable indicator of humid conditions. The relatively high diversity of spores in the Arizona sample supports clear sedimentological evidence of wetter conditions on the western edge of the formation (Demko et al. 2004; Turner and Peterson 2004), but a single datum is not sufficient to draw conclusions regarding a possible east-to-west gradient of increasing palynofloral diversity. The most northerly samples, those from the Dinosaur National Monument and Wyoming, display the highest

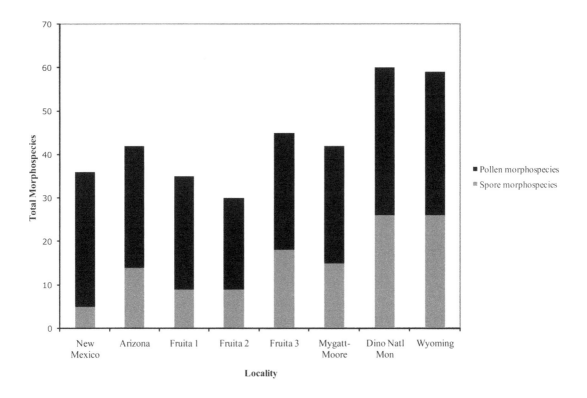

Fig. 13.6. Relative spore and pollen morphospecies richness from localities 1–8 (see Fig. 13.1).

overall species diversity and the highest abundance and diversity of spores. Coals appear in the Morrison Formation just north of the Wyoming site, a clear indication of humid conditions. Of course, purely local factors such as facies type also play a role in determining relative palynomorph abundance and diversity, which likely explains the variation observed in closely spaced samples such as the three Fruita samples, as well as the somewhat deviant sample from the Mygatt-Moore Quarry.

At first glance, our pollen analysis would appear to contradict the hypothesis that the vegetation of the Morrison consisted of an open landscape dominated by ferns with trees restricted to river and lake margins (Parrish et al. 2004). However, it would be a mistake to translate these data directly into a particular plant community type without also considering the sedimentary and taphonomic biases that affect palynological samples. Pollen and spores are drawn from a variety of both intrabasinal and extrabasinal sources. The proportion of source plants that end up in a pollen sample depend on a host of factors, both biotic (abundance and position on the landscape, size and stature of parent plant, its productivity, phenology, life cycle, pollination mode, spore/pollen preservation potential, openness of the landscape) and abiotic factors, such as facies type (swamp, lake, delta margin), degree of time averaging, and degree of oxidation and pH. Other things being equal, plants growing closer to the depositional site contribute more pollen/spores compared with plants growing at a greater distance, wind-pollinated plants contribute more than insect-pollinated plants, tall plants more than short-statured ones, long-lived plants more than ephemeral ones, common plants more than rare (Moore et al. 1991; Traverse 2007). Taking these factors into account, our palynological data could be interpreted as consistent with the hypothesis that conifers were confined to watercourses and ferns dominated in better-drained upland areas of the landscape: conifers would be far more palynologically productive, and ferns located distal to depositional sites would be disproportionately underrepresented in the palynological record.

Complicating matters further is the fact that plant fossils tend to represent the wetter end of the landscape spectrum because they are most likely to be preserved in dysaerobic, water-saturated, and generally acidic environments. The plant megafossil record, in particular, is confined to a narrow subset of quite local and generally wetland environments (Spicer 1989; Burnham et al. 1992), thereby conveying the impression that regional plant communities are more hygrophilous than they really are (Demko et al. 1998; DiMichele and Gastaldo 2008). This bias is a ready, if ad hoc, explanation for the mesic to hygrophilous appearance of much of the Morrison megafloras (Tidwell 1990; Tidwell and Medlyn 1992, 1993). Although palynomorphs can potentially sample a broader range of the regional vegetation than megafossils due to their small size and aerial habit, they are still subject to much of the same local flora bias as displayed by megafossils, especially their preservation in dysaerobic sediments (Farley and Dilcher 1986).

Ecological Tolerances of Key Groups

So are we back to our starting point, unable to distinguish between two contrasting views of the Morrison landscape? Consideration of ecological preferences of components of the flora may help tease apart plants tolerant of better-drained, more xeric environments (the "upland" element) from more mesic/hygrophilous streamside elements. Bryophytes and ferns in particular face certain fundamental physiological constraints that affect their ability to function away from water. For example, they require water for reproduction, possess a fragile, independent gametophyte phase, and have poor control of evapotranspiration—all traits that put them at a disadvantage in drier conditions (Page 2002). Although such limitations can serve as a general guide, numerous examples of dry-adapted bryophytes and pteridophytes exist among living species, and there is no reason to believe that similarly adapted species did not evolve in the past.

A safer approach is to examine ecological affinities for plants that were approximately contemporaneous with Morrison plants by using evidence of the sedimentological context in which megafossils are preserved. For example, plants preserved in coals indicate a preference for marsh or swamp habitats, preservation in mineral soils suggests occupancy of better-drained sites, and association with evaporites and fusain suggests some degree of xerophily. The most environmentally informative groups may be many of the fern families reflected in the palynofloral record. Generally speaking, many fossil fern groups throughout their history are associated with the same kinds of hydric conditions observed today in their extant relatives (Skog 2001). Among the most reliable indicators of hydric conditions is the family Osmundaceae, often found in coals in the Early and Middle Jurassic of Europe (van Konijnenburg-van Cittert 2002). Members of the tree fern family Cyatheaceae have a poor fossil record, but they appear to have occupied habitats similar to those occupied today, i.e., relatively wet and stable habitats (van Konijnenburg-van Cittert 2002). These two families comprise the most common elements of the Morrison palynoflora (*Cyathidites* = Cyatheaceae in part, *Todites*, *Baculatisporites*, *Osmundacidites* = Osmundaceae), suggesting that they may represent hygrophilous elements of the local (i.e., stream or lake margin) community.

Other fern families in the Morrison flora include representatives that appear to have been tolerant of more open and disturbed sites, from both sedimentological context and evidence of xeromorphy. For example, the Dipteridaceae (represented by *Dictyophyllidites* and *Cibotiumspora jurienensis* in the palynoflora) include the possibly brackish water–adapted fern *Hausmannia dichotoma* (van Konijnenburg-van Cittert and van der Burgh 1996); *Hausmannia* also occurs as fusain in Middle Jurassic flora of Scotland (Bateman et al. 2000). *Cibotiumspora jurienensis* has been recovered from the sporangia of the recently described species *Hausmannia morinii*, which may have grown in disturbed habitats (Stockey et al. 2006). The Gleicheniaceae and Matoniaceae appear to include many of the most important xeric ferns throughout the Mesozoic and Cenozoic, including the highly xeromorphic fern *Weichselia*. Representatives of these groups frequently occur as fusain in the Lower Cretaceous Wealden beds (Watson and Alvin 1996). Putatively xerophilous ferns occur but tend to be quite rare in Morrison floras, which can be interpreted in two ways: such groups may be present but distal from depositional sites, or they may be rare because regional conditions are humid.

The Schizaeaceae also include members that were tolerant of open and disturbed habitats in the Cretaceous (Collinson 1996, 2002). For example, an early Maastrichtian flora preserved in situ by volcanic ash in Wyoming, U.S.A., includes an *Anemia* type fern growing on mineral soil within an open fern-dominated landscape (Wing et al. 1993). Jurassic representatives of this family may have been more typically wetland plants, however (van Konijnenburg-van Cittert 2002).

Evidence for paleoecology of common conifers families in the Morrison is more ambiguous, except for the Cheirolepidiaceae, a well-established xeric indicator (Vakhrameev 1991; Watson and Alvin 1996). Today, both the Podocarpaceae and Araucariaceae characterize relatively humid and frost-free environments, but there are hints that this may not have applied to all fossil representatives. For example, podocarpaceous megafossils occur as fusain in the Tendaguru Formation of Tanzania (Suess and Schultka 2006), and the association of araucariaceous pollen with

Classopollis in Jurassic palynofloras combined with sedimentological evidence of aridity suggests xerophily (e.g., Barrón et al. 2006). However, direct evidence on paleoecology of Jurassic representatives of the Podocarpaeae and Araucariaceae is sparse. Similarly, the scant data available suggest that Jurassic members of the Cupressaceae may have occupied more mesic environments; however, taxodiaceous wood preserved as fusain has been reported from the Dinosaur Beds of Tendaguru, Tanzania, a formation believed to have been deposited in a semitropical, seasonal climate similar to the Morrison Formation (Suess and Schultka 2001). Given the wide range of habitats occupied by extant members of the Cupressaceae, from swamps (*Taxodium*) to deserts (*Juniperus*), xerophily in extinct members is plausible, but there is little direct evidence for it among Jurassic representatives.

Bryophytes, because of their low stature and consequent inability to place spores high into the airstream, represent primarily local vegetation. The Sphagnaceae in particular, represented by *Stereisporites* in the Morrison Formation, most likely inhabited waterlogged habitats (and this is one group whose morphology suggests there has been no significant trend toward adaptation to xeric environments over time). Bryophytes in the Morrison palynofloras most likely represent local and hygrophilous components of the vegetation.

Sedimentology

Careful examination of the sedimentology and taphonomy of a given site is an essential source of data for landscape reconstruction and may help distinguish distal from local elements in the flora. Most of the samples from this analysis were most likely derived from pond sediments, judging from the fine-grained lithology, fissility, and medium to dark gray Munsell color number (Table 13.1). This is in agreement with the observation that most palynomorphs in the Morrison are derived from pond environments (Dunagan and Turner 2004). Little further is known about the sedimentological context of most of these samples, with the exception of the Fruita and Mygatt-Moore samples, as discussed below.

Detailed mapping of the three-dimensional exposures at the Fruita Paleontological Area have revealed a complex of low sinuosity channels, levees, paleosols, spring-fed ponds, and ephemeral alkaline marshes (Kirkland 2006). The Fruita 1 sample was collected from a fine-grained grayish unit interpreted as a small pond/marsh (dubbed the *Ceratosaurus* Pond), whereas Fruita 2 and Fruita 3 were recovered from different levels of a site inferred to be an ephemeral marsh and crevasse splay environment (Kirkland 2006). Fruita 1 and Fruita 2 display high abundance of a few pollen types, especially *Exesipollenites*. In addition, araucariaceous pollen, which is subdominant in these samples, often occurs in clumps, suggesting that the parent trees of these pollen types were growing nearby and dropping their pollen grains directly into the water. In contrast, the

Fruita 3 sample, which was collected about a meter above Fruita 2 and in slightly coarser lithology, displays higher diversity and somewhat greater evenness of distribution of taxa. This sample was collected from within a crevasse splay deposit, which would have included palynomorphs washed in from greater distances as well as components of the local flora. The Fruita 3 sample contains a number of rare taxa that do not appear in any of the other samples studied, such as *Equisetosporites* cf. *E. volutus* and several species of spores. These forms could represent hinterland vegetation, or at least vegetation growing at some point distant from the site of deposition. In any case, Fruita 2 and 3 appear to represent more ephemeral sites compared with Fruita 1, which may represent the more stable and mature components of the flora growing along more permanent water sources.

The Mygatt-Moore sample can also be placed within a detailed sedimentological context. It was collected near the top of a tetrapod-bearing unit containing many abraded bones and abundant plant debris, which has been interpreted as a permanent water hole (Kirkland and Carpenter 1994; Chin and Kirkland 1998) or as an ephemeral overbank deposit (Foster 2007b). The palynoflorule is distinct from all others examined in its high abundance of podocarpaceous pollen, especially one morphospecies (*Podocarpidites* cf. *epistratus*, Fig. 13.3C; Plate 13C). In addition, as noted above, it contains significant quantities of spores from *Crybelosporites vectensis* as well as probable microspores of Isoetaceae, along with rare acritarchs, all signaling an unusual aquatic environment. Braman in Tidwell et al. (1998) also notes the presence of abundant ferns and conifers, although no identifications were presented in that paper.

The Mygatt-Moore locality is unusual in preserving recognizable plant fossils in the same unit as animals, including plants preserved in nodules that have been interpreted as coprolites by Chin and Kirkland (1998) (however, see comments questioning this interpretation in Sander et al. 2010). Tidwell et al. (1998) report the presence of *Equisetum*, Bennettitales (*Otozamites*), Cycadales (*Jensenispermum*), and ginkgophytes (*Ginkgo*, *Czekanowskia*), along with podocarpaceous and pinaceous permineralized wood and *Brachyphyllum rechtenii* shoots (?Cheirolepidiaceae). Growth rings in the woods vary from distinct to incomplete to absent (Tidwell et al. 1998). Rare fern fragments are also present (Kirkland and Carpenter 1994). All of the plant fossils were probably subjected to some degree of transport, although faunal trampling could also have contributed to their fragmentary condition. Many of the components of the megaflora are not reflected in the palynoflora, such as the smooth monosulcate pollen produced by cycads, bennettitaleans, and ginkgophytes. Conversely, tree ferns, the Osmundaceae, and presumed Isoetaceae are not reflected in the megaflora, although taphonomic processes and low preservation potential could account for their absence. The presence of podocarpaceous and pinaceous wood does mirror the abundant bisaccate pollen present in the palynoflora. Interestingly, these elements in the megaflora are interpreted as transported (Tidwell et al.

1998); however, the abundance of podocarpaceous pollen suggests that some of this wood may have been local. Varying development of growth rings in the fossil woods may reflect fluctuation in water supply: trees growing in drier uplands may experience seasonal fluctuations, whereas trees growing near a permanent water supply would not (Tidwell et al. 1998).

A similar bone bed containing both plant and animal fossils is described by Gee and Tidwell (2010) from the Howe Ranch near Greybull, Montana, just a few tens of kilometers north of our Wyoming site. The plant megafossils found there consist almost entirely of a single, newly described species of *Araucaria*, whereas the palynoflora consists of a diversity of bisaccate pollen grains, *Callialasporites*, and spores (Gee 2006). The taphonomy and palynology of this site are currently under study, but it clearly represents another valuable source for understanding the taphonomic context of the plant and animal fossil record.

The disparity between taxonomic makeup of the megafloras and palynofloras in the Morrison palynological record is striking and not confined to the Mygatt-Moore locality. Megafloras tend to contain conifers and ferns, as reflected in the spore/pollen record, but the Bennettitales, Cycadales, and ginkgophytes are common as well (Ash and Tidwell 1998), whereas the smooth, monosulcate pollen grains that these plants produced are quite rare in the palynofloral record. The paucity of smooth monosulcates may be related in part to phenology and life cycle of certain members of these groups. At least some extinct cycads, like their modern counterparts, probably relied on insect pollination (Klavins et al. 2005), and certain Bennettitales with flowerlike and bisexual strobili almost certainly did (Friis et al. 2006) (although Osborn and Taylor [2010] found no evidence of palynivory in *Cycadeoidea*). From what is known about their morphology, the Bennettitales and extinct Cycadales were relatively short, shrubby plants, although ginkgophytes may have been arborescent. Low-statured and insect-pollinated plants would be expected to contribute proportionally less to a pollen sample. Although at least one ginkgophyte has been reported to have a bisexual strobilus (Shipunov and Sokoloff 2003), a reliable indicator of insect pollination, most ginkgophytes appear to have been wind-pollinated trees. The rarity of smooth monosulcates in the pollen record suggests to us that the iconic cycad/bennettitalean/ginkgophyte representatives of the Morrison megaflora were restricted to streamside and lakeside habitats, which would account for their common appearance in the megafossil record. A mixture of conifers and ferns probably occupied better-drained and more upland areas of the landscape, in addition to stream margins. In this scenario, the megafossil record indeed signals a more hygrophilous flora than existed regionally. Furthermore, it seems likely that certain persistent but rare components of the Morrison palynoflora and other Jurassic floras, such as *Cerebropollenites mesozoicus* and *Equisetosporites*, whose parent plants are unknown in the megafloral record, represent extrabasinal or hinterland elements that are known only from their more readily dispersed pollen. These could represent components of a more xeric adapted, hinterland flora.

Time averaging is another factor that potentially accounts for the disparity between megafossil and palynofloral diversity (Behrensmeyer 1982). Megafossils are more likely to represent single snapshots in time, perhaps a season's leaf fall or a single storm event, whereas palynofloras, unless sampled at millimeter-level resolution, generally represent attritional accumulations of plant matter over significant intervals of time, perhaps tens or hundreds of years. Such attritional accumulations even out short fluctuations caused by yearly or decadal changes in temperature or rainfall and present a cumulative view of the vegetation over time, which is likely to increase the numbers of taxa encountered in a given sample.

Disparity between megafossil and palynofossil diversity is in itself probably not a reliable indicator of aridity, contrary to the assertion by Parrish et al. (2004). In fact, this disparity between palynofloral and megafloral diversity is commonly observed in a variety of environments throughout geological history. It likely stems from the broader pool of sampled vegetation and greater temporal range reflected in the palynological record, despite its inferior taxonomic resolution relative to megafloras. The notably poor plant fossil record throughout the Morrison does support the strong sedimentological evidence for generally semiarid or seasonal climates. Most of the sediments were probably too well drained, oxygenated, or alkaline to allow significant preservation of plant material (DiMichele and Gastaldo 2008). It is no coincidence that the Morrison possesses an excellent vertebrate fossil record and a poor plant record; the conditions that make the formation such a good place to find vertebrate fossils, well-drained and mildly alkaline sediments, are those that destroy plant fossils. Plants are best preserved in acidic environments that readily destroy vertebrate fossils. The absence of plant fossils in the Morrison should not be considered evidence that plants were largely absent during Morrison times.

Plant fossils, especially pollen, generally occur in dysaerobic, water-saturated environments; thus, the plant fossil record preserves the wetter parts of the landscape. This may be true on a temporal as well as a landscape-level scale—that is, water availability may have varied not just on a seasonal basis but on hundred-year to millennial cycles. Morrison palynofloras and megafossils may reflect those intervals during which more water was available, whether the result of higher rainfall, changes in groundwater availability, or other reasons. This could also help explain the discrepancy between the plant record and the sedimentological record. Further work is needed to document this, through intensive analysis of the context of plant fossil occurrences within a detailed sedimentological framework.

We see no contradiction between the presence of large dinosaur herbivores and seasonally arid conditions. As more recent analogs of the dinosaurian megafauna, the high-altitude northern hemisphere Pleistocene megafauna occupied open taiga and tundra, and the modern African megafauna occupy savanna rather than dense forests. Large herbivores subsist more readily on low-quality forage than small herbivores and can migrate greater distances in search of food and water, both advantages during times of drought (Engelmann et al. 2004). Recent studies of the energy content of modern representatives of typical Jurassic plants show that certain taxa, especially *Araucaria*, *Ginkgo*, and *Equisetum*, provide high-quality browse, comparable to that of angiosperms, which is sufficient to support communities of large herbivorous dinosaurs (Hummel et al. 2008; Sander et al. 2010). In contrast, cycads, tree ferns, and the Podocarpaceae proved to have much lower energy content in comparison with angiosperms (Hummel et al. 2008). This suggests that dinosaurs may not have fared so well in a fern- or cycad-dominated landscape and lends credence to our view that conifers probably comprised the dominant component of Morrison floras.

It seems implausible to us that conifers would be confined to watercourses during Morrison times, with ferns the exclusive or predominant tenants of the well-drained hinterlands. Conifers appear to have originated in well-drained uplands and expanded and radiated as climates warmed and dried out (Lyons and Darrah 1989), and they possess numerous physiological and morphological traits that suggest drought tolerance, such as small leaves, thick cuticle and sunken stomata, and narrow tracheids. Of course, as they radiated during the Mesozoic, they expanded into a broad range of ecological niches, including wetlands, but there is no evidence to suggest that by the Late Jurassic they had forsaken their Permo-Carboniferous dry land origins. Cheirolepidiaceous conifers probably constituted an important component of dry land floras; the reduction in frequency of *Classopollis* in the more northerly localities may reflect the fact that they were growing at some distance from the site of deposition. The association of the Araucariaceae with the Cheirolepidiaceae in other contexts suggests that they, too, represented upland components; on the other hand, they appear to have been growing close to water sources in the Fruita localities, so they may have tolerated a variety of environments. The Podocarpaceae are rather abundant and diverse throughout, so they may also have included elements that were tolerant of better-drained conditions. The (presumed) Pinaceae become conspicuously more abundant in the northern parts of the Morrison, which suggests that they preferred wetter conditions. Certain groups of ferns undoubtedly did occupy better-drained upland regions and likely included representatives of the archaic fern families Gleicheniaceae, Matoniaceae, and Dipteridaceae. A significant percentage of the spore component in the palynofloras, however, likely represents stream bank and pond margin vegetation; ferns restricted to hinterland areas would be expected to be quite rare in palynofloras for the reasons set out above. Much more data, especially detailed analysis of the sedimentological context of palynofloras and megafloras, are required to raise this discussion beyond the realm of speculation.

Conclusions

We must emphasize that eight scattered pollen samples are far from sufficient to reconstruct the regional vegetation in the geographically and temporally extensive Morrison Formation. Our data do support a south-to-north gradient of decreasing aridity. The palynofloras show strong conifer dominance, with subordinate ferns; other groups are poorly represented, in striking contrast to the megafloral record. The species richness in our samples is quite high, supporting the observation by Litwin et al. (1998) of a rich and diverse flora, especially among ferns. The diverse palynofloras and disparate megafloras together suggest a heterogeneous landscape mosaic throughout the Morrison. How much of the regional vegetation is reflected in these palynofloras, in contrast to stream bank or lakeside flora, remains to be determined. Our data may be consistent with an open, fern-dominated landscape; nevertheless, we think that the plant data currently available are insufficient to make this the preferred hypothesis, and that paleoecological considerations as well as the available data from the palynological record suggest that conifers were regionally dominant. The task ahead requires a detailed analysis of the sedimentological context of megafloral and palynofloral remains, at the level of the analysis by Kirkland (2006), to untangle facies biases and local from regional components of the vegetation.

Acknowledgments

We acknowledge the editorial assistance of Carole Gee and Al Traverse. C.L.H. thanks the following people: Mary McGann and Linda Wright, for preparing two of the samples in Jim Doyle's 1980 palynology class at the University of California, Davis, and sparking a lifelong interest in the Morrison; the Museum of Northern Arizona, George Callison of California State Long Beach, and Dan Chure of the Dinosaur National Monument for providing or allowing the collection of samples; George Engelmann for inviting me to participate in the 1988 Morrison Formation Project Development workshop and to collect samples; and Erik Kvale of Devon Energy for showing me the Wyoming site. Fieldwork in Wyoming was supported by a Scholarly Studies grant to Conrad Labandeira. This research was supported in part by the Intramural Research Program of the National Institutes of Health, National Library of Medicine. This is publication 104 of the Evolution of Terrestrial Ecosystems Consortium (National Museum of Natural History, Smithsonian Institution).

N.B.R. acknowledges Subramanian Swaminathan, Delta State University, for help with preparing the locality map, Keith E. Riding for help with graphics software, and DSU undergraduate students Laura Wiggins, Michelle Byrd, and Arquette McCoy for help with photography. I would also like to express my deep appreciation for Ted Delevoryas, to whom this contribution is dedicated. Ted was an inspiration—always upbeat, kind, and supportive as a major professor. At my very first meeting with him as a prospective graduate student, knowing my interest in dinosaur–plant interactions, he encouraged me to study the flora of the Upper Cretaceous Aguja Formation of Texas in conjunction with the vertebrate analyses that were being compiled at the time. Our Morrison study is appropriate in this context, given Ted's general interest in Mesozoic floras of North America (Delevoryas 1959, 1960), and it brings me full circle back to his encouragement to study the plant world that dinosaurs inhabited.

References Cited

Alvin, K. L. 1982. Cheirolepidiaceae: biology, structure and palaeoecology. *Review of Paleobotany and Palynology* 37: 71–98.

Archangelsky, S., and A. Archangelsky. 2006. Putative Early Cretaceous pteridaceous spores from the offshore Austral Basin in Patagonia, Argentina. *Cretaceous Research* 27: 473–486.

Archangelsky, S., and G. M. Del Fueyo. 2010. Endemism of early cretaceous conifers in Western Gondwana. In C. T. Gee (ed.), *Plants in Mesozoic Time: Morphological Innovations, Phylogeny, Ecosystems*, pp. 247–268. Bloomington: Indiana University Press.

Ash, S. R. 1994. First occurrence of *Czekanowskia* (Gymnospermae, Czekanowskiales) in the United States. *Review of Paleobotany and Palynology* 81: 129–140.

Ash, S. R., and W. D. Tidwell. 1998. Plant megafossils from the Brushy Basin member of the Morrison Formation near Montezuma Creek Trading Post, southeastern Utah. *Modern Geology* 22: 321–339.

Backhouse, J. 1988. Late Jurassic and Early Cretaceous palynology of the Perth Basin, Western Australia. *Geological Survey of Western Australia* 135: 1–232.

Balme, B. E. 1995. Fossil in situ spores and pollen grains: an annotated catalogue. *Review of Palaeobotany and Palynology* 87: 81–323.

Barrón, E., J. J. Gómez, A. Goy, and A. P. Pieren. 2006. The Triassic–Jurassic boundary in Asturias (northern Spain): palynological characterisation and facies. *Review of Palaeobotany and Palynology* 138: 187–208.

Bateman, R. M., N. Morton, and B. L. Dower. 2000. Early Middle Jurassic plant communities in northwest Scotland: paleoecological and paleoclimatic significance. In R. L. Hall and P. L. Smith (eds.), *Advances in Jurassic Research 2000; Proceedings of the Fifth International Symposium on the Jurassic System, Georesearch Forum* 6, pp. 501–511. Zurich: Trans Tech Publications.

Batten, D. J., and R. J. Dutta. 1997. Ultrastructure of exine of gymnospermous pollen grains from Jurassic and basal Cretaceous deposits in northwest Europe and implications for botanical relationships. *Review of Paleobotany and Palynology* 99: 25–54.

Behrensmeyer, A. K. 1982. Time resolution in fluvial vertebrate assemblages. *Paleobiology* 8: 211–227.

Bell, T. E. 1986. Deposition and diagenesis of the Brushy Basin Member and upper part of the Westwater Canyon Member of the Morrison Formation, San Juan Basin, New Mexico. In C. E. Turner-Peterson and N. S. Fishman (eds.), *A Basin Analysis Case Study: The Morrison Formation, Grants Uranium Region, New Mexico*, pp. 77–91. *American Association of Petroleum Geologists Studies in Geology* 22.

Brenner, G. J. 1963. The spores and pollen of the Potomac Group of Maryland. *Maryland Department of Geology, Mines and Water Resources Bulletin* 27: 1–215.

Burden, E. T., and L. V. Hills. 1989. Illustrated key to genera of Lower Cretaceous terrestrial palynomorphs excluding megaspores of western Canada. *American Association of Stratigraphic Palynologists Contribution Series* 21: 1–146.

Burnham, R. J., S. L. Wing, and G. G. Parker. 1992. The reflection of deciduous
  forest communities in leaf litter: implications for autochthonous litter
  assemblages from the fossil record. *Paleobiology* 18: 30–49.
Chin, K., and J. I. Kirkland. 1998. Probable herbivore coprolites from the upper
  Jurassic Mygatt-Moore Quarry, western Colorado. *Modern Geology* 23:
  249–275.
Chure, D. J., and G. E. Engelmann. 1989. The fauna of the Morrison Formation
  in Dinosaur National Monument. In J. J. Flynn (ed.), *Mesozoic/Cenozoic
  Vertebrate Paleontology: Classic Localities, Contemporary Approaches*, pp.
  8–14. 28th International Geological Congress Field Trip Guidebook T322.
  Washington, D.C.: American Geophysical Union.
Chure, D. J., R. J. Litwin, S. T. Hasiotis, E. Evanoff, and K. Carpenter. 2006.
  The fauna and flora of the Morrison Formation. In J. R. Foster and S. G.
  Lucas (eds.), *Paleonotology and Geology of the Upper Jurassic Morrison
  Formation*, pp. 233–249. *New Mexico Museum of Natural History and
  Science Bulletin* 36.
Collinson, M. E. 1996. "What use are fossil ferns?"—20 years on: with a review of
  the fossil history of extant pteridophyte families and genera. In J. M. Camus,
  M. Gibby, and R. J. Johns (eds.), *Pteridology in Perspective*, pp. 349–394.
  London: Royal Botanic Gardens, Kew.
———. 2002. The ecology of Cainozoic ferns. *Review of Paleobotany and
  Palynology* 119: 51–68.
Condon, S. M., and F. Peterson. 1986. Stratigraphy of Middle and Upper Jurassic
  Rocks of the San Juan Basin: historical perspective, current ideas and
  remaining problems. In C. E. Turner-Peterson and N. S. Fishman (eds.),
  *A Basin Analysis Case Study: The Morrison Formation, Grants Uranium
  Region, New Mexico*, pp. 7–26. *AAPG Studies in Geology* 22, American
  Association of Petroleum Geologists, Tulsa, Oklahoma.
Couper, R. A. 1958. British Mesozoic microspores and pollen grains. A systematic
  and stratigraphic study. *Palaeontographica*, Abt. B, 103: 75–179.
Delevoryas, T. 1959. Investigations of North American cycadeoids: *Monanthesia.
  American Journal of Botany* 46: 657–666.
———. 1960. Investigations of North American cycadeoids: trunks form
  Wyoming. *American Journal of Botany* 47: 778–786.
Demko, T. M., and J. T. Parrish. 1998. Paleoclimatic setting of the Upper Jurassic
  Morrison Formation. *Modern Geology* 22: 283–296.
Demko, T. M., R. F. Dubiel, and J. T. Parrish. 1998. Plant taphonomy in incised
  valleys: implications for interpreting paleoclimate from fossil plants. *Geology*
  26: 1119–1122.
Demko, T. M., B. S. Currie, and K. A. Nicoll. 2004. Regional paleoclimatic and
  stratigraphic implications of paleosols and fluvial/overbank architecture
  in the Morrison Formation, Upper Jurassic, Western Interior, U.S.A.
  Sedimentary Geology 167: 115–135.
Dettmann, M. E. 1963. Upper Mesozoic microfloras from southeastern Australia.
  Proceedings of the Royal Society of Victoria 771: 1–148.
DiMichele, W. A., and R. A. Gastaldo. 2008. Plant paleoecology in deep time.
  Annals of the Missouri Botanical Garden 95: 144–198.
Dodson, P., A. K. Behrensmeyer, R. T. Bakker, and J. S. McIntosh. 1980.
  Taphonomy and paleoecology of the dinosaur beds of the Jurassic Morrison
  Formation. *Paleobiology* 6: 208–232.
Doerhoefer, G., and G. Norris. 1977. Discrimination and correlation of highest
  Jurassic and lowest Cretaceous terrestrial palynofloras in northwestern
  Europe. *Palynology* 1: 79–93.
Dunagan, S. P., and C. E. Turner. 2004. Regional paleohydrologic and
  paleoclimatic settings of wetland/lacustrine depositional systems in the

Morrison Formation (Upper Jurassic), Western Interior, U.S.A. *Sedimentary Geology* 167: 269–296.

Engelmann, G. F., and G. Callison. 1998. Mammalian faunas of the Morrison Formation. *Modern Geology* 23: 343–380.

Engelmann, G., D. J. Chure, and A. R. Fiorillo. 2004. The implications of a dry climate for the paleoecology of the fauna of the Upper Jurassic Morrison Formation. *Sedimentary Geology* 167: 97–308.

Farley, M. B., and D. L. Dilcher. 1986. Correlation between miospores and depositional environments of the Dakota Formation (mid-Cretaceous) of north-central Kansas and adjacent Nebraska, U.S.A. *Palynology* 10: 117–113.

Fensome, R. A. 1983. Miospores from the Jurassic–Cretaceous boundary beds, Aklavik Range, Northwest Territories, Canada. Ph.D. diss., University of Saskatchewan, Canada.

———. 1987. Taxonomy and biostratigraphy of schizaealean spores from the Jurassic–Cretaceous boundary beds of the Aklavik Range, District of Mackenzie. *Paleontologia Canadiana* 4: 1–49.

Filatoff, J. 1975. Jurassic palynology of the Perth Basin, Western Australia. *Palaeontographica*, Abt. B, 154: 1–113.

Foster, J. R. 2007a. *Jurassic West, The Dinosaurs of the Morrison Formation and Their World.* Bloomington: Indiana University Press.

———. 2007b. Taphonomy of the Mygatt-Moore Quarry, a large dinosaur bonebed in the Upper Jurassic Morrison Formation of Western Colorado. *Geological Society of America Abstracts with Programs* 39 (6): 400.

Francis, J. E. 1984. The seasonal environment of the Purbeck (Upper Jurassic) fossil forests. *Palaeogeography, Palaeoclimatology, Palaeoecology* 48: 285–307.

Friis, E. M., K. R. Pedersen, and P. R. Crane. 2006. Cretaceous angiosperm flowers: innovation and evolution in plant reproduction. *Palaeogeography, Palaeoclimatology, Palaeoecology* 232: 251–293.

Furer, L. C., E. P. Kvale, and D. W. Engelhardt. 1997. Early Cretaceous hiatus much longer than previously reported. In E. B. Campen (ed.), *Bighorn Basin: 50 Years on the Frontier—Evolution of the Geology of the Bighorn Basin.* 1997 Field Trip and Symposium. Yellowstone Bighorn Research Association, pp. 47–56.

Gee, C. T. 2006. A whole-plant *Araucaria* from the Late Jurassic Morrison Formation of the Howe Ranch, Wyoming, U.S.A., and implications for sauropod dinosaur feeding ecology. *Abstracts, BOTANY 2006* (Annual meeting of the Botanical Society of America), California State University–Chico, abstract 449. http://www.2006.botanyconference.org/engine/search/index.php?func=detail&aid=449.

Gee, C. T., and W. D. Tidwell. 2010. A mosaic of characters in a new whole-plant *Araucaria, A. delevoryasii* Gee sp. nov., from the Late Jurassic Morrison Formation of Wyoming, U.S.A. In C. T. Gee (ed.), *Plants in Mesozoic Time: Morphological Innovations, Phylogeny, Ecosystems,* pp. 67–94. Bloomington: Indiana University Press.

Hallam, A. 1994. Jurassic climates as inferred from the sedimentary and fossil record. In J. R. L. Allen, B. J. Hoskins, B. W. Sellwood, R. A. Spicer, and P. J. Valdes (eds.), *Palaeoclimates and Their Modeling With Special Reference to the Mesozoic Era,* pp. 79–88. London: Chapman & Hall.

Hotton, C. L. 1986. Palynology of the Morrison Formation [abstract]. *Proceedings of the Fourth North American Paleontological Convention,* p. 20A.

———. 2003. Palynology of the Cretaceous–Tertiary boundary in central Montana: evidence for extraterrestrial impact as a cause of the terminal Cretaceous extinctions. In J. Hartman, K. R. Johnson, and D. J. Nichols (eds.), *The Hell Creek Formation and the Cretaceous–Tertiary Boundary in* the Northern Great Plains—An Integrated Continental Record of the End of the Cretaceous. *Geological Society of America Special Paper* 361: 473–501.

Hummel, J., C. T. Gee, K.-H. Südekum, P. M. Sander, G. Nogge, and M. Clauss. 2008. In vitro digestibility of fern and gymnosperm foliage: implications for sauropod feeding ecology and diet selection. *Proceedings of the Royal Society B* 275: 1015–1021.

Kirkland, J. I. 2006. Fruita Paleontological Area (Upper Jurassic, Morrison Formation), Western Colorado: an example of terrestrial taphofacies. In J. R. Foster and S. G. Lucas (eds.), *Paleontology and Geology of the Upper Jurassic Morrison Formation*, pp. 67–95. Bulletin of the New Mexico Museum of Natural History and Science 36.

Kirkland, J. I., and K. Carpenter. 1994. North America's first pre-Cretaceous ankylosaur (Dinosauria) from the Upper Jurassic Morrison Formation of western Colorado. *Brigham Young University Geological Studies* 40: 25–42.

Klavins, S. D., D. W. Kellogg, M. Krings, E. L. Taylor, and T. N. Taylor. 2005. Coprolites in a Middle Triassic cycad pollen cone: evidence for insect pollination in early cycads? *Evolutionary Ecology Research* 7: 479–488.

Kowallis, B. J., E. H. Christiansen, A. L. Deino, F. Peterson, C. E. Turner, M. J. Kunk, and J. D. Obradovich. 1998. The age of the Morrison Formation. *Modern Geology* 22: 235–260.

Kvale, E. P., G. D. Johnson, D. L. Mickelson, K. Keller, L. C. Furer, and A. W. Archer. 2001. Middle Jurassic (Bajocian and Bathonian) dinosaur megatracksites, Bighorn Basin, Wyoming, U.S.A. *Palaios* 16: 233–254.

Litwin, R. J., C. E. Turner, and F. Peterson. 1998. Palynological evidence on the age of the Morrison Formation, Western Interior U.S. *Modern Geology* 22: 297–319.

Lupia, R., H. Schneider, G. M. Moeser, K. M. Pryer, and P. R. Crane. 2000. Marsileaceae sporocarps and spores from the Late Cretaceous of Georgia, U.S.A. *International Journal of Plant Sciences* 161: 975–988.

Lyons, P. C., and W. C. Darrah. 1989. Earliest conifers of North America upland and/or paleoclimatic indicators. *Palaios* 4: 480–486.

Moore, P. D., J. A. Webb, and M. E. Collinson. 1991. *Pollen Analysis*, 2nd ed. Oxford: Blackwell Scientific Publications.

Norris, G. 1969. Miospores from the Purbeck beds and marine upper Jurassic of southern England. *Palaeontology* 12: 574–620.

———. 1970. Palynology of the Jurassic–Cretaceous boundary in southern England. *Geoscience and Man* 1: 57–65.

Osborn, J. M., and M. L. Taylor. 2010. Pollen and coprolite structure in *Cycadeoidea* (Bennettitales): implications for understanding pollination and mating systems in Mesozoic cycadeoids. In C. T. Gee (ed.), *Plants in Mesozoic Time: Morphological Innovations, Phylogeny, Ecosystems*, pp. 35–49. Bloomington: Indiana University Press.

Page, C. N. 2002. Ecological strategies in fern evolution: a neopteridological overview. *Review of Palaeobotany and Palynology* 119: 1–33.

Parrish, J. T. 1993. Mesozoic climates of the Colorado Plateau. *Museum of Northern Arizona Bulletin* 59: 1–11.

Parrish, J. T., F. Peterson, and C. E. Turner. 2004. Jurassic "savannah"—plant taphonomy and climate of the Morrison Formation Upper Jurassic, Western U.S.A. *Sedimentary Geology* 167: 137–162.

Peterson, F. 1994. Sand dunes, sabkhas, streams and shallow seas: Jurassic paleogeography in the southern part of the Western Interior Basin. In M. V. Caputo, J. A. Peterson, and K. J. Franczyk (eds.), *Mesozoic Systems of the Rocky Mountain Region*, U.S.A., pp. 233–272. Denver: Rocky Mountain Section SEPM (Society for Sedimentary Geology).

Pocock, S. A. J. 1962. Microfloral analysis and age determinations of strata at the Jurassic–Cretaceous boundary in the western Canada plains. *Palaeontographica*, Abt. B, 3: 1–95.

———. 1970. Palynology of the Jurassic sediments of western Canada. Part 1: terrestrial species. *Palaeontographica*, Abt. B, 130: 73–136.

Ricketts, B. D., and A. R. Sweet. 1986. Stratigraphy, sedimentology and palynology of the Kootenay–Blairmore transition in southwestern Alberta and southeastern British Columbia. *Geological Survey of Canada Paper* 84-15: 1–41.

Sander, P. M., C. T. Gee, J. Hummel, and M. Clauss. 2010. Mesozoic plants and dinosaur herbivory. In C. T. Gee (ed.), *Plants in Mesozoic Time: Morphological Innovations, Phylogeny, Ecosystems*, pp. 331–359. Bloomington: Indiana University Press.

Schrank, E. 2004. A gymnosperm pollen not a dinoflagellate: a new combination for *Mendicodinium? quadratum* and description of a new pollen species from the Jurassic of Tanzania. *Review of Paleobotany and Palynology* 131: 301–309.

Shipunov, A. B., and B. B. Sokoloff. 2003. *Schweitzeria*, a new name for *Irania* Schweitzer (fossil Gymnospermae). *Byulleten Moskovnyi o-va Ispytateley Prirody. Otdelat' Biologicheskiy* 108 (5): 89–90. [In English.]

Singh, C. 1964. Microflora of the Lower Cretaceous Manville Group, east-central Alberta. *Alberta Research Council Bulletin* 15: 1–238.

———. 1971. Lower Cretaceous microfloras of the Peace River area, northwestern Alberta. *Alberta Research Council Bulletin* 28: 1–548.

Skog, J. E. 2001. Biogeography of Mesozoic leptosporangiate ferns related to extant ferns. *Brittonia* 53: 236–269.

Spicer, R. A. 1989. The formation and interpretation of plant fossil assemblages. *Advances in Botanical Research* 16: 96–191.

Stockey, R. A., G. W. Rothwell, and S. A. Little. 2006. Relationships among fossil and living Dipteridaceae: anatomically preserved *Hausmannia* from the Lower Cretaceous of Vancouver Island. *International Journal of Plant Sciences* 167: 649–663.

Suess, H., and S. Schultka. 2001. First record of *Glyptostroboxylon* from the Upper Jurassic of Tendaguru, Tanzania. *Botanical Journal of the Linnean Society* 135: 421–429.

———. 2006. Koniferenhoelzer (Fusite) aus dem Oberjura vom Tendaguru (Tansania, Ostafrika). *Palaeontographica*, Abt. B, 275: 133–165.

Tidwell, W. D. 1990. Preliminary report on the megafossil flora of the Upper Jurassic Morrison Formation. *Hunteria* 2: 1–12.

Tidwell, W. D., and D. A. Medlyn. 1992. Short shoots from the Upper Jurassic Morrison Formation, Utah, Wyoming, and Colorado, U.S.A. *Review of Palaeobotany and Palynology* 71: 219–238.

———. 1993. Conifer wood from the Upper Jurassic of Utah, U.S.A.—part II: *Araucarioxylon boodii* sp. nov. *The Palaeobotanist* 42: 70–77.

Tidwell, W. D., B. B. Britt, and S. R. Ash. 1998. Preliminary floral analysis of the Mygatt-Moore Quarry in the Jurassic Morrison Formation, west-central Colorado. *Modern Geology* 22: 341–378.

Traverse, A. 2007. *Paleopalynology*, 2nd ed. Dordrecht: Springer.

Trujillo, K. C. 2003. Stratigraphy and correlation of the Morrison Formation Late Jurassic–Early Cretaceous of the Western Interior, U.S.A., with emphasis on southeastern Wyoming. Ph.D. diss., University of Wyoming, Laramie.

Tschudy, R. H., B. D. Tschudy, S. D. van Loenen, and G. Doher. 1980. Illustrations of plant microfossils from the Morrison Formation I. Plant microfossils from the Brushy Basin Member. *Open File Report*, U.S. Geological Survey, report no. 81-35.

Tschudy, R. H., B. D. Tschudy, and S. D. van Loenen. 1981. Illustrations of plant microfossils from the Morrison Formation II. Plant microfossils from the Westwater Canyon Member. *Open File Report, U.S. Geological Survey*, report no. 81-1154.

―――. 1988a. Illustrations of plant microfossils from the Morrison Formation III. Plant microfossils from the Recapture Member. *Open File Report, U.S. Geological Survey*, report no. 88-234.

―――. 1988b. Illustrations of plant microfossils from the Morrison Formation IV. Plant microfossils from the Salt Wash Member. *Open File Report, U.S. Geological Survey*, report no. 88-235.

Turner, C. E., and N. S. Fishman. 1991. Jurassic Lake T'oo'dichi'. A large alkaline, saline lake, Morrison Formation, eastern Colorado Plateau. *Geological Society of America Bulletin* 103: 538–558.

Turner, C. E., and F. Peterson. 2004. Reconstruction of the Upper Jurassic Morrison Formation extinct ecosystem—a synthesis. *Sedimentary Geology* 167: 309–335.

Turner-Peterson, C. E. 1986. Fluvial sedimentology of a major uranium-bearing sandstone—a study of the Westwater Canyon Member of the Morrison Formation, San Juan Basin, New Mexico. In C. E. Turner-Peterson and N. S. Fishman (eds.), *A Basin Analysis Case Study: The Morrison Formation, Grants Uranium Region, New Mexico*, pp. 47–75. AAPG Studies in Geology 22.

Turner-Peterson, C. E., L. C. Gundersen, D. D. Francis, and W. M. Aubrey. 1980. Fluvio-lacustrine sequences in the Upper Jurassic Morrison Formation and the relationship of facies to tabular uranium ore deposits in the Poison Canyon area, Grants Mineral Belt, New Mexico. In C. E. Turner-Peterson (ed.), *Uranium in Sedimentary Rocks: Application of the Facies Concept to Exploration*, pp. 177–211. SEPM Rocky Mountain Section, Short Course Notes.

Tyler, N., and F. G. Ethridge. 1983. Depositional setting of the Salt Wash Member of the Morrison Formation, southwest Colorado. *Journal of Sedimentary Petrology* 53: 7–82.

Vakhrameev, V. A. 1991. *Jurassic and Cretaceous Floras and Climates of the Earth.* Cambridge: Cambridge University Press.

van Konijnenburg-van Cittert, J. H. A. 1971. In situ gymnosperm pollen from the Middle Jurassic of Yorkshire. *Acta Botanica Neerlandica* 20: 1–96.

―――. 1989. Dicksoniaceous spores in situ from the Jurassic of Yorkshire, England. *Review of Palaeobotany and Palynology* 61: 272–302.

―――. 1991. Diversification of spores in fossil and extant Schizaeaceae. In S. Blackmore and S. H. Barnes (eds.), *Pollen and Spores*, pp. 103–118. Oxford: Clarendon Press.

―――. 2002. Ecology of some Late Triassic to Early Cretaceous ferns in Eurasia. *Review of Palaeobotany and Palynology* 119: 113–124.

van Konijnenburg-van Cittert, J. H. A., and J. van der Burgh. 1996. Review of the Kimmeridgian flora of Sutherland, Scotland, with reference to the ecology and in situ pollen and spores. *Proceedings of the Geological Association* 107: 97–105.

Watson, J., and K. L. Alvin. 1996. An English Wealden floral list, with comments on possible environmental indicators. *Cretaceous Research* 17: 5–26.

Wing, S. L., L. J. Hickey, and C. C. Swisher. 1993. Implications of an exceptional fossil flora for Late Cretaceous vegetation. *Nature* 363: 342–344.

Appendix 13.1. List of Morphospecies Present (X) in Each Sample

Species	NM	AZ	FR 1	FR 2	FR 3	MYG	DNM	WY
Approximate sample size	960	960	1020	1000	950	950	1100	1060
BRYOPHYTES								
Notothyladaceae								
Nevesisporites vallatus de Jersey et Paten			X	X	X			
Nevesisporites spp. complex			X	X	X	X	X	X
Sphagnaceae								
Stereisporites cf. *S. antiquasporites* (Wilson et Webster) Dettmann						X	X	
Stereisporites sp.							X	
?BRYOPHYTES								
"*Retialetes*" sp.	X						X	X
LYCOPODIOPHYTA								
Lycopodiales								
Ceratosporites rotundiformis (Kara-Murza) Pocock					X	X		
Ceratosporites sp.								X
Lycopodiacidites cerniidites (Ross) Brenner			X*					
Lycopodiumsporites sp. cf. *L. pseudoreticulatus* (Couper) Evans		X						X
Retitriletes austroclavatidites (Cookson) Doering, Krutzsch, Mai et Schulz		X	X*	X	X		X	
Retitriletes sp. cf. *R. circolumenus* (Cookson et Dettmann) Backhouse					X			
Staplinisporites caminus (Balme) Pocock	X				X			
Isoetales								
?Isoetaceae microspores	X		X	X	X	X	X	
Selaginellales								
Heliosporites sp. (sensu Norris 1969)	X*							
Heliosporites sp. T	X	X	X	X			X	X
Heliosporites sp. H		X						X
?Heliosporites	X				X		X	X

Species	NM	AZ	FR 1	FR 2	FR 3	MYG	DNM	WY
SPHENOPHYTA								
Calamospora mesozoica Couper		X						
FILICALES								
Cyatheaceae/Dicksoniaceae								
Cyathidites minor Couper	X*	X	X	X	X	X	X	X
Dipteridaceae								
Cibotiumspora jurienensis (Balme) Filatoff		X		X	X	X	X	X
Dictyophyllidites equiexinus (Couper) Dettmann	X*		X		X		X	X
Dictyophyllidites harrisii Couper								X
Gleicheniaceae								
Gleicheniidites senonicus Ross			X*		X			X
Matoniaceae								
?Matonisporites sp.						X		
Osmundaceae								
Baculatisporites comaumensis (Cookson) Potonié								X
Osmundacidites wellmanii Couper		X*					X	X
Todisporites minor Couper			X*		X	X		X
?Polypodiaceae								
Polypodiidites sp.				X				
Pteridaceae								
Contignisporites glebulentus Dettmann								X
Crassitudisporites problematicus (Couper) Hiltmann								X

Appendix 13.1 (continued).

Species	NM	AZ	FR 1	FR 2	FR 3	MYG	DNM	WY
Schizaeaceae								
Cicatricosisporites sp.							X	X
Concavissimisporites sp. cf. C. crassatus (Delcourt et Sprumont) Delcourt, Dettmann et Hughes	X*		X*				X	X
Concavissimisporites montuosus (Doering) Fensome						X	X	X
Ischyosporites sp. cf. I. disjunctus Singh							X	X
Ischyosporites sp. cf. I. pseudoreticulatus (Couper) Doering		X*			X		X	
Pilosisporites sp.								X
FILICALES INCERTAE SEDIS								
Biretisporites potoniei Delcourt et Sprumont	X*					X		
Concavissimisporites sp. cf. C. punctatus (Delcourt et Sprumont) Brenner	X*					X	X	
Crybelosporites vectensis Kemp	X	X	X	X	X	X	X	X
Deltoidospora gradata (Malyavkina) Pocock					1		X	
Foveotriletes sp. cf. F. parviretus (Balme) Dettmann		X*						
Granulatisporites sp. cf. G. infimus Cornet et Traverse					X			X
Leptolepidites eastendensis Pocock		X						
Leptolepidites sp. cf. L. epacrornatus Norris		X						X
Leptolepidites proxigranulatus (Brenner) Doerhoefer	X*	X	X			X	X	X
Leptolepidites verrucatus Couper		X	X				X	
Leptolepidites sp. WY1								X
Neoraistrickia sp. AZ	X*	X*						
Obtusisporites canadensis Pocock	X*		X*			X		
Tripartina sp. cf. T. variabilis Malyavkina		X	X*					
Undulatisporites undulapolus Brenner		X				X	X	
Undulatisporites sp.							X	
Verrucosisporites densus (Bolkh.) Pocock							X	
Verrucosisporites sp. HT			X*		X		X	
Verrucate/echinate trilete spore							X	

Appendix 13.1 (continued).

Species	NM	AZ	FR 1	FR 2	FR 3	MYG	DNM	WY
SEED PLANTS								
Coniferales								
Araucariaceae								
Araucariacites australis Cookson	X	X	X	X	X		X	X
Callialasporites dampieri (Balme) Dev	X		X				X	X
Callialasporites sp. cf. C. dampieri		X			X		X	
Callialasporites microvelatus Schulz		X			X		X	
Callialasporites sp. cf. C. microvelatus		X		X				
Callialasporites sp. cf. C. segmentatus (Balme) S.K. Srivastava	X	X	X		X	X	X	X
Callialasporites trilobatus (Balme) Dev		X	X		X		X	
Callialasporites turbatus (Balme) Schulz	X	X	X	X	X	X	X	X
Callialasporites sp. cf. C. turbatus					X			
Callialasporites sp. indet.	X					X	X	X
Cheirolepidiaceae								
Classopollis classoides Pflug	X	X	X	X		X	X	X
Classopollis itunensis Pocock	X	X						
Classopollis minor Pocock						X		
Cupressaceae								
Exesipollenites tumulus Balme	X	X	X	X	X	X	X	X
Perinopollenites elatoides Couper	X	X		X			X	X
Inaperturopollenites dubius (Potonié et Venitz) Thomson et Pflug		X		X				X
Inaperturopollenites spp. indet.		X	X	X			X	X
Podocarpaceae								
Microcachryidites sp.			X*					
Parvisaccites rugulatus Brenner	X				X	X	X	X
Podocarpidites sp. cf. P. biformis Rouse	X			X	X	X	X	X
Podocarpidites convolutus Pocock	X	X*						
Podocarpidites sp. cf. P. ellipticus	X	X	X		X	X	X	X
Podocarpidites epistratus Brenner	X	X	X	X	X	X	X	X

Appendix 13.1 (continued).

Species	NM	AZ	FR 1	FR 2	FR 3	MYG	DNM	WY
Podocarpidites sp. cf. *P. epistratus*	X		X	X	X	X	X	X
Podocarpidites sp. cf. *P. granulatus* Singh			X*					
Podocarpidites sp. cf. *P. herbstii* Berger		X						
Podocarpidites sp. cf. *P. multesimus* (Bolkh.) Pocock	X	X*						X
Podocarpidites naumovii Singh								X
Podocarpidites sp. cf. *P. ornatus* Pocock	X	X		X		X		X
"*Podocarpidites*" *rousei* Pocock	X*							X
Podocarpidites sp. FR3					X			
Pristinuspollenites sp. M15	X						X	X
Pristinuspollenites sp. M16	X	X			X	X	X	X
Bisaccates incertae sedis (mostly Pinaceae?)								
Alisporites bilateralis Rouse fide Singh	X	X	X	X	X	X	X	X
Alisporites grandis (Cookson) Dettmann	X	X	X	X	X	X	X	X
Alisporites microsaccus (Couper) Pocock	X				X	X	X	X
Alisporites sp. cf. *A. similis* (Balme) Dettmann sensu Pocock (1970)	X	X	X	X		X	X	X
Alisporites sp. M01		X	X	X	X	X	X	X
Alisporites sp. L								
Cedripites sp. cf. *C. canadensis* Pocock	X	X*	X*		X	X	X	X
Parvisaccites sp. M06		X		X		X	X	
Piceispollenites sp. E	X	X	X		X	X	X	X
Pityosporites divulgatus (Bolkh.1956) Pocock	X	X	X	X				X
Pityosporites sp. M27	X		X		X	X	X	
Protoconiferus funarius Pocock	X	X	X			X	X	
Protopicea exilioides Pocock					X	X	X	X
Pseudowalchia ovalis Pocock					X	X	X	X
Pseudowalchia sp.						X	X	
Vitreisporites jurassicus Pocock								X
Walchites minor Pocock	X	X	X	X	X	X	X	

Appendix 13.1 (continued).

Species	NM	AZ	FR 1	FR 2	FR 3	MYG	DNM	WY
Coniferales incertae sedis								
Cerebropollenites mesozoicus (Couper) Nilsson	X	X	X		X	X	X	X
Bennettitales/Cycadales/Ginkgoales								
Cycadopites nitidus Balme	X	X	X	X				X
Cycadopites follicularis Wilson et Webster		X	X				X	
Shanbeipollenites quadratus (Kumar) Schrank								X
Caytoniales								
Vitreisporites pallidus (Reissinger) Nilsson		X*	X	X			X	
Gnetales								
Equisetosporites sp. cf. E. multicostatus (Brenner) Norris	X		X					
Equisetosporites sp. cf. E. volutus (Stanley) Farabee et Canright					X			
Equisetosporites spp. indet.	X		X		X			
Eucommidites troedssonii			X*					

Note. Locality abbreviations: NM = Goat Mountain, New Mexico; AZ = Carrizo Mountains, Arizona; FR 1 = Fruita Paleontological Area 1, Colorado; FR 2 = Fruita Paleontological Area 2, Colorado; FR 3 = Fruita Paleontological Area 3, Colorado; MYG = Mygatt-Moore Quarry, Colorado; DNM = DNM96, Dinosaur National Monument, Utah; WY = Red Gulch dinosaur track site, Wyoming.

* Species found in a higher-specimen count (Hotton 1986) and mentioned in text, but not included in analyses in this study.

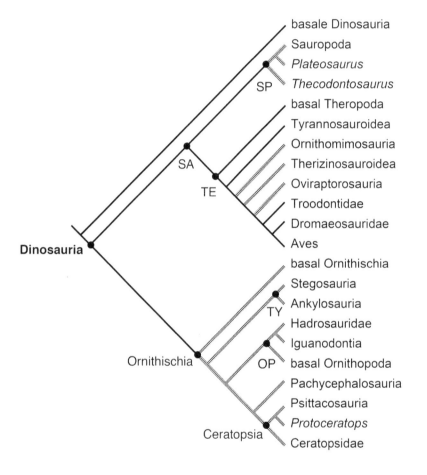

Fig. 14.1. Relationships of the major groups of dinosaurs. Bold lines indicate evolution of herbivory. Abbreviations of higher taxa: OP = Ornithopoda; SA = Saurischia; SP = Sauropodomorpha; TE = Theropoda; TY = Thyreophora.

MESOZOIC PLANTS AND DINOSAUR HERBIVORY

14

P. Martin Sander, Carole T. Gee, Jürgen Hummel, and Marcus Clauss

For most of their existence, herbivorous dinosaurs fed on a gymnosperm-dominated flora. Starting from a simple reptilian herbivory, ornithischian dinosaurs evolved complex chewing dentitions and mechanisms, while sauropodomorph dinosaurs retained the primitive condition of not chewing. Some advanced theropod dinosaurs evolved a bird-type herbivory with a toothless beak and a gastric mill. Dinosaur digestive tract remains, coprolites, and other trace fossils offer little evidence for dinosaur food preferences. Ferns, seed ferns, ginkgoes, and the Cheirolepidiaceae were previously viewed as the food plants favored by dinosaurs. However, animal nutrition science, comparative herbivore physiology, and digestive tract anatomy of modern herbivores suggest otherwise. In fermentation experiments, energy release is much greater in horsetails and most conifers, especially *Araucaria*, than in many ferns, cycads, and podocarp conifers. Sauropods as bulk feeders must have relied on plants that provided much biomass and regenerated foliage quickly, namely, conifers and ginkgoes. This argues against most ferns and cycads, which offer little biomass and energy.

During the Mesozoic, the flowering plants began their dominance of the global vegetation around the start of the Late Cretaceous, but the dinosaurs arose in the Late Triassic. Thus, for over 100 million years, herbivorous dinosaurs—the predominant terrestrial herbivores—fed on a flora consisting primarily of gymnosperms such as conifers, seed ferns, cycads, benettitaleans, and ginkgophytes. As a matter of necessity, dinosaurs evolved a variety of adaptations to feed on this preangiosperm flora. In light of new developments in our knowledge of dinosaur herbivory in the last 20 years, we present here an updated survey of evidence pertaining to the food ecology of dinosaurs and the plants that they fed on.

The morphology and anatomy of Mesozoic plants and the phytochemistry of living relatives have traditionally been used to infer herbivorous dinosaur dietary preferences. However, in the meantime, not only has new fossil evidence in the form of digestive tract contents, coprolites, and gastric mills been discovered, but also innovative studies on the inferred digestibility of the Mesozoic flora, tooth enamel microstructure, and allometric and scaling effects in the herbivore digestive tract offer new perspectives on dinosaur plant eaters and their food preferences.

Introduction

In this chapter, we synthesize this information to arrive at a more comprehensive, current, and better-supported understanding of dinosaurian herbivory.

Because it is impossible to conduct empirical feeding experiments on nonavian dinosaurs, all dietary information must come either from direct or indirect fossil evidence, such as "stomach" contents or coprolites, or through inferences based on indirect fossil evidence and data derived from field or laboratory experiments on birds or living representatives of the Mesozoic plant groups. Similarly, the functional analysis of dinosaur skeletons and dentitions in regard to herbivory (see recent summaries by Sues 2000) has contributed much in the way of indirect data.

Skeletal and Dental Evidence

Skeletal and dental evidence for dinosaur herbivory and particularly dietary preferences has received considerable attention by vertebrate paleontologists (for reviews, see Weishampel 1984; Galton 1986; Weishampel and Norman 1989; Weishampel and Jianu 2000; Fastovsky and Smith 2004; Weishampel et al. 2004; Fastovsky and Weishampel 2005). Herbivory and its specific expression among the dinosaurs is dependent to a major extent on phylogenetic relationships (Fig. 14.1). The sister group of the dinosaurs was carnivorous, while most dinosaurs are herbivorous. The entire clade Ornithischia, for example, is herbivorous. The Saurischia, with its two major clades, the Sauropodomorpha and the Theropoda, are both herbivorous and carnivorous (Fig. 14.1). Although all sauropodomorphs appear to have been plant eaters, theropods retained the plesiomorphic feeding habit of carnivory (Figs. 14.1, 14.2), with the exception of some derived forms on the line to birds and among birds themselves.

Sauropodomorphs

In the basal sauropodomorphs ("prosauropods") of the Late Triassic and Early Jurassic, the dentition consists of numerous, coarsely denticulate, leaf-shaped teeth (Fig. 14.2). This dentition is the generalized type that occurs in many groups of fossil reptiles and in modern herbivorous reptiles (herbivorous lizards) and indicates that the animals simply bit or stripped leaves from branches but did not chew them (Galton 1986; Barrett 2000).

The basal sauropodomorphs are the first high browsers in the fossil record that were able to reach plant parts higher than 1 m off the ground (Galton 1986, 1990). The lack of oral processing (e.g., chewing) was retained by the later sauropods, which achieved gigantic body sizes. Although there is considerable diversity in sauropod tooth morphology (Fig. 14.2), there is no evidence that any sauropods possessed fleshy cheeks or chewed their food (Upchurch and Barrett 2000; Barrett and Upchurch 2005). The large particles of plant tissue that resulted from this type of ingestion must have been fermented in specialized regions of the gut, probably in the hindgut (Farlow 1987; Hummel et al. 2008b). This large

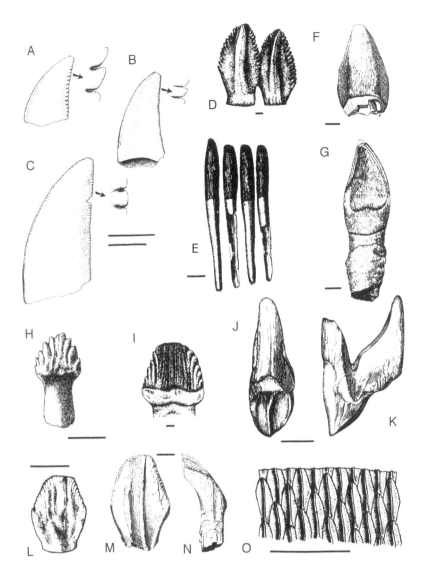

Fig. 14.2. Tooth morphology is the best indicator of dietary preference among dinosaurs. (A–C) Carnivore teeth are laterally compressed, recurved cones. They bear anterior and posterior cutting edges with fine serrations. (A) *Troodon*. (B) *Saurornitholestes*. (C) Plesiomorphic large theropod. (D–G) Sauropodomorph teeth. (D) *Plateosaurus* teeth are of the generalized reptilian herbivore type. (E) Sauropod teeth range from slender and pencil shaped (*Diplodocus*) to more massive and spatula shaped, with obvious wear facets (F, G, the brachiosaurid *Oplosaurus*). (H, I) Armored dinosaur (thyreophoran) teeth are simple structures with coarse denticles. (H) The ankylosaurus *Edmontonia*. (I) The stegosaur *Stegosaurus*. (J, K) Isolated upper tooth of *Triceratops* in lateral (J) and anterior (K) views. This tooth would have been part of a tooth battery and would have experienced heavy wear. (L–O) Ornithopod teeth all show extensive wear and in advanced forms are tightly integrated into tooth batteries. (L) *Dryosaurus*. (M) Isolated lower tooth of *Iguanodon* in medial view and (N) in lateral view. (O) Part of the tooth battery in a hadrosaur lower jaw in medial view. Shorter scale bar for A–C (entire teeth) = 5 mm; longer scale bar for A–C (close-ups of serrations) = 1 mm. Scale bar = 1 mm (D), 1 cm (E–K, M, N), 5 mm (L), 5 cm (O). *From Sander (1997)*.

particle size was offset by the large retention times that were possible in very large animals (Hummel et al. 2008a; Hummel and Clauss, in press; Clauss et al., in press).

Recent work by Wings and Sander (2007) indicates that it is unlikely that sauropods had a gastric mill. Although gastroliths are occasionally found with sauropod skeletons, comparison with modern birds with gastric mills, such as ostriches, suggests that the number of gastroliths associated with these sauropod individuals would have been wholly insufficient in number. Furthermore, the high polish found on sauropod gastroliths is inconsistent with the abrasive environment of a gastric mill (Wings and Sander 2007).

It should also be noted that any kind of mechanism leading to the comminution of plant tissue, by either oral processing or a gastric mill, would have limited the rate of food intake (Hummel and Clauss, in press). Limits on the rate of food intake, in turn, would have been disadvantageous to the evolution of gigantic body size of sauropods (Sander and Clauss 2008). This is suggested by the extraordinary amount of time modern elephants spend on feeding, although they do digest their food rather inefficiently (Clauss and Hummel 2005).

One of the most prominent characters of sauropods is their extremely long neck. It is thought that this long neck evolved as an adaptation for high browsing in trees and for increasing lateral feeding range (Dodson 1990; Christian and Dzemski 2007; Dzemski and Christian 2007). The long neck must have allowed sauropods to exploit the full spectrum of plant heights, from low-growing plants to midsized shrubs to large trees.

However, neck posture varied in sauropods, and there are opposing views on how some sauropod groups such as the diplodocids and brachiosaurids were able to hold their long necks. Some argue for a head position from low to medium height even in *Brachiosaurus* on the basis of modeling of the mobility of the joints in the neck (e.g., Stevens and Parrish 2005) and cardiovascular constraints (Seymour and Lillywhite 2000). Others argue, on the basis of the comparison with recent long-necked herbivores (camels and ostriches), that a wide range of movements and a habitually more erect neck posture was possible (Dzemski and Christian 2007; Christian and Dzemski 2007; Taylor et al. 2009). This would have resulted in a larger, three-dimensional "feeding envelope" allowing the animals to take up more energy from a single feeding station without moving the body.

Theropods

The theropods are generally known as meat-eating dinosaurs, but herbivory evolved repeatedly in nonavian theropods and in birds (Figs. 14.1, 14.2). This includes the enigmatic therizinosaurs (Weishampel and Jianu 2000; Clark et al. 2004), which have jaws, dentitions, and teeth that are similar in shape to those of prosauropods, suggesting a similar diet and feeding style. Two other lineages of derived theropods, the ornithomimosaurs and oviraptorosaurs, were toothless (or nearly so). The small head on the long flexible neck in the ornithomimosaurs would have been suitable for feeding on small vertebrates and invertebrates, as well as on plant matter (Kobayashi et al. 1999; Ji et al. 2003; Makovicky et al. 2004; Osmólska et al. 2004; Barrett 2005). However, some ornithomimosaurs clearly were filter feeders (Makovicky et al. 2004).

The jaws of advanced oviraptorosaurs appear adapted to high bite forces and were possibly used to feed on hard-shelled seeds, not on eggs as their name would suggest. The eggs on which the oviraptorosaurs were believed to be feeding turned out to be their own eggs (Norell et al. 1994). However, a skeleton of a young dromaeosaurid in one of these nests may indicate at least occasional carnivory in oviraptorosaurs (Norell et al. 1994). Similarly, tall and short jaws, like those in oviraptorosaurs, have

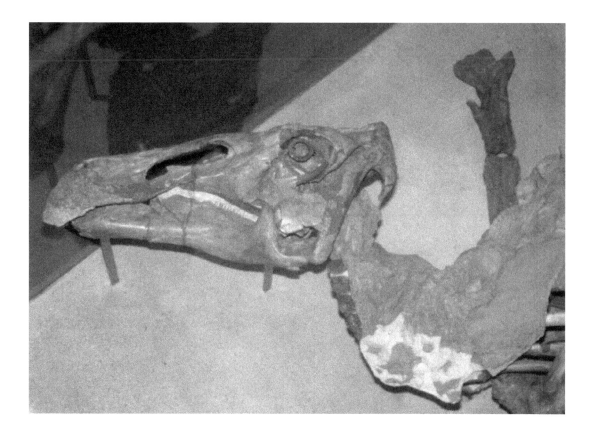

Fig. 14.3. Skull of a mummified carcass of *Edmontosaurus annectens* from the Late Cretaceous of Wyoming. Note the well-preserved, broad beak separated by a gap (diastema) from the inset tooth row. This carcass once contained matted plant remains in its abdominal region; this specimen is on display at the Senckenberg Museum in Frankfurt, Germany. *Photo by P.M.S.*

been observed in the dromornithid birds of the late Tertiary and Pleistocene of Australia, which are primarily interpreted as herbivores, but also as scavengers (Murray and Vickers-Rich 2004). Primitive oviraptorosaurs such as *Caudipteryx* show evidence of a well-developed gastric mill. The mass and shape of the gastric mill found with all known skeletons of *Caudipteryx* closely resembles that of modern granivorous birds (Wings and Sander 2007) and present strong evidence for herbivory.

Ornithischians

In the Ornithischia, the first adaption to herbivory was a horny beak, a so-called rhamphotheca; most ornithischians had lost their anterior teeth and replaced them with this horny beak (Fig. 14.3). Concurrent with this adaptation, the tooth rows were set inward from the margin of the jaw, which implies the development of muscular cheeks (Galton 1986; Weishampel and Norman 1989; Weishampel and Jianu 2000). This primitive pattern was retained by the armored dinosaurs (*Thyreophora*), the stegosaurs and ankylosaurs, which must have fed low to the ground. This would have meant a diet of low-growing ferns and fern allies (Weishampel and Norman 1989; Weishampel and Jianu 2000).

Basal ornithischians such as *Lesothosaurus* had fairly simple, denticulate teeth that show only limited wear, while the ornithopodan and

Fig. 14.4. Enamel microstructure of dinosaur teeth as seen by the scanning electron microscope. The outer enamel surface is at or beyond the top of all images except in (B) and (G). (A, B) Carnivore (indeterminate tyrannosaurid) enamel showing well-organized columnar enamel in (A) longitudinal and (B) tangential section. (C–E) Enamel of nonchewing herbivores such as that of the sauropodomorphs *Plateosaurus* (C, longitudinal section) and *Diplodocus* (D, cross section) and the ankylosaur *Palaeoscincus* (E, longitudinal section) shows a rather poorly organized microstructure. (F–H) The most complex enamel microstructures are seen in chewing ornithischian dinosaurs such as the advanced ornithopod *Iguanodon* (F, longitudinal section) and an indeterminate ceratopsid (H, oblique section cut from the inside of the tooth). (G) Globular micromorphology of the hadrosaur enamel surface. Scale bars = 10 μm (A–E), 30 μm (F–H). *From Sander (1997).*

ceratopsian clades both independently evolved grinding teeth (Fig. 14.2) (Norman and Weishampel 1985, 1991; Weishampel and Norman 1989; Sander 1997; Weishampel and Jianu 2000). A gradual increase in mastication is well documented by the evolution of ever more complex tooth batteries in these groups, especially in the ornithopods. The most elaborate of these tooth batteries is seen in the most derived ornithopods, the hadrosaurs (Figs. 14.2, 14.3). On the basis of their body size and facultatively bipedal stance, large ornithopods such as iguanodontians and hadrosaurs would have been able to feed on plant parts growing several meters above the ground (Weishampel and Norman 1989; Tiffney 1997; Weishampel and Jianu 2000).

In ornithopod dinosaurs, a gradual increase in body size accompanied the increasing complexity of tooth batteries and led to elongate horselike skulls in the larger ornithischians (Fig. 14.3) (Norman and Weishampel 2004). An equivalent trend occurred later in mammalian herbivores, in particular in the horses (MacFadden 1994). The explanation behind the evolutionary coupling of a more complex tooth battery and an elongate skull is allometric growth, because the grinding surface of the teeth increases by the second power while the volume of the animal's body size increases by the third power. Although metabolic rate increases by two-thirds or three-fourths power (Clauss et al. 2008), this still results in a strongly positive allometry of the dental and mastication process.

In the ceratopsian lineage, a rather similar evolutionary trend is evident (Weishampel and Norman 1989; Weishampel and Jianu 2000; You and Dodson 2004). Even the earliest ceratopsians show occlusion (secondary wear facets). In the psittatosaurus (small bipedal dinosaurs from the Lower Cretaceous of China), the development of a gastric mill aided the tooth batteries in the comminution of plant material (Wings and Sander 2007). There is no evidence in the ceratopsians proper for a gastric mill, but elaborate tooth batteries are present in neoceratopsians (Dodson et al. 2004), which similarly resulted in extremely large skulls in the very large forms. Even the large neoceratopsians must been low browsers because of their obligatory quadrupedal stance, short forelegs, and short neck.

In both the ornithopods and the derived ceratopsians, the tooth battery shows a single occlusal surface, which was kept fully functional by the continual replacement of teeth (Sander 1997). In hadrosaurs, up to four replacement teeth were present at any one time in a single tooth position (Fig. 14.2). Their solution to the problem of abrasive plant material is different from those found by the mammals; the mammals evolved high tooth crowns in their cheek teeth, as exemplified by the horses, or ever-growing teeth, such as in a few rodents and some South American ungulates (Rensberger 2000).

However, hadrosaurs differ from neoceratopsians in their specific mechanism of mastication, which led early on to the recognition that the elaborate tooth batteries of the two groups evolved convergently (Ostrom

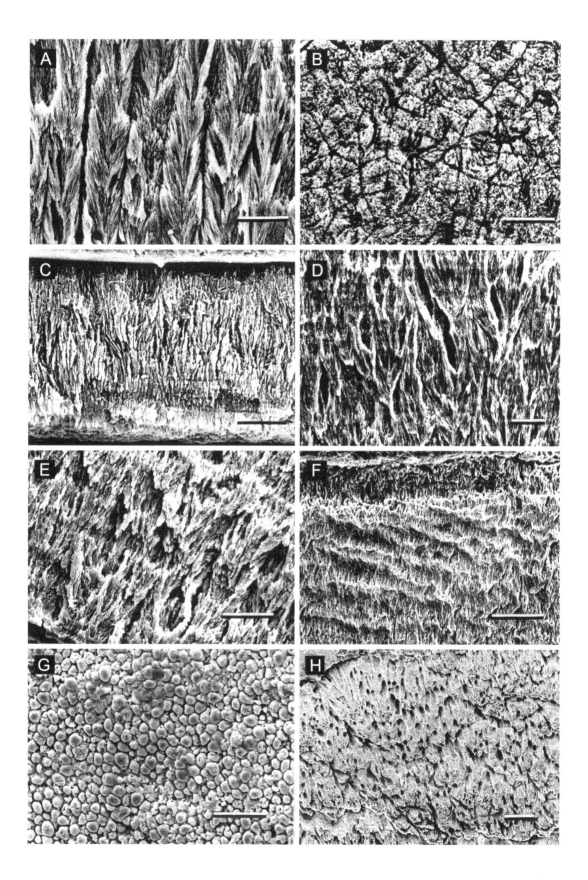

1964a, 1964b). In hadrosaurs, grinding of food in the tooth battery was achieved by a slight rotation of the jaw elements around their long axis (Norman and Weishampel 1985). Specifically, the grinding surface of the lower jaw faced upward and outward and that of the lower jaw downward and inward, and the upper tooth-bearing bone, the maxilla, was not tightly connected to the rest of the skull. Thus, when the jaws closed, it swung outward and produced a grinding movement. The grinding surfaces of the neoceratopsians, on the other hand, stood nearly vertical and sheared past each other as the jaws closed (Ostrom 1964a; Sander 1997; Dodson et al. 2004).

Evidence from Tooth Enamel Microstructure

The microstructure of dinosaurian tooth enamel has been extensively studied with the scanning electron microscope, and distinctive differences among carnivores, nonchewing herbivores, and chewing herbivores are apparent (Sander 1997, 1999, 2000; Hwang 2005). The recurved, laterally flattened teeth of theropod dinosaurs have well-organized columnar enamel (Fig. 14.4A, B), while those sauropodomorph teeth studied (e.g., *Plateosaurus, Diplodocus*) have poorly organized columnar enamel that is thin (Sander 1999; Hwang 2005).

Both advanced ornithopods (iguanodontians and hadrosaurs) and ceratopsians show a more complex enamel microstructure (Fig. 14.4F, H) and unusual globular enamel surface morphologies (Fig. 14.4G) that are correlated with their elaborate tooth batteries. Differences in enamel microstructure and surface morphology between the two taxa are consistent with the convergent origin of ornithopod and ceratopsian tooth batteries. The wavy enamel of advanced ornithopods seems to have evolved to increase abrasion resistance (Sander 1997, 1999, 2000), while the functional significance of ceratopsid columnar enamel is not clear.

Evidence from Digestive Tract Remains

Reports of fossil digestive tract remains, or "stomach" contents, are not uncommon in the literature, although most of them tend to be descriptive or anecdotal in nature.

Sauropods

In the Late Jurassic Morrison Formation of North America, plant remains were found in the abdominal region of a sauropod that contained "an amount of carbonaceous material . . . old stems, bits of leaves, other plant matter" during Barnum Brown's well-known excavation of the Howe Quarry in north-central Wyoming in 1934 (Bird 1985: 63). Judging from this account, it seems improbable that this occurrence was looked at as anything more than a curiosity, or that any of the fossil plant material was conserved or formally studied.

Fig. 14.5. The fossiliferous deposit (arrow) from which Stokes's (1964) "fossilized stomach contents of a sauropod dinosaur" came; site relocated by S. R. Ash and W. D. Tidwell. *Photo courtesy of S. R. Ash.*

Putative stomach contents were noted again in the Morrison Formation on the Howe Ranch nearly half a century later. During a second set of excavations by Hans-Jakob Siber in the original Howe Quarry in 1990–1991 (cf. Gee and Tidwell 2010), patches of compressed coalified plant matter were observed among the sauropod bones and interpreted as stomach contents (Michelis 2004). One of these patches of plant material contained a single smooth pebble that was thought to represent a gastrolith. This occurrence of stomach contents was not, however, found in the abdominal cavity of a skeleton, as in the case in 1934, or even in association with a fossil carcass because the Howe Quarry was nearly played out in 1990 and only isolated ribs and vertebrae were found. Incidentally, Brown, in his 1934 dig, had found a group of similar pebbles that he took to be gastroliths, but these stomach stones were not associated with any fossil plants (Bird 1985; Michelis 2004). As with 1934 dig, the 1990–1991 putative stomach content remains from Howe Quarry have not received any formal study, although a short description is provided by Michelis (2004).

In the Brushy Basin Member of the Morrison Formation in Utah, a 40-kg heavy, fossiliferous mass in the shape of a flattened ovoid was heralded as the fossilized stomach contents of a sauropod (Stokes 1964). The rock mass contained short lengths of woody branches and bone and tooth fragments, and was found among the dispersed skeletal remains of a large sauropod. This was sensational news that made it into *Science*, because it pointed to an omnivorous food habit in sauropods.

In May 1992, the original collecting site was relocated by Sidney Ash and Don Tidwell with help of one of the original discoverers. They found that the source of the ovoid mass was a fossiliferous, elongate lens of predominately calcareous sediment about 30 m in length (Fig. 14.5). This deposit contained a mix of plant material, small teeth, and bones (Fig. 14.6A, B; Plate 15A, B) similar to that described from the supposed stomach contents. Larger bones, such as sauropod vertebrae, were also scattered along the ledge where the deposit cropped out (W. D. Tidwell, personal communication to C.T.G., 2008). The fossil plants consisted primarily of 3–4-cm-long stems of the enigmatic gymnosperm

Fig. 14.6. A sample of Stokes's (1964) "fossilized stomach contents of a sauropod dinosaur" collected by S. R. Ash and W. D. Tidwell. (A) Top view of a rock sample from the fossiliferous deposit in Figure 14.5. Note the slender bone fragments (arrows) and larger, flatter bone fragment (b), all presumably of sauropod origin, on the left. (B) Lateral view of cut and polished section of the rock sample in (A). Also note the poorly sorted clasts in this matrix-supported calcareous mass. *Photos courtesy of S. R. Ash.*

A

b

2 cm

B

2 cm

Hermatophyton, but also included some fragments of conifer wood, a few seeds and leaves, and small bits of bone of a sauropod, probably *Camarasaurus* (Ash 1993: 4). These fossil plant remains and bones are embedded as clasts in the calcareous matrix; some of these dinosaur bones measured up to 50 cm in length.

Hence, the stomach contents described by Stokes (1964) are not an example of the digestive tract remains of a sauropod but are likely "merely a small fragment of a peculiar conglomerate that may have accumulated on the floor of a small pond during the Late Jurassic" (Ash 1993). Facies analysis and interpretation of the deposit is currently in progress (S. R. Ash, personal communication to C.T.G., 2008).

Basal Birds

In contrast to the disappointing lack of convincing evidence in the sauropods, an unequivocal example of fossil gut remains is known among herbivorous theropods, more specifically in a basal bird, *Jeholornis prima*, from the Early Cretaceous of China (Zhou and Zhang 2002). *Jeholornis prima* was a large and long-tailed bird and is known from a partially articulated skull and a nearly complete postcranial skeleton. In its abdominal area, there are clear impressions of dozens of round, lightly and longitudinally striated seeds about 1 cm in diameter, which have been assigned to the form genus *Carpolithus*. This is an unequivocal instance of a Cretaceous bird with stomach contents and shows that granivory, or seed eating, had evolved in the theropod line by the Early Cretaceous. This is also suggested by the gastric mill seen in the derived nonavian theropods of the clades Ornithomimosauria and Oviraptorosauria (Wings and Sander 2007). The intact nature of the seeds in *Jeholornis* indicates that the bird swallowed the seeds whole and would have processed them up into smaller particles in its gizzard, while the great number of seeds suggests that it possessed a large crop (Zhou and Zhang 2002). Furthermore, we note that the plant material in this bird's digestive tract consists of only one species of plant (*Carpolithus* sp.) and of only one type of plant part (seeds).

Hadrosaurs

Hadrosaurs with putative stomach contents have been discovered three times in Late Cretaceous dinosaur faunas in North America (Taggart and Cross 1997). The first two were mummified specimens of *Edmontosaurus annectens* (cf. Currie et al. 1995), which were collected by Charles H. Sternberg and his sons, George and Charles M., in Wyoming (Sternberg 1909, 1917). The first specimen, deposited in the collections of the American Museum of Natural History in New York, contained carbonized stomach contents, but the plant material was apparently not described (cf. Currie et al. 1995). The second specimen, collected from the same region in 1910, was sent to the Senckenberg Museum in Frankfurt, Germany, where it can still be seen on display today (Fig. 14.3). This mummified carcass contained a balled-up mass of plant material, which was macerated and studied by Richard Kräusel, who identified needles of a common Cretaceous conifer that he called *Cunninghamites elegans*, twigs of conifers and dicots, and numerous small seeds or fruits (Kräusel 1922). Maceration of the plant matter apparently prevented conservation of the digestive tract remains that would enable a critical assessment of its authenticity (Currie et al. 1995; Chin 1997).

A third hadrosaur with putative stomach contents was recovered from the Late Cretaceous Dinosaur Park Formation in Alberta, Canada (Currie et al. 1995). This specimen was identified as *Corythosaurus casuarius* and is on display in the Dinosaur Hall of the Royal Tyrrell Museum in Drumheller, Alberta, Canada; the plant-filled mass found in its abdominal region is deposited in the collections of this museum (Fig. 14.7A, B). Like the other two specimens, this hadrosaur was mummified, and the gut contents of plant material were found within the skin of the animal. This plant material was made up of small pieces of wood and small twigs, seeds and seed pods, and small pieces of charcoal; there are twice as many conifer remains as angiosperm remains. Palynological study of the stomach contents revealed that a relatively diverse palynoflora of 30 taxa of spores, pollen, megaspores, fungal spores, acritarchs, dinocysts, and possible algae was present. As a result of their thorough investigation, Currie et al. (1995) posit that the most conservative conclusion is that the plant remains associated with the Canadian hadrosaur cannot be shown to represent the remains of its digestive tract. The diversity of the palynoflora, as well as the size uniformity of the plant megafossils and the occurrence of charcoal, dinocysts, and rare algae, suggest that the plant remains may have been accumulated and washed into the body after death by hydraulic processes.

Evidence from Coprolites

Coprolite evidence is more common than the fossil remains of digestive tracts, but it is more difficult to pin down the exact provenance of the fecal material. The producer of a specific type of coprolite must be inferred from the size of the coprolite relative to the size of the probable animal in the local fauna. Hence, most Mesozoic coprolites can only be generally ascribed to herbivorous dinosaurs and not to any one particular group.

Sauropods

Coprolites of titanosaur sauropods were reported from the Late Cretaceous Lameta Formation of central India (Matley 1939; Ghosh et al. 2003; Mohabey 2005; Prasad et al. 2005). There are four types of coprolites, which were all attributed to this group of giant sauropods, as titanosaur bones are the only herbivorous dinosaur remains in these deposits. The plant matter in these coprolites consists mostly of gymnospermous remains, such as cuticles and woody tissue, but also includes bacterial colonies, fungal spores, and algal remains, as well as phytoliths of palms and a variety of grasses. As with the Mygatt-Moore coprolites (see below), such a diverse assemblage of plant parts would also reflect a generalist feeding habit. Nevertheless, there is some doubt in this case, too, because it is questionable if the coprolites from this locality that were examined by one of us (P.M.S.) truly pertain to titanosaurs, owing to their relatively small size (less than 10 cm in diameter). The small size of the coprolites would not be inconsistent with sauropod body size if sauropods had produced small fecal pellets like some modern herbivores (e.g., horses, deer, rabbits). However, these putative sauropod fecal pellets should then

be grouped in clusters and should not occur as solitary nodules, as they do in the Lameta Formation (cf. Matley 1939; A. Sahni, personal communication to P.M.S., 2006). Just as in the case of dinosaur eggs (Norell et al. 1994), caution is necessary in regard to the assignment of specific coprolites to specific dinosaur groups, and only unequivocal associations (e.g., a fecal mass inside a dinosaur skeleton) would provide a firm basis for the identification of the producer.

Hadrosaurs

Another case of putative herbivorous dinosaur coprolites filled with fungally decayed wood has recently been reported from the Late Cretaceous Two Medicine Formation of Montana (Chin 2007). The find consists of 18 large, irregularly shaped masses with a volume of up to 7 liters. These nodules are composed of up to 85% conifer wood, either in the form of wood clasts or smaller fragments. Striking is the occurrence of disaggregated tracheids and of fossil hyphae, both of which are evidence of fungal activity in the wood. The preponderance of wood fragments and lack of small-diameter twigs suggest that the coprolite producers would have intentionally fed on the fungally degraded wood and that they did not accidentally ingest the rotten wood while browsing on fresh leaves, twigs, and small branches. Because ingestion appears to have been intentional, Chin (2007) argued that the dinosaurs ate the rotten wood to exploit the nutrient resources in the wood released by fungal attack. The coprolites occur in a series of strata ranging from 74 to 80 million years old, which represents a reoccurring, perhaps seasonal, pattern of decayed wood consumption, and were most likely produced by the hadrosaur *Maiasaura*, which is the most common dinosaur at the coprolite localities and was large enough to have produced these gigantic coprolites (Chin 2007).

Fig. 14.7. Putative stomach contents of the hadrosaur *Corythosaurus casuarius* from the Late Cretaceous Dinosaur Park Formation in Alberta, Canada. TMP 1980.040.0001, Collections of the Royal Tyrrell Museum, Drumheller, Alberta, Canada. (A) The larger of the two blocks of matrix recovered from within the hadrosaur. (B) Close-up of the block in (A) showing the fragments of plant material in the matrix. *Photos by Nadine Pajor and Thomas Breuer.*

If these nodules do indeed represent coprolites, an alternative suggestion is that the fungus developed in the feces after excretion. This is commonly observed by two of us (M.C. and J.H.) in herbivore feces during feeding trials, if the feces are not immediately frozen after collection. Moreover, it seems unlikely that a herbivore would regularly and willingly ingest fungally decayed wood.

Unspecified Herbivorous Dinosaurs

In the Middle Jurassic of England, numerous coprolite pellets with the leaf remains of various gymnosperms were studied from localities in North Yorkshire famous for their rich and diverse fossil flora (Hill 1976). More than 250 pellets were found clustered into random groups, which were spread over an area of about one square meter and dispersed over several bedding planes. Most of the individual pellets measured 10–13 mm in diameter and resembled the pelleted form of sheep's dung. Maceration of the coprolites yielded predominantly the thick, resinous leaf cuticle of the bennettitalean *Ptilophyllum pectinoides*, although there were also a few cuticle fragments of *Solenites* or *Czekanowskia*, *Nilssoniopteris vittata*, and *Williamsonia*—all gymnosperms now extinct. On the basis of the correlation between their size and possible producers in the local fauna, these pellets were attributed to a herbivorous dinosaur about the size of a sheep or large rabbit, such as an ornithopod dinosaur (Hill 1976).

Coprolites of herbivorous dinosaurs also occur in the Late Jurassic Morrison Formation of North America. In western Colorado, for example, plant-bearing nodules of round to irregular shapes were discovered in the Mygatt-Moore Quarry (Chin and Kirkland 1998). These nodules range in size from 5 to 15.5 cm in diameter and contain a variety of plant parts from a number of plant groups, such as gymnosperm seeds, fern sporangia, cycadophyte fronds and petioles, and conifer wood and cuticle (*Brachyphyllum*), as well as volcanic ash. The identification of these nodules as herbivore coprolites was made on the bases of the inclusion of plant parts and on their distinctive shape in the mudstone matrix of the quarry.

The combination of a diverse assemblage of plant remains and ingested volcanic ash is believed to indicate a broad diet on the part of the herbivore. The size of the coprolites suggests that they may have been produced by small herbivorous dinosaurs, for example, the Mygatt-Moore nodosaur *Mymoorapelta*, or perhaps by juvenile dinosaurs of a larger species passing through this area during Jurassic times. Some doubt has been cast on the fecal origin of these nodules. It has been pointed out that the well-preserved, intact fern sporangia still retaining spores and attached to their original frond, cycad fronds and petioles, and cuticle and seeds with their fleshy outer layers still attached found in these nodules would not have survived the digestive system of a herbivorous dinosaur. Passing through the dinosaur, the material should have been macerated, rather than remaining in pristine condition (W. D. Tidwell, personal communication to C.T.G., 2008). Similarly, it is questionable whether an animal would willingly feed on volcanic ash.

Assessment of Stomach Contents and Coprolites

On the basis of our survey, it appears that the reports of dinosaur stomach contents and coprolites are variable in their authenticity. The authenticity of the stomach contents in sauropods, for example, is not at all well supported, whereas the report of stomach contents in a seed-eating Cretaceous bird (Zhou and Zhang 2002) is well documented. Ascertaining the authenticity of coprolites and assigning them to a specific dinosaur producer seems to be equally as difficult.

One characteristic that seems to typify the better-substantiated cases of stomach contents and coprolites is their monotypic nature in regard to plant parts and generally monospecific composition in regard to plant species. Examples include the *Carpolithus* seeds in the gut of the Cretaceous bird *Jeholornis prima* (Zhou and Zhang 2002), or the plenitude of bennettitalean leaf cuticle of *Ptilophyllum pectinoides* in the Yorkshire dinosaur pellets (Hill 1976). This would be consistent with a feeding behavior in which an animal would exploit just one plant part of a single species, depending on what is available and ripe at the time. Stomach contents would thus be a snapshot of the dietary intake of the animal at one moment in time—that is, what the animal was eating immediately before death and burial—and are less likely to represent a sampling of its overall diet. Accumulations resembling stomach contents or coprolites that bear a wide variety of plant groups and plant parts might have very well come together as a result of taphonomic processes and not through ingestion by an animal.

Evidence from Tracks

Dinosaur tracks are direct evidence of behavior and as such have the potential to record feeding in dinosaurs. Only one such case appears to have been documented in the fossil record, however. It consists of ornithopod tracks with toes pointing toward the bases of large trees preserved in the roof shales of a coal mine in Utah, U.S.A. (Balsley and Parker 1983; Carpenter 1992; Taggart and Cross 1997). The roof shales and tree bases represent a fossilized forest floor on which large ornithopods, presumably some kind of hadrosaur, congregated and walked up to the trees to browse. The trees pertain to large taxodiaceous conifers, suggesting that the ornithopods fed on the foliage of these conifers (Taggart and Cross 1997).

The Plant Perspective

In regard to the issue of dinosaur herbivory, paleobotanists and vertebrate paleontologists have mainly been interested in dinosaur–plant interactions and possible coevolution between the two groups, particularly during the rise of the angiosperms during the Cretaceous (e.g., Krassilov 1981; Coe et al. 1987; Wing and Tiffney 1987a, 1987b; Weishampel and Norman 1989; Taggart and Cross 1997; Tiffney 1997; Weishampel and Jianu 2000; Leckey 2004; Butler et al. 2009). Nearly all of these workers also tackled the question of dinosaur food preferences from the perspective of fossil plants.

A focus of these studies have been the paleobotanical and sauropod data from the Morrison Formation. The general consensus is that the sauropod dinosaurs of the Morrison were low-browsing fern grazers (Taggart and Cross 1997) in open expanses that were prairielike (Taggart and Cross 1997) or that were dotted with small stands or groves of tree ferns and conifers (Coe et al. 1987; Rees et al. 2004). These studies also generally agree that the giant sauropods fed on large volumes of low-quality fodder (Coe et al. 1987; Wing and Tiffney 1987b), although the nutritional content of potential food sources was not quantified or documented in any of these studies.

In 1997, Tiffney took a slightly different tack and assessed specific Mesozoic plant groups as potential fodder on the basis of a comparative survey of growth forms, morphological structures, and chemical defenses in living relatives. He concluded that most Mesozoic gymnosperms would have been poor nutrition for the herbivorous dinosaurs, owing to their tough bark, hard and often spiny foliage, and wealth of indigestible chemical compounds. Tiffney also postulated that plant tissues with the fewest structural and chemical defenses would have included cycadophyte pinnae, *Ginkgo* leaves, and possibly also the foliage of the Cheirolepidiaceae, an extinct conifer group. Even more succulent, as well as less heavily armed with chemical defenses, would have been angiosperm leaves, which would have become a common source of food in the mid-Cretaceous, at the onset of the angiosperm radiation and diversification.

Leckey (2004), in a master's thesis, followed up on Tiffney's ideas by looking for correlations between members of the Cretaceous flora and herbivorous dinosaur faunas in North America, primarily to discern possible coevolutionary relationships between the angiosperms and herbivores. Several statistical tests—second-order, nearest-neighbor, and bivariate K analyses—were used to find the degree of overlap between dinosaur and plant groups. Although a weak correlation was found between the sauropods and either conifers or angiosperms, there was a significant degree of overlap between end-Cretaceous (Campanian and Maastrichtian) ornithischians and angiosperms. Similarly, Weishampel and Norman (1989), Weishampel and Jianu (2000), Fastovsky and Smith (2004), and Butler et al. (2009) were unable to find support for the hypothesis that the rise of the angiosperms influenced dinosaur evolution to any major extent.

Hummel and Clauss (in press) have suggested that one potential difference between feeding on conifers versus angiosperms could be the respective biomass that can be cropped in one bite, with higher returns per bite associated with conifer foliage. It would have thus been more advantageous in terms of bite efficiency for the sauropods to have consumed conifer foliage. Chewing herbivores, such as ornithopods, would have been more limited in their cropping efficiency by the time necessary for mastication; higher bite volumes would therefore not have necessarily translated into higher intake rates for them, at least not to the same extent as for nonmasticating herbivores such as sauropods. The end-Cretaceous association of ornithopods and angiosperms found by Leckey (2004) is interpreted here as a secondary effect of ornithopods choosing habitats where they faced less competition from sauropods.

The study of herbivory in modern animals, particularly mammals, is a large and economically important field that draws on the expertise of not only zoologists but also ecologists, animal nutritionists, and veterinarians. It is particularly the perspective of the last three that has many new and exciting insights to offer to the study of dinosaur herbivory, as will become apparent in the sections that follow.

Evidence from Modern Herbivores

Bulk Feeders Versus Selective Feeders

In comparison with extant ungulate herbivores (the modern ecological analog of the herbivorous dinosaurs), it becomes clear that the diet of large herbivorous dinosaurs must have consisted of vegetative plant parts (leaves, stems, bark), all of which are rich in cellulose and lignin and are therefore difficult to digest (Hummel and Clauss, in press). Food of higher quality, such as fruits and seeds, were probably eaten whenever available, but the high quantitative requirements of large herbivores—the need to eat *a lot*—as well as their comparatively large mouthparts, result in the necessity to forage rather unselectively (Demment and Van Soest 1985), a strategy known as bulk feeding. This does not mean that large herbivores do not prefer high-quality food; there is just no way for them to obtain it in large enough amounts. Even in the giraffe, which is less of a bulk feeder than most other extant megaherbivores, seed pods (3%) or flowers (5%) make up minor proportions of the overall dry matter intake (Pellew 1984).

Digestive Tract and Scaling Effects

Because they consist largely of cellulose, plant cell walls can only be digested with the help of symbiotic microbes. Given the broad distribution of a significant symbiotic gut flora, among the great majority of extant herbivores, the favorable growing conditions of the animal gut for microbes, the relative ease of acquiring a bacterial gut flora (Van Soest 1994), and the tremendously long existence of cellulolytic anaerobic bacteria (Van Soest 1994), the use of gut microbes in the digestion of the Mesozoic leaves can be taken as the by far most likely possibility (Farlow 1987). Any scenario without symbiotic microbes implies an extreme intake level on the part of the herbivore, which is unlikely for large dinosaurs.

There is some uncertainty about the location of a fermentation chamber within the dinosaur gut. The option of either hindgut fermentation, such as that in horses or elephants, or foregut fermentation, such as in ruminants or hippos, are both plausible, although the hindgut generally seems to be the more likely possibility, especially in large, nonchewing forms (Farlow 1987; Hummel and Clauss, in press).

Overall gut capacity (mass of gut contents) has been shown to scale with body mass $(BM)^{1.0}$, while energy requirements scale with $BM^{0.75}$ (reviewed in Clauss et al. 2008). From this relation, it can be concluded that the relative digestive capacity of herbivores increases with BM to some extent because more space is available in the gut per unit energy requirement. For example, Illius and Gordon (1992) considered ingesta retention time (the time that the ingested food stays in the gut) to scale with $BM^{0.25}$, i.e., to increase with increasing BM in mammalian herbivores. However, a recent reevaluation could not establish an increase in retention time above a BM threshold of 1–10 kg (Clauss et al. 2007), and raises some doubts about a simple relation of BM and retention time in mammals. Elephants are the best example for a megaherbivore with a comparatively fast passage time (Clauss et al. 2003). Apart from highly probable, differential adaptations to different forage types and their varying fermentation characteristics in mammals (Hummel et al. 2006), the influence of mastication efficiency in mammals must not be forgotten (Clauss and Hummel 2005). Actually, among large mammalian herbivores, there appears to be a trade-off between mastication efficiency and ingesta retention that overrules the effects of BM on ingesta retention (Clauss et al. in press).

Therefore, when extrapolating from mammals to herbivorous dinosaurs, one would simply expect the same variety in digestive strategies among those dinosaurs that had evolved a chewing mechanism (i.e., ornithopods and ceratopsians). In particular, it appears likely that these forms had comparatively high metabolic rates and could have posed considerable competition for one another, and especially for similar-sized, nonchewing herbivores (e.g., ankylosaurs or stegosaurs). Sauropods, however, are special. They neither chewed nor ground their food in a gizzard, and thus they most likely digested material of comparatively large particle size. Although experiments by Bjorndal et al. (1990) have shown that the rate of fermentation of large plant particles is much lower than that of finely comminuted ones, this may not have represented a problem to sauropod digestion (Hummel et al. 2008b). This is because the extraordinarily large body size of sauropods most likely allowed for a strong allometric increase in their gut capacity relative to energy requirements, as well as for an efficiency in digestion that would have precluded any competition from chewing herbivores (Hummel and Clauss, in press). In other words, only with a nonchewing form of food intake, which was made possible by scaling effects of the digestive tract and long retention times, could the sauropods have attained the huge sizes they achieved during the Mesozoic, particularly in the Late Jurassic and Late Cretaceous (Sander and Clauss 2008).

Little if anything is known about the dietary quality of potential food plants of herbivorous dinosaurs, which mainly consisted of nonangiosperm plants. Evaluating extant relatives as proxies for potential dinosaur food plants obviously has the shortcoming of disregarding any changes that may have occurred in the plants during their long evolutionary history since the Mesozoic. However, the gross morphology and ecological preferences between some fossil and living plants, for example, *Equisetum* and *Araucaria*, are so similar that the extant generic names are applied to the fossil forms (e.g., Gould 1968; Gee 1989; Stockey 1994 and references therein; Gee and Tidwell 2010), implying that little change, structural, ecological, or otherwise, has occurred (Behrensmeyer et al. 1992). Furthermore, there is good correlation between the structural variation of certain biochemical compounds and the evolutionary status of these plant taxa (Swain 1974).

Although diet quality has many facets—including energy content, protein content, and the content of certain elements (e.g., phosphorus)—the digestibility of the plant material and thus energy contained in it was considered to be the best overall parameter by Owen-Smith (1988). Hummel et al. (2008a) followed this rationale by testing the nearest living relatives of potential dinosaur food plants for their energy content by means of an in vitro fermentation method (Menke et al. 1979) commonly used in feed evaluation for extant herbivores such as ruminants. Specifically, Hummel et al. (2008a) evaluated the degradability of ferns, horsetails, conifers, and other gymnosperms in response to herbivore hindgut fermentation (Fig. 14.8). Although Weaver (1983) already had evaluated the energy content of the nearest living relatives of dinosaur food plants, it should be noted that her approach of evaluating energy content from an analysis of gross energy (i.e., combustion energy) was fairly imprecise for estimating the actual energy available to an herbivore. For example, the gross energy of a piece of wood, which is hardly digestible to a herbivore, is similar to that of young leaves, which are relatively well digestible to the same herbivore.

In the in vitro experiments, the Mesozoic plant relatives performed moderately to very well and yielded energy levels comparable to extant dicot leaf species in many instances (Fig. 14.8; Plate 16; Table 14.1). *Araucaria*, a widespread conifer during the Mesozoic, has an intriguing pattern of fermentation behavior; its fermentation starts off rather slowly, but then quickly speeds up after 30 hours. This type of fermentation behavior would potentially favor a long retention time in the hindgut of a herbivore, such as that of a gigantic sauropod.

Experimental Evaluation of Potential Food Plants

Fig. 14.8. Fermentative behavior of potential dinosaur food plants compared with that of angiosperms. Gas production in the Hohenheim gas test is plotted versus fermentation time. DM = dry matter; the mean and standard error of the mean of each data point are indicated. (A) Various gymnosperms compared to angiosperms. Note that *Ginkgo* and some conifer families (Cephalotaxaceae, Taxodiaceae, Pinaceae, and Taxaceae) performed at the level of dicot leaf browse, whereas podocarp conifers and cycads fared poorly. (B) Ferns compared with angiosperms. Note the great variability among ferns, including the poor performance of the tree fern *Dicksonia*. (C) The Araucariaceae and horsetails (*Equisetum* spp.) compared with angiosperms. Note that horsetails even surpass grasses and that araucarias outperform dicot leaf browse after 72 hours. *Modified from Hummel et al. (2008a).*

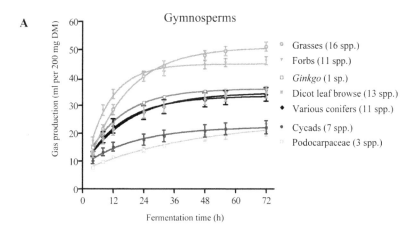

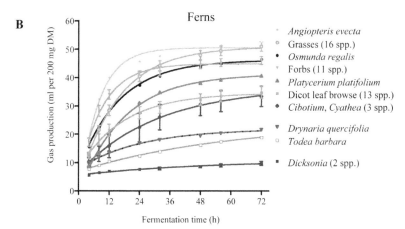

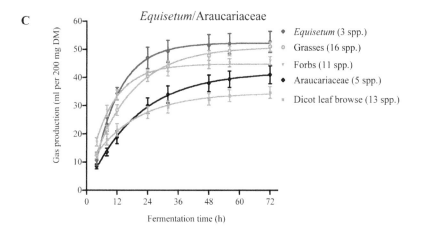

Sample Type (number of spp.)	ME (Gp 72 h) (MJ/ kg DM)	ME (Weaver, 1983) (MJ/kg DM)	Crude Protein (% DM)	NDF (% DM)
Equisetum (3)	11.6 (10.8–12.9)	5.3	11.7	48.4
Grasses (16)	11.3 (9.3–13.6)		15.3	62.8
Forbs (11)	10.4 (9.1–11.9)		19.8	37.8
Araucariaceae (5)	9.4 (8.0–11.6)	7.0	4.4	65.2
Ginkgo (1)	8.6	6.7	15.6	27.5
Various conifers (13)	8.3 (6.3–10.8)	7.0 (6.4–7.5)	10.0	51.3
Various ferns (9)	7.7 (4.7–11.7)	6.6 (5.4–7.4)	11.5	62.8
Dicot leaf browse (13)	7.5 (5.5–10.0)		20.7	43.2
Tree ferns (5)	6.4 (3.6–9.3)	6.9 (6.6–7.2)	11.3	63.6
Cycads (7)	6.1 (4.4–7.7)	7.6 (7.1–8.6)	11.4	65.3
Podocarpaceae (3)	5.9 (5.0–6.1)		6.6	62.3

Table 14.1. Nutrient and Metabolizable Energy (ME) Content of Potential Dinosaur Food Plants

Note. The ME content based on Weaver's (1983) data was obtained by multiplying gross energy with a factor of 0.5 (digestible energy, according to Weaver 1983) and consequently with a factor of 0.76 (according to Robbins 1993). DM = dry matter; NDF = neutral detergent fiber; Gp 72 h = results from the 72-hour trials. Source. Modified from Hummel et al. (2008a).

The horsetails (*Equisetum* spp.) were another plant group that performed surprisingly well. Interestingly, despite the high content of abrasive silica, living horsetails are indeed fed on by a few extant herbivores such as geese and waterfowl (Gee in press).

Protein levels of most plants investigated were at acceptable levels for herbivores (Table 14.1), although they were found to be low in *Araucaria* (Hummel et al. 2008a), suggesting that a pure *Araucaria* diet would have been insufficient for a large sauropod and that other plants must have been taken in as well.

In summary, the results of the in vitro fermentation experiments of Hummel et al. (2008a) indicate that, on the basis of the nutritive analysis of their nearest living relatives, the Mesozoic flora would have offered solid nutrition for the herbivorous dinosaurs. In vitro fermentation experiments and a comprehensive evaluation of the dietary quality of recent relatives of potential dinosaur food plants invites further exploration.

Synthesis and Conclusions

Because herbivory is the major pathway of energy flow in terrestrial ecosystems from the Late Paleozoic to the present day, the study of dinosaurian herbivory is necessary for understanding the evolution of terrestrial ecosystems. In this chapter, we have reviewed the various lines of evidence for dinosaur herbivory, starting with the most tangible evidence, the fossil teeth (including their enamel microstructure) and bones of the dinosaurs themselves. Long recognized and more tantalizing is the evidence provided by the rare finds of fossilized digestive tract contents, permineralized feces, and dinosaur tracks. The major impetus for writing this review, however, did not come from new discoveries in this area, but rather from new data derived from the fields of ecology and animal nutrition, which provide a wholly different angle on the subject.

It turns out that the herbivory of advanced ornithischians dinosaurs (ornithopods and ceratopsians) resembles that of modern-day herbivores (Norman and Weishampel 1985; Weishampel and Norman 1989). More precisely, herbivory in mammals evolved convergently in many ways to that of advanced ornithischian dinosaurs, after these dinosaurs went extinct at the end of the Cretaceous. The evolutionary origin of herbivory in ceratopsian and ornithopod dinosaurs began within the early dinosaurs that did not chew, but simply bit off and swallowed plant matter (Fastovsky and Smith 2004), much as living herbivorous reptiles such as iguana lizards and turtles do today. This plant matter was then digested by symbiotic microbes in special fermentation chambers in their gut (Farlow 1987; Hummel et al. 2008a). Most basal ornithischians dinosaurs (armored dinosaurs and pachycephalosaurs) were more highly evolved than this and must have engaged in some sort of oral processing of their food, which was aided by the presence of fleshy cheeks (Norman and Weishampel 1985; Fastovsky and Smith 2004). Full-fledged mastication evolved twice among ornithischians, in the ceratopsians and the ornithopods (Norman and Weishampel 1985; Weishampel and Norman 1989). Saurischians, on the other hand, modified primitive reptilian herbivory to a much lesser extent but were still able to evolve some very large body sizes with it, surpassing the largest mammalian herbivores and the largest ornithischian dinosaurs by an order of magnitude in BM (Sander and Clauss 2008). In other words, the sauropods show that chewing is not a prerequisite for evolving gigantic body size. Advanced herbivorous theropods apparently used gastric mills of the kind still in existence in birds today (Wings and Sander 2007). This is what can be gleaned from the paleontological evidence.

The organisms that bore the brunt of herbivory, namely, terrestrial plants growing during Mesozoic, have not yet yielded much direct information on dinosaurian herbivory. Evaluation of their attractiveness as fodder to the specific groups of dinosaurs has been limited because fossil evidence of dinosaur bite or gnaw marks on leaves, branches, bark, or reproductive structures has not been documented; this is in direct contrast to insect herbivory, in which evidence of arthropod-mediated damage occurs on plant parts throughout the last 420 million years, even on some of the oldest land plants of the Late Silurian (Labandeira 2006). In the better-substantiated cases of fossil gut contents and coprolites, it should be noted that a monotypic nature is characteristic in regard to plant parts and generally monospecific composition in regard to plant species.

Paleobotanical studies have attempted to infer interaction between plants and dinosaurs by focusing on patterns of co-occurrence with reference to the possible coevolution of dinosaurs and angiosperms in the mid-Cretaceous. Despite several attempts (Coe et al. 1987; Taggart and Cross 1997; Tiffney 1997; Weishampel and Jianu 2000; Leckey 2004; Butler et al. 2009), clear support for this intriguing hypothesis has not come forth, and the dinosaurs show remarkably little response to this major event in the evolution of terrestrial ecosystems.

As most potential food plants of the dinosaurs have living relatives, they represent fertile ground for comparative studies. Comparisons based on growth habit, morphology, and chemistry of the plant tissues have led to the prevailing view that soft-leaved and chemically poorly defended plant parts such as the foliage of the ferns, seed ferns, and ginkgoes must have played a major role in dinosaur nutrition (Coe et al. 1987; Tiffney 1997).

Recent fermentation experiments on the living relatives of the Mesozoic flora (Hummel et al. 2008a) suggest that horsetails, *Angiopteris* and *Osmunda* ferns, araucarias and other nonpodocarpaceous conifers, and ginkgoes would have been good to excellent sources of food energy to herbivorous dinosaurs, while cycads, podocarpaceous conifers, and dicksoniaceous tree ferns would have been the least attractive.

Ecological considerations, namely, that large dinosaurs must have been bulk feeders, also indicate that the plants most commonly consumed by dinosaurs must have offered large amounts of biomass and were able to regenerate quickly (Gee, in press; Hummel and Clauss, in press). This again points to conifers as major food plants in a preangiosperm world (see also Hotton and Baghai-Riding 2010), while cycads must have played a minor role in the diet of herbivorous dinosaurs as a result of the small amounts of nonnutritious biomass that they offer, coupled with a slow rate of regeneration (Gee, in press). We have shown that comparison with digestive tract anatomy and digestive physiology of modern herbivores sheds much light on the subject of dinosaur herbivory, particularly that of sauropods. Isometric scaling of gut capacity with body size means that the sauropods were able to enjoy the benefits of primitive reptilian herbivory (i.e., higher food uptake rate as a result of the lack of chewing) without suffering from the negative effects (i.e., slow digestion) (Sander and Clauss 2008; Hummel and Clauss, in press).

Thus, the application of methods from animal nutrition science and animal ecology results in a greatly improved picture of dinosaurian herbivory, especially in regard to the dominant herbivores of the Mesozoic, the giant sauropods, mainly because unlike other dinosaur groups, the herbivory of these giant dinosaurs lacks a modern analog. The study of dinosaurian herbivory helps to improve our understanding of mammalian herbivory in modern ecosystems and thus transforms paleontology from an academic study of fossil organisms to a valuable tool to be used in understanding contemporary life on earth.

Acknowledgments

P.M.S. and C.T.G. dedicate this paper to Ted Delevoryas, whose paleobotany lab at U.T. Austin brought them together 28 years ago and awakened their lifelong interest and enthusiasm for fossil plants. We thank Sidney R. Ash and William D. Tidwell for photos, loan of material, and helpful discussions in regard to Stokes's material; Nadine Pajor and Thomas Breuer for the photos in Figure 14.7, and James Gardner and Donald Brinkman for facilitating access to this material at the Royal Tyrrell Museum; and Dorothea Kranz and Georg Oleschinski for technical support in the creation of Figures 14.2 and 14.4. This is contribution no.

58 of the DFG Research Unit 533 on the Biology of the Sauropods: The Evolution of Gigantism.

References Cited

Ash, S. R. 1993. Current work. *The Morrison Times* 2: 4.

Balsley, J. K., and L. R. Parker. 1983. Cretaceous wave-dominated delta, barrier islands, and submarine fan depositional systems: Book Cliffs, eastern-central Utah. *American Association of Petroleum Geologists Field Guide*.

Barrett, P. M. 2000. Prosauropod dinosaurs and iguanas: speculations on the diets of extinct reptiles. In H.-D. Sues, *Evolution of Herbivory in Terrestrial Vertebrates: Perspectives from the Fossil Record*, pp. 42–78. Cambridge: Cambridge University Press.

———. 2005. The diet of ostrich dinosaurs (Theropoda: Ornithomimosauria). *Palaeontology* 48: 347–358.

Barrett, P. M., and P. Upchurch. 2005. Sauropod diversity through time: possible macroevolutionary and paleoecological implications. In K. A. Curry-Rogers and J. A. Wilson (eds.), *Sauropod Evolution and Paleobiology*, pp. 125–156. Berkeley: University of California Press.

Behrensmeyer, A. K., J. D. Damuth, W. A. DiMichele, and R. Potts. 1992. *Terrestrial Ecosystems through Time: Evolutionary Paleoecology of Terrestrial Plants and Animals*. Chicago: University of Chicago Press.

Bird, R. T. 1985. *Bones for Barnum Brown. Adventures of a Dinosaur Hunter*. Ed. V. T. Schreiber. Fort Worth: Texas Christian University Press.

Butler, R. J., P. M. Barrett, P. Kenrick, and M. G. Penn. 2009. Diversity patterns amongst herbivorous dinosaurs and plants during the Cretaceous: implications for hypotheses of dinosaur/angiosperm co-evolution. *Journal of Evolutionary Biology* 22: 446–459.

Bjorndal, K. A., A. B. Bolten, and J. E. Moore. 1990. Digestive fermentation in herbivores: effect of food particle size. *Physiological Zoology* 63: 710–721.

Carpenter, K. 1992. Behavior of hadrosaurs as interpreted from footprints from the "Mesaverde" Group (Campanian) of Colorado, Utah, and Wyoming. *Contributions to Geology, University of Wyoming* 29: 81–96.

Chin, K. 1997. What did dinosaurs eat? Coprolites and other direct evidence of dinosaur diets. In J. O. Farlow and M. K. Brett-Surman (eds.), *The Complete Dinosaur*, pp. 371–382. Bloomington: Indiana University Press.

———. 2007. The paleobiological implications of herbivorous dinosaur coprolites from the Upper Cretaceous Two Medicine Formation of Montana: why eat wood? *Palaios* 22: 554–566.

Chin, K., and J. I. Kirkland. 1998. Probable herbivore coprolites from the Upper Jurassic Mygatt-Moore Quarry, western Colorado. *Modern Geology* 23: 249–275.

Christian, A., and G. Dzemski. 2007. Reconstruction of the cervical skeleton posture of *Brachiosaurus brancai* Janensch, 1914 by an analysis of the intervertebral stress along the neck and a comparison with the results of different approaches. *Fossil Record* 10: 38–49.

Clark, J. M., T. Maryansky, and R. Barsbold. 2004. Therizinosauroidea. In D. B. Weishampel, P. Dodson, and H. Osmólska (eds.), *The Dinosauria*, 2nd ed., pp. 151–164. Berkeley: University of California Press.

Clauss, M., and J. Hummel. 2005. The digestive performance of mammalian herbivores: why big may not be that much better. *Mammal Review* 35: 174–187.

Clauss, M., R. Frey, B. Kiefer, M. Lechner-Doll, W. Loehlein, C. Polster, G. E. Rössner, and W. J. Streich. 2003. The maximum attainable body size of herbivorous mammals: morphophysiological constraints on foregut, and adaptations of hindgut fermenters. *Oecologia* 136: 14–27.

Clauss, M., A. Schwarm, S. Ortmann, W. J. Streich, and J. Hummel. 2007. A case of non-scaling in mammalian physiology? Body size, digestive capacity, food intake, and ingesta passage in mammalian herbivores. *Comparative Biochemistry and Physiology*, Part A, 148: 249–265.

Clauss, M., J. Hummel, W. J. Streich, and K.-H. Südekum. 2008. Mammalian metabolic rate scaling to 2/3 or 3/4 depends on the presence of gut contents. *Evolutionary Ecology Research* 10: 153–154.

Clauss, M., C. Nunn, J. Fritz, and J. Hummel. 2009. Evidence for a tradeoff between retention time and chewing efficiency in large mammalian herbivores. *Comparative Biochemistry and Physiology A* 154: 376–382.

Coe, M. J., D. L. Dilcher, J. O. Farlow, D. H. Jarzen, and D. A. Russell. 1987. Dinosaurs and land plants. In E. M. Friis, W. G. Chaloner, and P. R. Crane (eds.), *The Origins of Angiosperms and Their Biological Consequences*, pp. 225–258. Cambridge: Cambridge University Press.

Currie, P. J., E. B. Koppelhus, and A. F. Muhammad. 1995. "Stomach" contents of a hadrosaur from the Dinosaur Park Formation (Campanian, Upper Cretaceous) of Alberta, Canada. In A. Sun and Y. Wang (eds.), *Sixth Symposium on Mesozoic Terrestrial Ecosystems and Biota, Short Papers*, pp. 111–114. Beijing: China Ocean Press.

Demment, M. W., and P. J. Van Soest. 1985. A nutritional explanation for body-size patterns of ruminant and nonruminant herbivores. *American Naturalist* 125: 641–672.

Dodson, P. 1990. Sauropod paleoecology. In D. B. Weishampel, P. Dodson, and H. Osmólska (eds.), *The Dinosauria*, pp. 402–407. Berkeley: University of California Press.

Dodson, P., C. A. Forster, and S. D. Sampson. 2004. Ceratopsidae. In D. B. Weishampel, P. Dodson, and H. Osmólska (eds.), *The Dinosauria*, 2nd ed., pp. 494–516. Berkeley: University of California Press.

Dzemski, G., and A. Christian. 2007. Flexibility along the neck of the ostrich (*Struthio camelus*) and consequences for the reconstruction of dinosaurs with extreme neck length. *Journal of Morphology* 268: 707–714.

Farlow, J. O. 1987. Speculations about the diet and digestive physiology of herbivorous dinosaurs. *Paleobiology* 13: 60–72.

Fastovsky, D. E., and J. B. Smith. 2004. Dinosaur paleoecology. In D. B. Weishampel, P. Dodson, and H. Osmólska (eds.), *The Dinosauria*, pp. 614–626, 2nd ed. Berkeley: University of California Press.

Fastovsky, D. E., and D. B. Weishampel. 2005. *The Evolution and Extinction of the Dinosaurs*, 2nd ed. Cambridge: Cambridge University Press.

Galton, P. M. 1986. Herbivorous adaptations of the Late Triassic and Early Jurassic dinosaurs. In K. Padian (ed.), *The Beginning of the Age of Dinosaurs*, pp. 203–221. Cambridge: Cambridge University Press.

———. 1990. Basal Sauropodomorpha–Prosauropoda. In D. B. Weishampel, P. Dodson, and H. Osmólska (eds.), *The Dinosauria*, pp. 320–344. Berkeley: University of California Press.

Gee, C. T. 1989. Revision of the Late Jurassic/Early Cretaceous flora from Hope Bay, Antarctica. *Palaeontographica*, Abt. B, 213: 149–214.

———. In press. Dietary options for the sauropod dinosaurs from an integrated botanical and paleobotanical perspective. In N. Klein, K. Remes, C. T. Gee, and P. M. Sander (eds.), *Biology of the Sauropod Dinosaurs: Understanding the Life of Giants*. Bloomington: Indiana University Press.

Gee, C. T., and W. D. Tidwell. 2010. A mosaic of characters in a new whole-plant *Araucaria*, *A. delevoryasii* Gee sp. nov., from the Late Jurassic Morrison Formation of Wyoming, U.S.A. In C. T. Gee (ed.), *Plants in*

Mesozoic Time: Morphological Innovations, Phylogeny, Ecosystems, pp. 67–94. Bloomington: Indiana University Press.

Ghosh, P., S. K. Bhattacharya, A. Sahni, R. K. Kar, D. M. Mohabey, and K. Ambwani. 2003. Dinosaur coprolites from the Late Cretaceous (Maastrichtian) Lameta Formation of India: isotopic and other markers suggesting a C₃ plant diet. *Cretaceous Research* 24: 743–750.

Gould, R. E. 1968. Morphology of *Equisetum laterale* Phillips, 1829, and *E. bryanii* sp. nov. from the Mesozoic of south-eastern Queensland. *Australian Journal of Botany* 16: 153–176.

Hill, C. R. 1976. Coprolites of *Ptilophyllum* cuticles from the Middle Jurassic of North Yorkshire. *Bulletin of the British Museum (Natural History), Geology* 27: 289–293.

Hotton, C. L., and N. L. Baghai-Riding. 2010. Palynological evidence for conifer dominance within a heterogeneous landscape in the Late Jurassic Morrison Formation, U.S.A. In C. T. Gee (ed.), *Plants in Mesozoic Time: Morphological Innovations, Phylogeny, Ecosystems*, pp. 295–328. Bloomington: Indiana University Press.

Hummel, J., and M. Clauss. In press. Feeding and digestive physiology. In N. Klein, K. Remes, C. T. Gee, and P. M. Sander (eds.), *Biology of the Sauropod Dinosaurs: Understanding the Life of Giants.* Bloomington: Indiana University Press.

Hummel, J., K.-H. Südekum, W. J. Streich, and M. Clauss. 2006. Forage fermentation patterns and their implications for herbivore ingesta retention times. *Functional Ecology* 20: 989–1002.

Hummel, J., C. T. Gee, K.-H. Südekum, P. M. Sander, G. Nogge, and M. Clauss. 2008a. In vitro digestibility of fern and gymnosperm foliage: implications for sauropod feeding ecology and diet selection. *Proceedings of the Royal Society B* 275: 1015–1021.

Hummel, J., K.-H. Südekum, and M. Clauss. 2008b. Relevance of particle size for the digestion of different plant taxa and implications for herbivorous dinosaurs. *Proceedings of the Comparative Nutrition Society Symposia* 7: 106–110.

Hwang, S. H. 2005. Phylogenetic patterns of enamel microstructure in dinosaur teeth. *Journal of Morphology* 266: 208–240.

Illius, A. W., and I. J. Gordon. 1992. Modelling the nutritional ecology of ungulate herbivores: evolution of body size and competitive interactions. *Oecologia* 89: 428–434.

Ji, Q., M. A. Norell, P. J. Makovicky, K. Q. Gao, S. A. Ji, and C. Yuan. 2003. An early ostrich dinosaur and implications for ornithomimosaur phylogeny. *American Museum Novitates* 3420: 1–19.

Kobayashi, Y., J.-C. Lü, Z.-M. Dong, R. Barsbold, Y. Azuma, and Y. Tomida. 1999. Herbivorous diet in an ornithomimid dinosaur. *Nature* 402: 480–481.

Krassilov, V. A. 1981. Changes of Mesozoic vegetation and the extinction of dinosaurs. *Palaeogeography, Palaeoclimatology, Palaeoecology* 34: 207–224.

Kräusel, R. 1922. Die Nahrung von *Trachoodon. Palaeontologische Zeitschrift* 4: 80.

Labandeira, C. 2006. Silurian to Triassic plant and hexapod clades and their associations: new data, a review, and interpretations. *Arthropod Systematics and Phylogeny* 64: 53–94.

Leckey, E. H. 2004. Co-evolution in herbivorous dinosaurs and land plants: the use of fossil locality data in an examination of spatial co-occurrence. M.S. thesis, University of California, Santa Barbara.

MacFadden, B. J. 1994. *Fossil Horses: Systematics, Paleobiology, and Evolution of the Family Equidae.* Cambridge: Cambrige University Press.

Makovicky, P. J., K. Kobayashi, and P. J. Currie. 2004. Ornithomimosauria. In D. B. Weishampel, P. Dodson, and H. Osmólska (eds.), *The Dinosauria*, 2nd ed., pp. 137–150. Berkeley: University of California Press.

Matley, C. A. 1939. The coprolites of Pijdura, Central Provinces. *Records of the Geological Survey of India* 74: 535–547.

Menke, K. H., L. Raab, A. Salewski, H. Steingass, D. Fritz, and W. Schneider. 1979. The estimation of the digestibility and metabolizable energy content of ruminant feeding stuffs from the gas production when they are incubated with rumen liquor in vitro. *Journal of Agricultural Science* 93: 217–222.

Michelis, I. 2004. Vergleichende Taphonomie des Howe Quarry's (Morrison-Formation, Oberer Jura), Bighorn County, Wyoming, U.S.A. Ph.D. diss., University of Bonn, Germany.

Mohabey, D. M. 2005. Late Cretaceous (Maastrichtian) nests, eggs, and dung mass (coprolites) of sauropods (titanosaurs) from India, pp. 466–489. In V. Tidwell and K. Carpenter (eds.), *Thunder-Lizards: The Sauropodomorph Dinosaurs*. Bloomington: Indiana University Press.

Murray, P. F., and P. Vickers-Rich. 2004. *Magnificent Mihirungs: The Colossal Flightless Birds of the Australian Dreamtime*. Bloomington: Indiana University Press.

Norman, D. B., and D. B. Weishampel. 1985. Ornithopod feeding mechanisms: their bearing on the evolution of herbivory. *American Naturalist* 126: 151–164.

———. 1991. Feeding mechanisms in some small herbivorous dinosaurs: processes and patterns. In J. M. V. Rayner and R. J. Wootton (eds.), *Biomechanics in Evolution*, pp. 161–181. Cambridge: Cambridge University Press.

———. 2004. Basal Ornithischia. In D. B. Weishampel, P. Dodson, and H. Osmólska (eds.), *The Dinosauria*, 2nd ed., pp. 325–334. Berkeley: University of California Press.

Norell, M. A., J. M. Clark, D. Dashzeveg, R. Barsbold, M. L. Chiappe, A. R. Davidson, M. C. McKenna, A. Perle, and M. J. Novacek. 1994. A theropod dinosaur embryo and the affinities of the Flaming Cliffs dinosaur eggs. *Science* 266: 779–782.

Osmólska, H., P. J. Currie, and R. Barsbold. 2004. Oviraptorosauria. In D. B. Weishampel, P. Dodson, and H. Osmólska (eds.), *The Dinosauria*, 2nd ed., pp. 165–183. Berkeley: University of California Press.

Ostrom, J. H. 1964a. A functional analysis of jaw mechanics in the dinosaur *Triceratops*. *Postilla* 88: 1–35.

———. 1964b. A reconsideration of the paleoecology of hadrosaurian dinosaurs. *American Journal of Science* 262: 975–997.

Owen-Smith, N. 1988. *Megaherbivores*. Cambridge: Cambridge University Press.

Pellew, R. A. 1984. The feeding ecology of a selective browser, the giraffe (*Giraffa camelopardalis tippelskirchi*). *Journal of Zoology* 202: 57–81.

Prasad, V., C. A. E. Strömberg, H. Alimohammadian, and A. Sahni. 2005. Dinosaur coprolites and the early evolution of grasses and grazers. *Science* 310: 1177–1180.

Rees, P. M., C. R. Noto, J. M. Parrish, and J. T. Parrish. 2004. Late Jurassic climates, vegetation, and dinosaur distributions. *Journal of Geology* 112: 643–653.

Rensberger, J. M. 2000. Dental constraints in the early evolution of mammalian herbivory. In H.-D. Sues (ed.), *Evolution of Herbivory in Terrestrial Vertebrates: Perspectives from the Fossil Record*, pp. 144–167. Cambridge: Cambridge University Press.

Robbins, C. T. 1993. *Wildlife Feeding and Nutrition*. San Diego: Academic Press.

Sander, P. M. 1997. Teeth and jaws. In P. J. Currie and K. Padian (eds.), *Encyclopedia of Dinosaurs*, pp. 717–725. San Diego: Academic Press.

———. 1999. The microstructure of reptilian tooth enamel: terminology, function, and phylogeny. *Münchner geowissenschaftliche Abhandlungen* 38: 102.

———. 2000. Prismless enamel in amniotes: terminology, function, and evolution. In M. F. Teaford, M. W. J. Ferguson, and M. M. Smith (eds.), *Development, Function and Evolution of Teeth*, pp. 92–106. New York: Cambridge University Press.

Sander, P. M., and M. Clauss. 2008. Sauropod gigantism. *Science* 322: 200–201.

Seymour, R. S., and H. B. Lillywhite. 2000. Hearts, neck posture and metabolic intensity of sauropod dinosaurs. *Proceedings of the Royal Society B* 267: 1883–1887.

Sternberg, C. H. 1909. *The Life of a Fossil Hunter.* Henry Holt. Reprint, Bloomington: Indiana University Press, 1990.

———. 1917. *Hunting Dinosaurs in the Bad Lands of the Red Deer River, Alberta, Canada.* 3rd ed. published in 1985. Edmonton: NeWest Press.

Stevens, K. A., and J. M. Parrish. 2005. Neck posture, dentition, and feeding strategies in Jurassic sauropod dinosaurs. In V. Tidwell and K. Carpenter (eds.), *Thunder-Lizards: The Sauropodomorph Dinosaurs*, pp. 212–232. Bloomington: Indiana University Press.

Stockey, R. A. 1994. Mesozoic Araucariaceae: morphology and systematic relationships. *Journal of Plant Research* 107: 493–502.

Stokes, W. L. 1964. Fossilized stomach contents of a sauropod dinosaur. *Science* 143: 576–577.

Sues, H.-D., ed. 2000. *Evolution of Herbivory in Terrestrial Vertebrates: Perspectives from the Fossil Record.* Cambridge: Cambridge University Press.

Swain, T. 1974. Biochemical evolution of plants. In M. Florkin and E. H. Stotz (eds.), *Comprehensive Biochemistry, Molecular Evolution*, vol. 29, part A, pp. 125–302. Amsterdam: Elsevier.

Taggart, R. E., and A. T. Cross. 1997. The relationship between land plant diversity and productivity and patterns of dinosaur herbivory. In D. L. Wolberg, E. Stump, and G. D. Rosenberg (eds.), *Dinofest International. Proceedings of a Symposium Sponsored by Arizona State University*, pp. 403–416. Philadelphia: Academy of Natural Sciences.

Taylor, M. A., M. J. Wedel, and D. Naish. 2009. Head and neck posture in sauropod dinosaurs inferred from extant animals. *Acta Palaeontologica Polonica* 54: 213–220.

Tiffney, B. H. 1997. Land plants as food and habitat in the Age of Dinosaurs. In J. O. Farlow and M. K. Brett-Surman (eds.), *The Complete Dinosaur*, pp. 352–370. Bloomington: Indiana University Press.

Upchurch, P., and P. M. Barrett. 2000. The evolution of sauropod feeding mechanisms. In H.-D. Sues (ed.), *Evolution of Herbivory in Terrestrial Vertebrates: Perspectives from the Fossil Record*, pp. 79–122. Cambridge: Cambridge University Press.

Van Soest, P. J. 1994. *Nutritional Ecology of the Ruminant*, 2nd ed. Ithaca, N.Y.: Cornell University Press.

Weaver, J. C. 1983. The improbable endotherm: the energetics of the sauropod dinosaur *Brachiosaurus. Paleobiology* 9: 173–182.

Weishampel, D. B. 1984. The evolution of jaw mechanics in ornithopod dinosaurs. *Advances in Anatomy, Embryology and Cell Biology* 87: 1–109.

Weishampel, D. B., and C.-M. Jianu. 2000. Plant-eaters and ghost lineages: dinosaurian herbivory revisted. In H.-D. Sues (ed.), *Evolution of Herbivory in Terrestrial Vertebrates: Perspectives from the Fossil Record*, pp. 123–143. Cambridge: Cambridge University Press.

Weishampel, D. B., and D. B. Norman. 1989. Vertebrate herbivory in the Mesozoic: jaws, plants, and the evolutionary metrics. In J. O. Farlow (ed.),

Paleobiology of the Dinosaurs: Geological Society of America Special Paper 238: 87–1000.

Weishampel, D. B., P. Dodson, and H. Osmólska, eds. 2004. *The Dinosauria*, 2nd ed. Berkeley: University of California Press.

Wing, S. L., and B. H. Tiffney. 1987a. The reciprocal interaction of angiosperm evolution and tetrapod herbivory. *Review of Palaeobotany and Palynology* 50: 179–210.

———. 1987b. Interactions of angiosperms and herbivorous tetrapods through time. In E. M. Friis, W. G. Chaloner, and P. R. Crane (eds.), *The Origins of Angiosperms and Their Biological Consequences*, pp. 203–224. Cambridge: Cambridge University Press.

Wings, O., and P. M. Sander. 2007. No gastric mill in sauropod dinosaurs: new evidence from analysis of gastrolith mass and function in ostriches. *Proceedings of the Royal Society B* 274: 635–640.

You, H., and P. Dodson. 2004. Basal Ceratopsia. In D. B. Weishampel, P. Dodson, and H. Osmólska (eds.), *The Dinosauria*, 2nd ed., pp. 478–493. Berkeley: University of California Press.

Zhou, Z., and F. Zhang. 2002. A long-tailed, seed-eating bird from the Early Cretaceous of China. *Nature* 418: 405–409.

CONTRIBUTORS

Sergio Archangelsky, División Paleobotánica, Museo Argentino de Ciencias Naturales "B. Rivadavia," CONICET, Avenida Angel Gallardo 470, Buenos Aires (C1405DJR) Argentina; sarcang@fibertel.com.ar

Analía Artabe, MLP-CONICET, Paseo del Bosque s/n, La Plata, Argentina; aartabe@fcnym.unlp.edu.ar

Sidney R. Ash, Department of Earth and Planetary Sciences, Northrop Hall, University of New Mexico, Albuquerque, New Mexico 87131-1116, U.S.A.; sidash@aol.com

Nina L. Baghai-Riding, Division of Biological and Physical Sciences, Delta State University, Cleveland, Mississippi 38733, U.S.A.; nbaghai@deltastate.edu

Brooks B. Britt, Museum of Paleontology and Department of Geological Sciences, Brigham Young University, Provo, Utah 84602, U.S.A.; brooks_britt@byu.edu

Marcus Clauss, Clinic for Zoo Animals, Exotic Pets and Wildlife, Vetsuisse Faculty, University of Zurich, Winterthurerstr. 260, CH-8057 Zurich, Switzerland; mclauss@vetclinics.unizh.ch

William L. Crepet, Department of Plant Biology and L.H. Bailey Hortorium, 462 Mann Library, Cornell University, Ithaca, New York 14853, U.S.A.; wlc1@cornell.edu

N. Rubén Cúneo, MEF-CONICET, Avenida Fontana 140, Trelew, Argentina; rcuneo@mef.org.ar

Charles P. Daghlian, Dartmouth College, E. M. Facility, 7605 Remsen, Dartmouth College, Hanover, New Hampshire 03755, U.S.A.; Charles.P.Daghlian@Dartmouth.edu

Georgina M. Del Fueyo, División Paleobotánica, Museo Argentino de Ciencias Naturales "B. Rivadavia," CONICET, Avenida Angel Gallardo 470, Buenos Aires (C1405DJR) Argentina; georgidf@yahoo.com.ar

David L. Dilcher, Department of Biology, Indiana University, Bloomington, IN 47405, paleoleo@yahoo.com

Ignacio Escapa, MEF-CONICET, Avenida Fontana 140, Trelew, Argentina; iescapa@mef.org.ar

Carole T. Gee, Steinmann Institute, Division of Paleontology, University of Bonn, Nussallee 8, D-53115 Bonn, Germany; cgee@uni-bonn.de

Silvia Gnaedinger, Centro de Ecología Aplicada del Litoral (CE-COAL)—Consejo Nacional de Investigaciones Científicas y Técnicas (CONICET), Facultad de Ciencias Exactas y Naturales y Agrimensura, Universidad Nacional del Nordeste (UNNE), Casilla de Correo 128, CP. 3400 Corrientes, Argentina; scgnaed@hotmail.com

Carol L. Hotton, Department of Paleobiology, National Museum of Natural History, Smithsonian Institution, Washington, D.C. 20560, and National Center for Biological Information, National Library of Medicine, National Institutes of Health, Bethesda, Maryland 20892-6510, U.S.A.; hotton@ncbi.nlm.nih.gov

Jürgen Hummel, Institute of Animal Science, University of Bonn, Endenicher Allee 15, D-53115 Bonn, Germany; jhum@itw.uni-bonn.de

Nancy Kerk, Department of Molecular, Cellular and Developmental Biology, Yale University, New Haven, Connecticut 06520, U.S.A.; nancy.kerk@yale.edu

Michael Krings, Bayerische Staatssammlung für Paläontologie und Geologie und GeoBio-Center[LMU], D-80333 Munich, Germany; m.krings@lrz.uni-muenchen.de

Jeffrey M. Osborn, School of Science, The College of New Jersey, Ewing, New Jersey 08628, U.S.A.; josborn@tcnj.edu

Gar W. Rothwell, Department of Environmental and Plant Biology, Ohio University, Athens, Ohio 45701, U.S.A.; rothwell@ohio.edu

P. Martin Sander, Steinmann Institute, Division of Paleontology, University of Bonn, Nussallee 8, D-53115 Bonn, Germany; martin.sander@uni-bonn.de

Andrew B. Schwendemann, Department of Ecology and Evolutionary Biology, and Natural History Museum and Biodiversity Research Center, University of Kansas, Lawrence, Kansas 66045-7534, U.S.A.; aschwend@ku.edu

Dennis W. Stevenson, New York Botanical Garden, 2900 Southern Blvd., Bronx, New York 10458, U.S.A.; dws@nybg.org

Ruth A. Stockey, Department of Biological Sciences, University of Alberta, Edmonton, AB, T6G 2E9, Canada; ruth.stockey@ualberta.ca

Ian Sussex, Department of Molecular, Cellular and Developmental Biology, Yale University, New Haven, Connecticut 06520, U.S.A.; ian.sussex@yale.edu

David Winship Taylor, Department of Biology, Indiana University Southeast, 4201 Grant Line Rd., New Albany, Indiana 47150, U.S.A.; dwtaylo2@ius.edu

Edith L. Taylor, Department of Ecology and Evolutionary Biology, and Natural History Museum and Biodiversity Research Center, University of Kansas, Lawrence, Kansas 66045-7534, U.S.A.; etaylor@ku.edu

Mackenzie L. Taylor, Department of Ecology and Evolutionary Biology, University of Tennessee, Knoxville, Tennessee 37996, U.S.A.; mtaylo37@utk.edu

Thomas N. Taylor, Department of Ecology and Evolutionary Biology, and Natural History Museum and Biodiversity Research Center, University of Kansas, Lawrence, Kansas 66045-7534, U.S.A.; tntaylor@ku.edu

William D. Tidwell, BYU Museum of Paleontology, Brigham Young University, Provo, Utah 84602, U.S.A.; william_tidwell@gmail.com

Liliana Villar de Seoane, MACN-CONICET, Avenida Angel Gallardo 470, Buenos Aires, Argentina; lvillar@macn.gov.ar

INDEX

Milton Keynes UK
Ingram Content Group UK Ltd.
UKHW050305230923
429204UK00004B/93